Infrastructure Development and Construction Management

T0136228

This is a comprehensive book on infrastructure development and construction management. It is written keeping in mind the curricula of construction management programmes in India and abroad.

It covers infrastructure development, the construction industry in India, financial analysis of the real estate industry in India, economic analysis of projects, tendering and bidding, contracts and contract management, FIDIC conditions of contract, construction disputes and claims, arbitration, conciliation and dispute resolution, international construction project exports and identifying, analysing and managing construction project risk. Thus, this book covers most of the construction management activities that are carried out at different stages of a construction project.

This is an essential book for students of construction management, construction professionals, academicians and researchers.

J.C. Edison is a Professor and Dean at the National Institute of Construction Management and Research (NICMAR), India. He has over 30 years' experience in the industry and academia. He joined the NICMAR in 1995. He has taught post-graduate students for over 25 years, and is engaged with guiding research scholars and management consultancy. Professor Edison has presented papers at national and international conferences including at Oxford and Harvard Universities. He has authored books on the construction industry, international contracting and project exports. His research interests include industrial economics, economics of construction and infrastructure development, project exports, project economics and real estate, and urban infrastructure.

Infrastructure Development and Construction Management

J.C. Edison

Routledge
Taylor & Francis Group

LONDON AND NEW YORK

First published 2021
by Routledge
2 Park Square, Milton Park, Abingdon, Oxon OX14 4RN

and by Routledge
52 Vanderbilt Avenue, New York, NY 10017

Routledge is an imprint of the Taylor & Francis Group, an informa business

British Library Cataloguing-in-Publication Data
A catalogue record for this book is available from the British Library

Library of Congress Cataloging-in-Publication Data
Names: Edison, J. C., author.
Title: Infrastructure development and construction management/J. C. Edison.
Description: Abingdon, Oxon; New York, NY: Routledge, 2021. |
Includes bibliographical references and index.
Identifiers: LCCN 2020017536 (print) | LCCN 2020017537 (ebook) |
ISBN 9780367518943 (hbk) | ISBN 9780367518929 (pbk) |
ISBN 9781003055624 (ebk)
Subjects: LCSH: Construction industry–Management. |
Infrastructure (Economics) | Economic development.
Classification: LCC HD9715.A2 E327 2021 (print) | LCC HD9715.A2
(ebook) | DDC 624.068–dc23
LC record available at https://lccn.loc.gov/2020017536
LC ebook record available at https://lccn.loc.gov/2020017537

ISBN: 978-0-367-51894-3 (hbk)
ISBN: 978-0-367-51892-9 (pbk)
ISBN: 978-1-003-05562-4 (ebk)

Typeset in Goudy
by Deanta Global Publishing Services, Chennai, India

Contents

Figures

Tables

Boxes

Preface

This book is intended for students and graduates of civil engineering and construction management. It contains different aspects of infrastructure development and construction management. The chapters were selected based on their relevance to the needs of students, construction professionals, academicians and researchers. This book is written keeping in mind the curricula of construction management programmes in India and abroad. It covers infrastructure development; an overview of construction and industry in India; financial analysis of the real estate industry in India; economic analysis of projects; tendering and bidding; contracts and contract management; the International Federation of Consulting Engineers (FIDIC) conditions of contract; construction disputes and claims; arbitration, conciliation and dispute resolution; international construction project exports; and identifying, analysing and managing construction project risk.

Chapter 1: Why is infrastructure development important? provides different aspects of infrastructure development and establishes linkages between infrastructure investment (capital expenditure) and gross domestic product (GDP). The study takes (i) gross domestic product as output; (ii) gross capital formation (GCF) as capital expenditure; (iii) compensation to employees, of both organised and unorganised sectors, at constant prices, as labour; and (iv) capital expenditure of both central and state governments is considered as investment in Indian infrastructure. The results of the study reveal that investment in capital expenditure, i.e. construction, has a significant and positive effect on the growth and development of India. However, the results of the study indicate that time also has a role in the determination of GDP.

Chapter 2: Construction industry in India, covers the critical role of the construction industry in economic development; its characteristics and structure; nature of investment; opportunities in the construction industry in India; present scenario of the construction industry; growth of the firms in the construction industry; and constraints and factors influencing construction productivity.

Chapter 3: Real estate industry in India: A financial analysis, examines the relationship between real estate sector investment and GDP through regression analysis; and the financial parameters of the Indian real estate industry. In addition, a financial analysis is carried out by taking financial aggregates, financial ratios, ratios pertaining to margins on income, returns on investments and efficiency ratios of the real estate industry in India.

Chapter 4: Economic analysis of projects, presents economic analysis (sector analysis, demand analysis, alternative analysis), benefit–cost analysis, investment decision criteria (benefit–cost ratio, the economic net present value [ENPV] and the economic internal rate of return [EIRR], effectiveness analysis), shadow pricing, risk and sensitivity analysis, sustainability analysis and environmental sustainability.

Chapter 5: Tendering and bidding, explains the contract and the tender; project procurement cycle; contractor selection; types of tendering processes; preparation of tender documents; bidding documents; the prequalification process; preparation of bids; submission of bids; processing of bids; and award of work.

Chapter 6: Contracts and contract management, presents the contract and its various aspects; contracts on the basis of validity, formation and performance; contract delivery methods; special purpose vehicle; role of the employer, contractor, architect and engineer in contract management; management of contracts; factors to be controlled in contract administration; contractor's reporting to the employer; and contract closure.

Chapter 7: FIDIC conditions of contract, presents clause-by-clause details of all 20 clauses of the 2006 Multilateral Development Bank (MDB) harmonised version of the FIDIC Conditions of Contracts for Construction Works.

Chapter 8: Construction disputes and claims, discusses the types of construction claims; claims between parties; management of claim; claim documentation; claim presentation; and negotiation and settlement of claims.

Chapter 9: Arbitration, conciliation and dispute resolution, deals with arbitration, conciliation and dispute resolution; the Arbitration and Conciliation Act, 1996; domestic arbitration; international commercial arbitration; principal causes of disputes and claims and their effect on the project; dispute settlement methods in the construction industry; arbitration as a dispute settlement mechanism; types of arbitration proceedings; advantages of arbitration; and the process of arbitration and settlement.

Chapter 10: International construction project exports, provides details of the global construction market and project exports from India; complexities of international contracting; factors in bidding for international projects; overseas contract tendering process; clearance of project export proposals; and requirements relating to completed projects. The steps to be taken by exporters on completion of contracts and preparation of the final report are explained and a case study is provided.

Chapter 11: Identifying, analysing and managing construction project risk, presents the types of risk in construction projects; common risks encountered by international construction projects; risk classification as per different phases of the project; risk management process in construction projects; common techniques used to identify risks in developing countries; project risk analysis techniques; responding to risk; options for dealing with risks; risk avoidance, reduction and transfer; risk sharing; and risk retention.

This book is a comprehensive textbook for students of engineering colleges and construction management institutes and essential reading on construction management for professionals, academicians and researchers.

I wish to express my gratitude to Professor (Dr) Mangesh G. Korgaonker, Director General NICMAR for extending all his support and allowing me time to write the book and to use the institute's resources.

I am indebted to my faculty colleagues Dr Pradeepta Kumar Samanta, Dr Harish Singla and Dr Amit Hiray for their editorial inputs and Mr A.R. Jadhav, senior librarian, NICMAR, Pune, for his help and cooperation in collecting the required literature to write this book.

My special thanks to Professor (Dr) Chandrakant Gokhale, Dean, NICMAR School of Construction Management for his constructive criticisms.

I thank Mr V.K.J. Rane, former managing director of Ircon International Limited, who has provided immense input on international construction contracting and project exports.

I extend my gratitude to the trustees and governors of NICMAR who are the major players in the Indian construction industry, especially Mr Ajit Gulabchand, chairman of NICMAR who inspired and enriched my thoughts through his scholarly lectures and speeches at NICMAR during the early days of my career. I believe that this is what has encouraged me to continue working in this industry-driven institute for a quarter of a century.

Finally, I thank my wife Anitha and my daughter Isabel for their constant support.

J.C. Edison

Abbreviations

ADB	Asian Development Bank
ADR	Alternative Disputes Resolution
AED	Africa Economic Digest
B/C ratio	Benefit–Cost Ratio
BCR	Benefit–Cost Ratio
BITs	Bilateral Investment Treaties
BOLT	Build Operate Lease and Transfer
BOO	Build Own Operate
BOOT	Build Own Operate Transfer
BOT	Build Operate Transfer
BQ	Bill of Quantities
CAGR	Compound Annual Growth Rate
CBA	Cost Benefit Analysis
CER	Cost Effectiveness Ratio
CISG	Contracts for the International Sale of Goods
CMIE	Centre for Monitoring Indian Economy
CMP	Contract Management Plan
CPWD	Central Public Works Department
DART	Dispute Avoidance and Resolution Task Force
DBFO	Design Build Finance Operate
DRB	Dispute Review Board
ECD	Exchange Control Department
EEPC	Engineering Export Promotion Council
EIRR	Economic Internal Rate of Return
EMD	Earnest Money Deposit
ENPV	Economic Net Present Value
EOI	Expression of Interest
EPC	Engineering, Procurement and Construction
ESHS	Environmental, Social, Health and Safety
FIDIC	International Federation of Consulting Engineers
FIRR	Financial Internal Rate of Return
GC	General Conditions
GCF	Gross Capital Formation
GDP	Gross Domestic Product
GFCF	Gross Fixed Capital Formation

GFCFC	Gross Fixed Capital Formation from Construction
GOI	Government of India
GSE	Georgian State Electro-system
GSHP	Gujarat State Highway Project
GVA	Gross Value Added
HAM	Hybrid Annuity Model
IBEF	India Brand Equity Foundation
ICB	International Competitive Bidding
ICSID	International Centre for Settlement of Investment Disputes
ILO	International Labour Organization
Ind. AS	Indian Accounting Standard
IRC	Indian Roads Congress
IRR	Internal Rate of Return
ITB	Instructions to Bidders
KPI	Key Performance Indicator
kV	Kilovolt
LOA	Letter of Acceptance
LOI	Letter of Intent
LSTK	Lump Sum Turnkey
MDB	Multilateral Development Bank
MEED	Middle East Economic Digest
MNC	Multinational Corporation
NCAER	National Council of Applied Economic Research
NHAI	National Highways Authority of India
NIT	Notice Inviting Tender
NPV	Net Present Value
NSDC	National Skill Development Corporation
O&M	Operating and Maintenance
OECD	Organization for Economic Cooperation and Development
OPRC	Output and Performance Based Road Contract
P&E	Prior Period Income and Extraordinary Income
P&E&OI	Prior Period and Extraordinary Transactions and Excluding Other Income
PAT	Profit After Tax
PC	Particular Conditions
PEM	Project Exports & Service Exports
PEPCI	Project Export Promotion Council of India
PPP	Public Private Partnership
PQ	Prequalification
Q-on-Q	Quarter-on-Quarter
RBI	Reserve Bank of India
RFP	Request for Proposal
RFQ	Request for Qualification
RTP	Request for Technical Proposals
SBD	Standard Bidding Document
SPV	Special Purpose Vehicle
UNCITRAL	United Nations Commission on International Trade Law
UNCTAD	United Nations Conference on Trade and Development

UNIDO	United Nations Industrial Development Organization
UNIDROIT	International Institute for the Unification of Private Law
WACC	Weighted Average Cost of Capital
WB	World Bank
Y-on-Y	Year-on-Year

1 Why is infrastructure development important?

1.1 Introduction

Construction activity is an integral part of an economy's infrastructure and industrial development. The construction industry is characterised by its fragmented and complex nature having a large number of players, with many backward and forward linkages across both public and private sectors and with diverse manufacturing and service industries. It generates substantial employment and provides a growth impetus to other sectors through backward and forward linkages (Government of India, 2002). Nearly 250 ancillary industries, for instance, cement, steel, brick, timber and building materials, depend on the growth of the construction industry. According to the Economic Survey 2017–18, by 2040, the cumulative figure for an infrastructure investment gap in India could reach around US\$526 billion(Government of India, 2018). The construction industry encompasses the whole spectrum of subsectors of the economy as well as infrastructure development, such as highways, roads, ports, railways, airports, power systems, irrigation and agriculture systems, telecommunication systems, hospitals, schools, townships, offices, houses and other buildings; urban infrastructure, including water supply, sewerage and drainage, industrial and mining; and rural infrastructure (Edison, 2016). Consequently, infrastructure development has become a key input for the socioeconomic development of an economy. In this context, the growth and profitability of the construction industry has become extremely important since it plays a major role in the development of infrastructure in India.

1.2 Role of infrastructure development in economic growth

Infrastructure can influence aggregate output both directly and indirectly. It influences gross domestic product (GDP) directly through its contribution and as an ancillary input in the production process of other sectors, and indirectly by reducing transaction and other costs through raising total factor productivity and consequently permitting optimum use of conventional productive resources. Several studies found a significant positive effect of infrastructure on output, productivity and growth rates.

The infrastructure capital of India is owned by both the public and the private sector. This century has witnessed the production, operation, maintenance and transfer of infrastructure capital by the private sector also. Consequently, it is difficult to identify and choose infrastructural capital investment in India. Generally, infrastructure output such as roads, highways, schools and hospitals are rarely sold in the market and their utilisation and benefits are complicated to measure. The current empirical study, therefore, has chosen the monetary values of ongoing and completed projects, provided by the Centre for Monitoring Indian Economy (CMIE), as infrastructure capital.

Infrastructure investment in India, as in other democratic countries, is not driven by the economy's growth requirements, such as the number of hospitals or hospital beds per population, sewer systems or the requirement for electricity in the growth of manufacturing, etc. It is done by assessing the genuine requirements of the economy. Often, political considerations offset economic concerns. The present empirical study is not addressing these factors. It attempts only to establish the association between infrastructure capital investment and the growth of the economy.

Development economists have considered the physical infrastructure as a precondition for industrialisation and economic development. The physical infrastructure consists of (i) economic infrastructure and (ii) social infrastructure. The economic infrastructure comprises telecommunications, roads, irrigation and electricity; the social infrastructure includes water supply, sewage systems, schools and hospitals (Murphy et al., 1989). The studies of Barro (1990) and Futagami, Morita and Shibata (1993) showed that physical infrastructure development enhances the long-term production and income levels of an economy in the macro-economic endogenous growth literature. Similarly, the empirical studies of Easterly and Rebelo (1993), Lipton and Ravallion (1995), Jimenez (1995), Canning and Bennathan (2000), Esfahani and Ramirez (2003), Canning and Pedroni (2008) and Calderón, Moral-Benito and Servén (2015) also show a relationship between physical infrastructure development and growth. In addition, a number of micro studies such as Van de Walle (1996), Lokshin and Yemtsov (2005), Jalan and Ravallion (2003), Jacoby (2000) and Gibson and Rozelle (2003) have revealed that the development of infrastructure is an indispensable component in poverty reduction. The positive effect of the quantity of infrastructure stocks and the quality of infrastructure services on growth was revealed in the study by Calderón (2009). A 10 per cent increase in infrastructure development contributes to a 1 per cent growth in the long term (2012). The study attempts to expand the Cobb–Douglas production function by adding infrastructure capital investment.

1.3 Role of infrastructure in economic growth

Infrastructure development enhances the economic growth of an economy in the long run through enhancing growth in other sectors, viz., agriculture, industry and services. Aschauer (1989) established that a sharp fall in productivity growth can almost simultaneously reduce public investment. The Aschauer study was the first to suggest that the shrinkage of productive public services in the United States may be decisive in explaining the overall decline in the rate of productivity growth in the economy. Aschauer's paper (1990) on 'Why is Infrastructure Important?' reviews a few of the probable ways in which infrastructure can be important. It considers, by implication, the validity of any case to increase investment in infrastructure facilities to enhance improvements in the economic performance of a country and the quality of life of its people. Mamatzakis' computations (2008) led to empirical evidence of the positive effects of public capital on output and established that infrastructure is a significant constituent of economic activity in Greece. His estimates revealed that an increase in public infrastructure diminishes costs in the majority of manufacturing industries, since it supports the growth in productivity of resources. Baldwin and Dixon (2008) investigated the investment in infrastructure in Canada, and outlined a taxonomy to define those assets that can be considered as infrastructure and can be used to assess the significance of different kinds of infrastructure investments and how Canada's infrastructure has evolved over the last four decades. An efficient infrastructure enhances quality of life, supports growth and is significant for national security. Peter Nijkamp (1986) validates the role of infrastructure in regional development. The study by Snieska and Draksaite (2007) confirms that the physical infrastructure is one of the indicators of the competitiveness of an economy. Martinkus and Lukasevicius (2008) consider infrastructure services and the physical infrastructure as major factors affecting the investment climate at the local level and augmenting the charm of a region.

The study by Bristow and Nellthorp (2007) reviewed the transport appraisal techniques in use in the European Union. It also analysed the impact of infrastructure on different aspects: output, inequality of income, economic growth, productivity of labour, regional competitiveness, the impact on the environment and well-being, and found strong agreement in the treatment of the various direct effects, where monetary valuation and inclusion in cost benefit analysis are common. Infrastructure investment is capable of stimulating organisational and management changes. Mattoon (2004) discussed the available tools to evaluate infrastructure decisions to help policymakers and the technical difficulties in calculating infrastructure benefits. The study also briefly discussed structural issues that might impede infrastructure decisions such as lack of a regional governance structure for funding infrastructure investments that cross local boundaries. The study concluded that infrastructure that benefits one jurisdiction may well harm another region. Public infrastructure has a vital input in private sector production, the geographic concentration of economic resources and wider and deeper markets for output and employment (2009). The study by Palei (2015) reveals that infrastructure influences national competitiveness.

The association between infrastructure and economic growth was investigated empirically by Sanchez-Robles (1998). This study found a positive impact of infrastructure expenditure on growth. In his study of 15 Indian states, Lall (1999) examined the role of public infrastructure investment on regional development and found that social infrastructure has a positive and significant impact on output whereas economic infrastructure does not. However, these results seem improbable in the light of other studies.

Agenor and Moreno-Dodson (2006) studied the relationship between the presence of infrastructure and education and health in the community, and established that infrastructure services are crucial to ensure the 'quality and availability of health and education', which provide a wealth effect to a great extent. Demetriades and Mamuneas (2000) found that social capital infrastructure has a considerable positive impact on earnings, the demand for private means of production and the delivery of products. In his study, Macdonald (2008) established that infrastructure is imperative for the private manufacturing sector. An interesting empirical analysis by Devarajan et al. (1996) suggests that excessive amounts of capital expenditure can become unproductive. The study by Zugravu and Sava (2014) reveals the positive impact of public capital on economic activities. Additionally, Easterly and Rebelo's (1993) study found that investment in transport and communication is consistently correlated with growth.

Government expenditure on infrastructure is the primary demand driver for the construction industry. Since adequate infrastructure is essential for sustained economic growth, infrastructure construction has gained significant importance over the past few years, mainly in the form of the development of roads, water supply and sanitation, irrigation, seaports and airports projects.

1.4 Indian infrastructure

It is well established that investment in physical infrastructure is a precondition for the growth and development of an economy. Investment in physical infrastructure spans several subsectors such as the industrial and mining sector, roads and highways, railways, seaports, airports, power systems, irrigation and agriculture systems, telecommunication systems, schools and hospitals, townships, offices, houses, commercial complexes and other buildings; the urban infrastructure, such as water supply, sewerage and drainage, sanitation; and the rural infrastructure. Infrastructure development has major linkages with the building material manufacturing industry, viz., cement, steel, bricks, tiles, sand and aggregates, fixtures and fittings, paints, chemicals, timber, aluminium, glass, plastics, mineral products and construction equipment, petrol and petro-products. Consequently, it is the basic input for socioeconomic development. Investment requirement in the infrastructure development sector in India over the next 5 years (2017–2022) is Rs. 31 trillion (US$454.83 billion), with 70 per cent of funds required for roads, power and urban infrastructure segments (India Brand Equity Foundation, 2008). The Indian construction industry employs over 40 million people for blue-collared jobs. Of these, about 80 per cent, i.e. 32 million, are employed in the infrastructure construction sector (National Skill Development Corporation, 2009). With India delaying the development of its infrastructure and the resultant slowdown in the economy, high growth in this sector is inevitable in the coming years.

1.5 Role of construction industry in socioeconomic development

Turin (1973) established that a stronger construction industry is present in developed countries than less-developed economies. His proposition was that the proportion of construction in national output would grow as an economy grows, up to a point, and would there after level out. A contradiction to this proposition is the study by Wells (1986) who states that the role of construction declines in an economy as the economy attains 'developed' status. Ofori (1990) noted the significance of construction in the national economy through a multifaceted set of inter-relationships. The study by Park (1989) has established the construction industry's extensive backward and forward linkages with other sectors of the economy. Similarly, the World Bank (1984) stated that the significance of the construction industry in an economy stems from its strong linkages with building materials and other sectors of the economy. The value added by construction is considerable if the backward linkages of construction, such as building materials and components, are taken into account (Dang and Low, 2015). In India, although the percentage share in the value of construction output shows a declining trend (Figure 1.1), it steadily increased from 2003–04 in terms of absolute values (Figure 1.2). Osei's (2013) study contends that the construction sector holds enormous potential for growth stimulation, the enhancement of project exports and employment generation.

1.6 Real estate and urban development in India

1.6.1 Real estate

The real estate sector in India has a steady demand supported growth due to population growth, a larger allocation to savings in real estate, the actualisation of mortgaging by buyers and bridging the gap of the existing housing shortfall (Mehta, 2007). The study by Vishwakarma (2013) finds a sign of a positive periodically collapsing bubble in the Indian

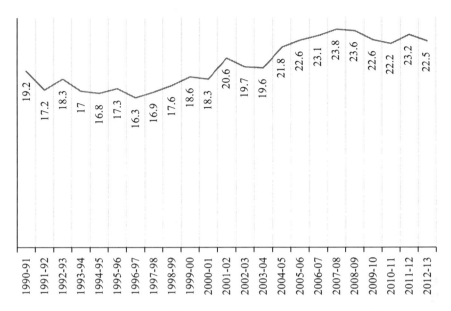

Figure 1.1 Value of the output of the construction sector as a percentage of GDP.

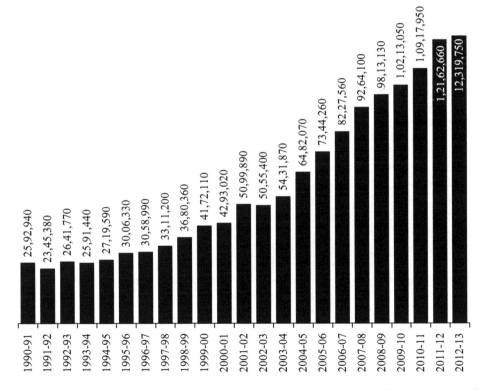

Figure 1.2 Value of the output of the construction sector at constant prices (base year 2004–05) in Rs. million.

real estate market. Even though 'real estate is long cycle prone business' (Kaiser, 1997), the size of the Indian real estate market is expected to reach US$1 trillion by 2030 (India Brand Equity Foundation, 2020). It is estimated that the demand for affordable housing will increase to 38 million housing units in 2030 from 19 million in 2012 (RBI Bulletin, January 2018). It is estimated that at the beginning of India's Twelfth Five-Year Plan (2012–17), the overall shortage of housing in India will be 18.78 million (Press Information Bureau, 2012).

National Housing Bank's (India) Report on Trend and Progress of Housing in India 2018 quotes the Population Division of the UN Department of Economic and Social Affairs' (UN DESA) report "World Urbanization Prospects – The 2018 Revision" and states that by 2050, it is projected that India will have added 416 million urban dwellers compared to 255 million in China (National Housing Bank, 2018). Currently, India has 40 million middle-class households. By 2025–26, the number of middle-class households in Indiais expected to reach 113.8 million, or 547 million people, an almost three-fold growth from current levels. As per the study, households with an annual income of between Rs. 3.4 lakh and Rs. 17 lakh (at 2009–10 price levels) fall into the middle-class category. According to the report, a typical middle-class household in India spends about 50 per cent of its total income on everyday expenses while the remainder is saved. This establishes the sustainability of the economy and the existence of huge numbers of middle-class households with a substantial disposable income and savings and affirms that there is immense potential for India's real estate market.

The real estate industry has considerable direct and indirect linkages with nearly 300 sectors, such as cement, steel, paints and building hardware, that contribute to capital formation and the generation of employment and income opportunities. Moreover, it also catalyses and stimulates economic growth. Consequently, investment in this sector can be considered a barometer of growth for the entire economy (Govt. of India, 2012).

1.6.2 Urban development

The rate of urban growth in India is very high compared to developed countries. Large cities are expanding primarily because of the continuous migration of the population. This urban growth is increasing the contribution of the urban sector to the economy. According to the Census of India 2011, about 31 per cent of the country's population, i.e. 377 million people, live in urban areas. In 1950–51, the contribution of the urban sector to India's GDP was only 29 per cent, increasing to 47 per cent in 1980–81 and it currently contributes 62 per cent to 63 per cent which is likely to increase to 75 per cent by 2021 (GOI, 2007a). Estimates indicate that the requirement for water for domestic use in both urban and rural areas is expected to almost double from the present level of about 5 per cent by 2050 (GOI, 2007b).

A major challenge of urbanisation is to provide basic services such as sanitation, water supply and basic housing to the urban population which is expected to increase to around 600 million within 20 years. Using 2006 data, the Water and Sanitation Programme of the World Bank suggests that the per capita economic cost of inadequate sanitation in India is Rs. 2,180 (GOI, 2013, p. 321). The Isher Ahluwalia Committee on Urban Infrastructure and Services[1] has estimated a total capital investment in urban infrastructure of about Rs. 390 million over the next 20 years (GOI, 2013, pp. 329–331). In the 2015–16 budget, the budgetary allocation for the Ministry of Urban Development was Rs. 160.542 billion including Rs. 60 million for smart cities and 500 habitations (GOI, 2015). The plan outlay of the 2016–17 budget for the Ministry of Urban Development is Rs. 211 billion (GOI, 2016)[2].

1.7 Project investments

Investment in projects is increasing in India, both implemented and completed, barring announcement of new investment projects. A reduction in the announcement of new investment projects has been noticed (Tables 1.1 and 1.2). Although the private sector recently took the lead in project announcements, most of these projects did not take off. During 2014–16, the private sector announced 2,856 investment projects worth Rs.11.337 trillion. On the contrary, 60 per cent of the Rs.7.42 trillion new investment projects announced by the government during the same period have already started.

Government sector project completion is greater than that of the private sector in India (Figure 1.3). A further examination of projects under implementation reveals that the share of private sector projects is much less than government projects (Figure 1.4). Low profit margins are inhibiting the private sector from staying away from project implementation. Announcements of new investment projects are showing a declining trend (Figure 1.5). Since the December 2014 quarter, there has been a steady and steep decline in new investment proposals (Table 1.3 and Figure 1.6). Another noteworthy fact is that the number of stalled projects is also steadily increasing (Table 1.4). The proportion of stalled projects as a percentage of total investments in outstanding projects is increasing (Figure 1.7).

Project commissioning in the real estate industry are likely to plunge in 2020–21. According to CMIE, during the year, projects worth Rs.686.2 billion are expected to be completed, 35 per cent lower as compared to the year 2019–20. Projects worth Rs.1056.4 billion were the corresponding figure for 2019–20. The drastic decline in project completions

Table 1.1 Spend and growth of new investment projects

Year	Government	Private sector	Total	Government	Private sector	Total
	(Rs. million)			% Growth		
1995-96	1,177,722	2,392,753	3,570,476	–	–	–
1996-97	1,194,226	1,500,943	2,695,169	1.4	−37.27	−24.52
1997-98	880,100	958,476	1,838,576	−26.3	−36.14	−31.78
1998-99	1,096,548	861,052	1,957,600	24.59	−10.16	6.47
1999-00	1,455,618	1,661,998	3,117,615	32.75	93.02	59.26
2000-01	1,269,640	1,205,968	2,475,608	−12.78	−27.44	−20.59
2001-02	1,039,102	797,043	1,836,145	−18.16	−33.91	−25.83
2002-03	1,060,560	1,010,731	2,071,291	2.07	26.81	12.81
2003-04	1,528,382	1,333,397	2,861,778	44.11	31.92	38.16
2004-05	1,335,559	2,802,954	4,138,513	−12.62	110.21	44.61
2005-06	2,401,379	6,111,574	8,512,953	79.8	118.04	105.7
2006-07	6,109,945	10,917,897	17,027,842	154.43	78.64	100.02
2007-08	6,112,381	13,758,162	19,870,543	0.04	26.01	16.69
2008-09	6,838,493	15,394,152	22,232,645	11.88	11.89	11.89
2009-10	7,292,161	8,859,027	16,151,188	6.63	−42.45	−27.35
2010-11	4,512,280	11,589,287	16,101,567	−38.12	30.82	−0.31
2011-12	4,025,502	5,760,222	9,785,723	−10.79	−50.3	−39.23
2012-13	1,838,763	3,331,453	5,170,216	−54.32	−42.16	−47.17
2013-14	3,246,012	2,709,341	5,955,354	76.53	−18.67	15.19
2014-15	4,831,459	5,702,109	10,533,568	48.84	110.46	76.88
2015-16	2,587,444	5,635,797	8,223,241	−46.45	−1.16	−21.93

Source: Economic Outlook Database, Centre for Monitoring Indian Economy, Mumbai, India. https://economicoutlook.cmie.com/ Accessed on 5 June 2017.

Table 1.2 Public and private investment projects announced and stalled (in Rs. million)

Year	Public investment projects	Private investment projects	Stalled projects	Year	Public investment projects	Private investment projects	Stalled projects
Jun-10	1,128,070	2,897,322	133,224	Dec-12	550,266	682,825	622,476
Sep-10	1,113,375	2,313,975	129,886	Mar-13	837,848	561,986	651,382
Dec-10	1,085,013	1,943,218	484,062	Jun-13	811,503	677,911	587,426
Mar-11	1,003,859	1,745,552	642,353	Sep-13	693,681	523,574	664,098
Jun-11	1,006,375	1,440,055	773,576	Dec-13	1,002,820	784,539	712,448
Sep-11	867,073	1,359,125	812,980	Mar-14	1,200,534	1,232,197	720,470
Dec-11	679,790	1,271,614	471,888	Jun-14	1,207,495	1,425,490	568,904
Mar-12	510,218	1,121,424	474,095	Sep-14	1,257,267	1,439,256	484,892
Jun-12	459,691	832,901	590,760	Dec-14	941,363	1,501,664	454,702
Sep-12	487,703	736,387	513,505	Mar-15	521,779	1,226,383	243,559

Source: Economic Outlook Database, Centre for Monitoring Indian Economy, Mumbai, India. https://economic outlook.cmie.com/ Accessed on 5 June 2015.

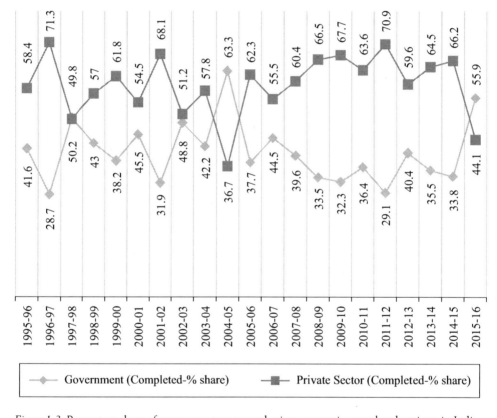

Figure 1.3 Percentage share of government sector and private sector in completed projects in India.

in the industry can be attributed to the extreme slowdown in the economic activities in India caused by the Covid-19 pandemic. Similarly, fresh investment proposal in the industry declined 18 per cent, i.e., Rs.582.5 billion during April–December 2019 as compared to Rs.709.4 billion in April–December 2018.[3] The quarterly aggregate order backlog of industrial and infrastructural firms reached Rs. 3861.2 billion at the end of March 2019 (Table 1.5). New order inflows for industrial and infrastructural construction have reduced

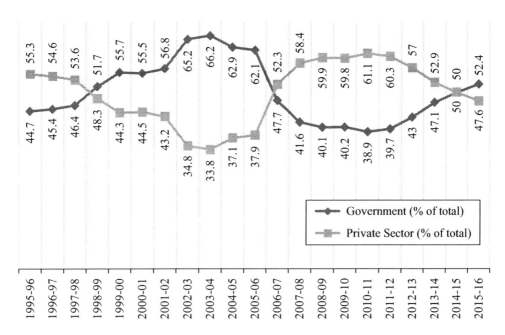

Figure 1.4 Percentage share of government sector and private sector in aggregate project investments in India.

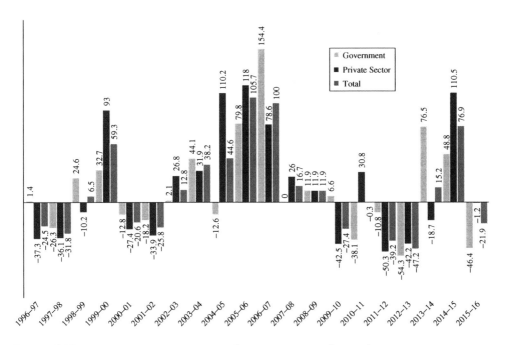

Figure 1.5 New investment projects announced – percentage growth over the previous year.

Table 1.3 Quarterly new investment announcements (Rs. billion)

Sr. No.	Quarter ended	Amount (Rs. billion)	Y-o-Y% change	Sr. No.	Quarter ended	Amount (Rs. billion)	Y-o-Y% change
1.	Jun-14	2,770	−18.29	12.	Mar-17	5,259	−9.23
2.	Sep-14	6,464	179.77	13.	Jun-17	3,115	−36.46
3.	Dec-14	6,602	180.21	14.	Sep-17	2,552	−33.63
4.	Mar-15	5,237	126.61	15.	Dec-17	2,796	−6.05
5.	Jun-15	4,199	51.6	16.	Mar-18	4,704	−10.56
6.	Sep-15	5,880	−9.04	17.	Jun-18	3,819	22.61
7.	Dec-15	4,646	−29.63	18.	Sep-18	2,307	−9.6
8.	Mar-16	5,794	10.65	19.	Dec-18	3,030	8.39
9.	Jun-16	4,902	16.72	20.	Mar-19	2,655	−43.56
10.	Sep-16	3,846	−34.6	21.	Jun-19	820	−78.53
11.	Dec-16	2,976	−35.95	22.	Sep-19	955	−58.6

Source: Economic Outlook Database, Centre for Monitoring Indian Economy, Mumbai, India. https://economic outlook.cmie.com/ Accessed on 24 October 2019.

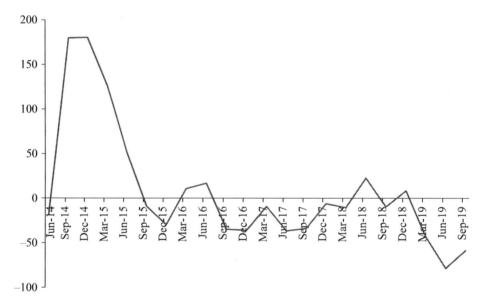

Figure 1.6 Quarterly new investment projects announced – percentage growth over the previous year.

in the recent quarters. The hopes of a revival in construction now depend of the recovery of the Indian economy.

1.8 Macro-economic aggregates

1.8.1 *Gross domestic product and the sectoral contributions*

Gross domestic product growth is considered an indicator of an economy's growth. Different sectors contribute to GDP and its growth such as agriculture, forestry, fishing, mining, quarrying, manufacturing, electricity, gas, water supply, construction, trade, hotels, transport, storage, communication, financing, insurance, real estate and business services,

Table 1.4 Implementation stalled projects (Rs. million):
June 2017 to September 2019

Investments	Implementation stalled
Jun-17	11,509,319
Sep-17	12,078,509
Dec-17	12,635,478
Mar-18	12,308,959
Jun-18	11,159,316
Sep-18	11,306,418
Dec-18	11,552,024
Mar-19	12,320,207
Jun-19	13,117,715
Sep-19	13,231,094

Source: Economic Outlook Database, Centre for Monitoring
Indian Economy, Mumbai, India. https://economicoutlook.cmie
.com/ Accessed on 24 October 2019.

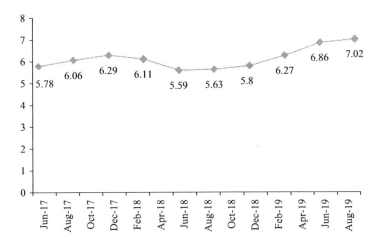

Figure 1.7 Stalled projects as a percentage of investments in outstanding projects.

Table 1.5 Order book position and new order inflows (in Rs. billion)

Quarter	Order book position	New order inflows
Dec-17	3,710.9	1,298.1
Mar-18	3,538.4	712.8
Jun-18	3,624.9	557.4
Sep-18	3,710.4	521.3
Dec-18	3,671.5	388.4
Mar-19	3,861.2	139.1

Source: Economic Outlook Database, Centre for Monitoring Indian Economy,
Mumbai, India. https://economicoutlook.cmie.com/ Accessed on 24 October
2019.

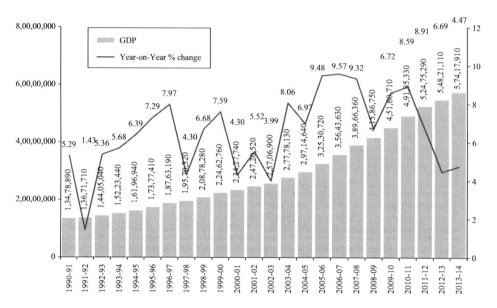

Figure 1.8 Growth of gross domestic product in India.

community, social and personal services. Estimates of 2013–14 show that the GDP of India was Rs. 57,417,910 million, up 4.74 per cent on the previous year (Figure 1.8). In 2013–14, the share of the service sector was the maximum (59.93 per cent) (Table 1.6 and Figure 1.9). The contribution of the industrial sector and the agricultural sector was 31.42 per cent and 17.88 per cent, respectively, in the same year. The construction sector's contribution to GDP was 8.22 per cent in 2013–14.

1.8.2 Gross value added

In India, the gross value added (GVA), at basic prices, grew by 6.29 per cent in 2013–14 up on the 5.43 per cent growth in the preceding year, as per the data released by the CMIE and the Central Statistics Office (CSO). The latest data released for 2014–15 and 2015–16 are 7.08 per cent and 7.19 per cent, respectively. Most of the sectors showed a decline in growth in 2015–16; for example, mining and quarrying from 10.8 per cent to 7.4 per cent; electricity, gas, water supply and other utility services from 8 per cent to 6.6 per cent; and the construction sector from 4.4 per cent to 3.9 per cent in the preceding year. The percentage share of construction in GVA has reduced from 2011–12 (Table 1.7 and Figure 1.10).

1.8.3 Gross fixed capital formation from construction

Gross fixed capital formation from construction (GFCFC), one of the indicators of investment activity, grew in 2012–13 but lost its share in GDP. Even the growth in GFCFC was meagre, 1.70 per cent in 2013–14, compared to growth of 10.30 per cent in the previous year (Figures 1.11, 1.12 and 1.13). The GFCFC in 2016–17 was Rs. 2284790 million increasing to Rs. 2797850 million in 2017–18. The GFCFC's share in total GFCF was 6.04 per cent

Table 1.6 Gross domestic product by economic activity at constant prices (base year 2004–05) (in Rs. million)

Year	GDP	Agriculture	Industry	Services	Year	GDP	Agriculture	Industry	Services
1990–91	13,478,890	3,979,710	3,723,590	5,734,650	2002–03	25,706,900	5,175,590	7,040,950	13,490,360
1991–92	13,671,710	3,902,010	3,736,340	6,003,660	2003–04	27,778,130	5,643,910	7,556,240	14,577,960
1992–93	14,405,040	4,161,530	3,856,480	6,345,490	2004–05	29,714,640	5,654,260	8,297,830	15,762,550
1993–94	15,223,440	4,299,810	4,068,470	6,813,510	2005–06	32,530,720	5,944,870	9,104,130	17,481,720
1994–95	16,196,940	4,502,580	4,441,220	7,211,400	2006–07	35,643,630	6,191,900	10,212,040	19,239,690
1995–96	17,377,410	4,471,270	4,942,630	7,940,400	2007–08	38,966,360	6,550,800	11,199,950	21,215,610
1996–97	18,763,190	4,914,840	5,258,640	8,538,420	2008–09	41,586,750	6,556,890	11,697,360	23,332,500
1997–98	19,570,320	4,789,330	5,469,650	9,300,900	2009–10	45,160,710	6,609,870	12,769,190	25,781,650
1998–99	20,878,280	5,092,030	5,696,560	10,071,380	2010–11	49,185,330	7,178,140	13,733,390	28,273,790
1999–00	22,462,760	5,227,950	6,036,310	11,198,480	2011–12	52,475,290	7,538,320	14,806,560	30,130,410
2000–01	23,427,740	5,227,550	6,400,430	11,799,760	2012–13	54,821,110	7,645,100	14,949,210	32,226,790
2001–02	24,720,520	5,541,570	6,567,380	12,611,580	2013–14	57,417,910	8,005,480	15,002,250	34,410,170

Source: Economic Outlook Database, Centre for Monitoring Indian Economy, Mumbai, India. https://economicoutlook.cmie.com/ Accessed on 5 June 2015.

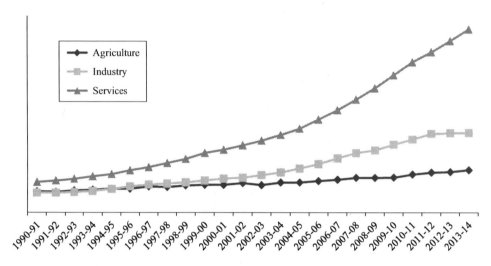

Figure 1.9 Gross domestic product by economic activity.

and 6.76 per cent, respectively. It appears that the contribution of the construction sector must increase considerably to allow an acceleration in GDP growth.

1.8.4 Employment

Infrastructure development is one of the largest sources of employment, both organised and unorganised, in the country. The majority of the manpower in the construction industry, unskilled and semi-skilled manual workers, are unorganised (ILO, 2001). The employment figures have shown a steady increase from 14.6 million in 1995 to more than double in 2005, i.e. from 31.46 million to 33 million personnel comprising engineers, technicians, foremen,

Table 1.7 Gross value added at constant basic prices: Construction (in Rs. million)

Year	GVA	GVA from construction	% share in GVA
2009–10	71,318,360	6,476,390	9.08
2010–11	77,045,140	6,870,710	8.92
2011–12	81,069,460	7,773,340	9.59
2012–13	85,462,750	7,800,500	9.13
2013–14	90,636,490	8,007,710	8.83
2014–15	97,121,330	8,352,290	8.6
2015–16	104,918,700	8,653,350	8.25
2016–17	113,189,720	9,177,540	8.11
2017–18	121,041,650	9,691,940	8.01
2018–19	129,069,365	10,539,010	8.17

Source: Economic Outlook Database, Centre for Monitoring Indian Economy, Mumbai, India. https://economicoutlook.cmie.com/ Accessed on 24 October 2019.

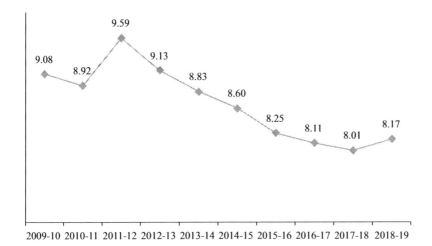

Figure 1.10 Percentage share of construction in GVA.

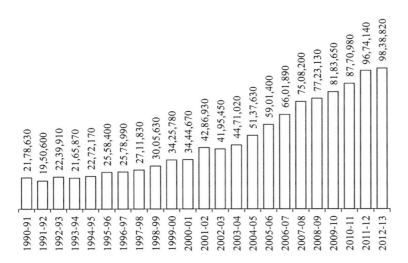

Figure 1.11 Gross fixed capital formation from construction at constant prices in Rs. million.

clerical staff and skilled and unskilled workers (GOI, 2007c). Larger investments in infrastructure have augmented the demand for construction and, consequently, the demand for construction engineers and technicians. In India, compensation to employees in the unorganised sector, as per the CMIE data, is almost similar to the organised sector (Table 1.8 and Figure 1.14). However, it is difficult to estimate the numbers employed in the unorganised sector; therefore, the real figure may be much more.

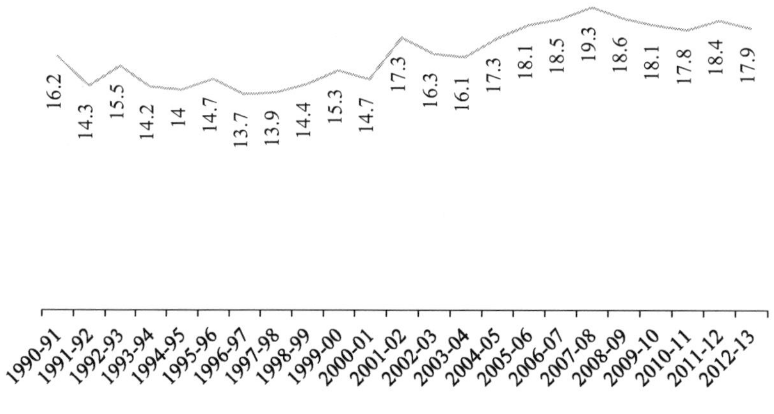

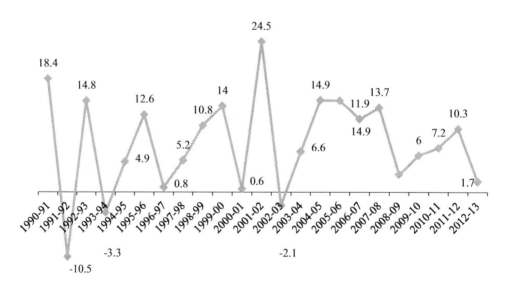

Figure 1.12 Gross fixed capital formation from construction as a percentage of GDP.

Figure 1.13 Percentage growth of gross fixed capital formation from construction.

1.8.5 Capital expenditure, compensation to employees and economic growth

The central and state government capital expenditure of India is showing an upward trend (Table 1.9 and Figure 1.15). On the contrary, the compensation to employees in the unorganised sector (percentage share of construction) is showing a downward trend (Table 1.10 and Figure 1.16). This may be due to the reduction in sales in the construction industry.

1.9 Data and the model

A review of the literature shows that most growth models are based on the Cobb–Douglas (1928) theory of production function. The current study attempted to statistically investigate

Table 1.8 Compensation to employees of organised and unorganised sector at current prices (base year 2004–05): 1990–91 to 2012–13 (in Rs. million)

Year	Organised	Unorganised	Total	Year	Organised	Unorganised	Total
1990–91	631,650	697,840	1,329,490	2002–03	3,433,470	2,865,360	6,298,830
1991–92	736,460	786,290	1,522,750	2003–04	3,735,300	2,757,500	6,492,800
1992–93	853,690	886,380	1,740,070	2004–05	4,135,990	3,266,150	7,402,140
1993–94	961,270	1,017,260	1,978,530	2005–06	4,513,440	3,566,890	8,080,330
1994–95	1,097,670	1,254,650	2,352,320	2006–07	4,946,590	4,185,170	9,131,760
1995–96	1,355,740	1,429,060	2,784,800	2007–08	5,598,630	4,960,200	10,558,830
1996–97	1,557,770	1,643,910	3,201,680	2008–09	7,240,270	6,034,700	13,274,970
1997–98	1,900,400	1,907,250	3,807,650	2009–10	9,149,570	6,785,280	15,934,850
1998–99	2,352,890	2,096,100	4,448,990	2010–11	10,586,750	8,065,570	18,652,320
1999–00	2,760,040	2,312,780	5,072,820	2011–12	11,862,360	9,814,940	21,677,300
2000–01	3,023,980	2,476,930	5,500,910	2012–13	13,632,240	10,735,430	24,367,670
2001–02	3,216,130	2,645,190	5,861,320	–	–	–	–

Source: Economic Outlook Database, Centre for Monitoring Indian Economy, Mumbai, India. https://economicoutlook.cmie.com/ Accessed on 5 June 2015.

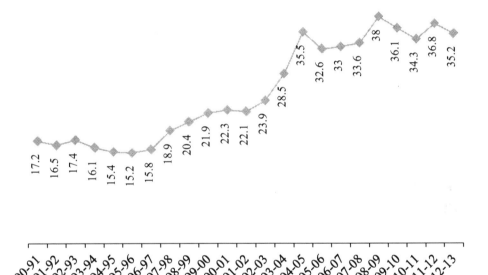

Figure 1.14 Compensation to employees of unorganised sector: Percentage share of construction.

the relationship between infrastructure and economic development. Therefore, in this section, the study used the preceding theoretical discussion to derive an empirical model to empirically assess the relationship between explanatory variables. The GDP[4](aggregate output) was used as the dependent variable because the GDP is the final result of all the economic activities in an economy at the aggregate level.

The study attempted to relate inputs and output at the national aggregate level by including investment in infrastructure as an additional factor along with capital and labour. Therefore, it is assumed that the output at the national aggregate level is a result of labour, capital and investment in infrastructure. The current study took (i) gross domestic product as output; (ii) gross capital formation (GCF) as capital expenditure; (iii) compensation to employees of both

Table 1.9 Central and state government capital expenditure in India: 1990–91 to 2013–14 (in Rs. billion)

Year	Total capital expenditure	Central government	State government	Year	Total capital expenditure	Central government	State government
1990–91	511	318	193	2002–03	1,595	745	850
1991–92	509	291	217	2003–04	2,596	1,091	1,504
1992–93	530	299	231	2004–05	2,658	1,139	1,519
1993–94	590	337	253	2005–06	1,936	664	1,272
1994–95	717	386	331	2006–07	2,264	688	1,576
1995–96	710	384	326	2007–08	2,985	1,182	1,803
1996–97	759	421	338	2008–09	2,993	902	2,091
1997–98	932	517	415	2009–10	3,391	1,127	2,265
1998–99	1,092	629	463	2010–11	3,927	1,566	2,361
1999–00	1,019	490	529	2011–12	4,446	1,586	2,860
2000–01	1,034	478	557	2012–13	4,792	1,669	3,123
2001–02	1,233	608	624	2013–14	5,263	1,877	3,387

Source: Economic Outlook Database, Centre for Monitoring Indian Economy, Mumbai, India. https://economicoutlook.cmie.com/ Accessed on 5 June 2015.

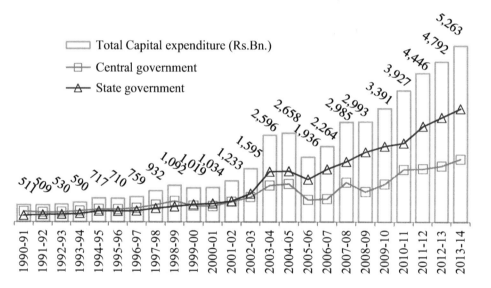

Figure 1.15 Central and state government capital expenditure in India.

organised and unorganised sectors at constant prices, as labour; and (iv) capital expenditure of both central and state governments was considered as investment in Indian infrastructure.

The production function of the present study can be written as:

$$\ln Y = \ln K + \ln L + \ln I$$

where:

Y is output (gross domestic product);
K is capital, i.e. gross capital formation;
L is labour, i.e. compensation to employees of organised and unorganised sectors at constant prices;

I is stock of infrastructure, i.e. capital expenditures of both central and state governments. All variables are in logs.

The relationship between government/capital expenditure and economic growth is a well-researched area. In most developing economies, governments intervene in economic activities to attain macro-economic objectives by using policies and strategies provided by both Keynesians and neoclassical economists (Usman, 2011). The effect of government development expenditure on economic growth in India was examined by Ranjan and Sharma (2008) who found a significant positive impact. Many researchers (Laudau, 1983; Folster and Henrekso, 2001; Akpan, 2005)examined the impact of government expenditure on economic growth with negative or insignificant results. The current study tested the relationship between the GDP and the capital expenditure of India by employing a log-linear equation and found that the impact of capital expenditure on GDP is positive and significant.

$$\ln GDP = x + \ln CAPEX + e$$

$$\ln GDP = (8.96437141936566) + (0.569530589075882) \times (\ln CAPEX)$$

The current study considered total capital expenditure (central and state government) as investment in infrastructure.

1.9.1 Multiple regression results

The equation relating gross domestic product (LnGDP), capital (LnCAPEX), gross capital formation (LnGCF) and labour (LnEmploy) is estimated as: LnGDP = (8.606) + (0.091) LnCAPEX + (0.207) LnGCF + (0.276) LnEmploy using the 24 observations. They-intercept, the estimated value of LnGDP when LnCAPEX is zero, is 8.606 with a standard error of 0.204. The slope, the estimated change in LnGDP per unit change in: (i) LnCAPEX is 0.091 with a standard error of 0.023; (ii) LnGCF is 0.207 with a standard error of 0.033; and (iii) LnEmploy is 0.276 with a standard error of 0.023. The value of R^2, the proportion of the variation in LnGDP that can be accounted for by variation in LnCAPEX, LnGCF and LnEmploy is 0.999. The correlation between LnGDP and LnCAPEX, LnGCF and LnEmploy is 0.157, 0.288 and 0.560, respectively. Time is significant at 5 per cent.

1.9.1.1 Coefficients

Model	Unstandardised coefficients		Sig.
	B	Std. error	
1 (Constant)	8.606	0.204	0*
CAPEXLn	0.091	0.023	0.001*
LnGDCF	0.207	0.033	0*
LnEmploy	0.276	0.023	0*
Time	0.014	0.007	0.049**
F		5334.972	.000.

a Dependent variable: GDPLn.
R^2 = 99.9%.
* Significant at 1%.
** Significant at 5%.

Table 1.10 Compensation to employees of organised and unorganised sector (base year 2004–05) (in Rs. million)

Year	Total	Organised	Unorganised	Year	Total	Organised	Unorganised
1990–91	1,329,490	631,650	697,840	2002–03	6,298,830	3,433,470	2,865,360
1991–92	1,522,750	736,460	786,290	2003–04	6,492,800	3,735,300	2,757,500
1992–93	1,740,070	853,690	886,380	2004–05	7,402,140	4,135,990	3,266,150
1993–94	1,978,530	961,270	1,017,260	2005–06	8,080,330	4,513,440	3,566,890
1994–95	2,352,320	1,097,670	1,254,650	2006–07	9,131,760	4,946,590	4,185,170
1995–96	2,784,800	1,355,740	1,429,060	2007–08	10,558,830	5,598,630	4,960,200
1996–97	3,201,680	1,557,770	1,643,910	2008–09	13,274,970	7,240,270	6,034,700
1997–98	3,807,650	1,900,400	1,907,250	2009–10	15,934,850	9,149,570	6,785,280
1998–99	4,448,990	2,352,890	2,096,100	2010–11	18,652,320	10,586,750	8,065,570
1999–00	5,072,820	2,760,040	2,312,780	2011–12	21,677,300	11,862,360	9,814,940
2000–01	5,500,910	3,023,980	2,476,930	2012–13	24,367,670	13,632,240	10,735,430
2001–02	5,861,320	3,216,130	2,645,190	–	–	–	–

Source: Economic Outlook Database, Centre for Monitoring Indian Economy, Mumbai, India. https://economicoutlook.cmie.com/ Accessed on 5 June 2015.

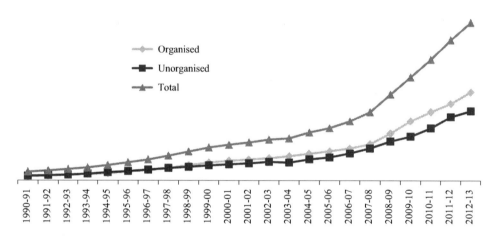

Figure 1.16 Compensation to employees of organised and unorganised sector: Base year 2004–05.

1.10 Conclusion

This chapter examined the relationship between investment in infrastructure and the growth of output, i.e. GDP, after providing an overview of the Indian infrastructure development sector. The foregoing results from the study established linkages between infrastructure investment (capital expenditure) and GDP. The analysis proved the association between GDP and capital expenditure, gross capital formation and labour. The results of the study revealed that investment in capital expenditure, i.e. construction, has a significant and positive effect on the growth and development of India. However, the results of the study also indicated that time plays a role in determining GDP. Therefore, we can infer that infrastructure is important for faster economic growth, social as well as economic performance, improved access to job markets and poverty diminution in an economy. Generally, the availability of good quality infrastructure raises the productivity levels of an economy, brings down the costs of the business enterprises and facilities to retain an adequate quality of life and increases the growth and development of the economy.

Notes

1 Isher Ahluwalia Committee on Urban Infrastructure and Services was prepared by the High Powered Expert Committee to estimate the investment requirement for urban infrastructure services by the Ministry of Urban Development, Government of India in 2008. The report, released in 2011, documents the nature of the urbanisation challenges facing India. Retrieved from: https ://icrier.org/pdf/FinalReport-hpec.pdf Accessed on 15 July 2016.
2 The budget 2016–17 gives an average assistance of over Rs. 80 lakh per gram panchayat and over Rs. 210 million per urban local body.
3 The real estate industry was already reeling under lower sales and tight liquidity in the preceding year. With the disruption in work due to lockdown, movement of labour, tight liquidity and breakdown of the supply chain, recovery in the sector would be a big challenge in terms of both sales and project completions. Due to liquidity crunch, subdued residential demand, tightening of lending norms for home loans by financial institutions and mass labour movement across the states will result in significant credit pressure on developers' and their cash flows and execution capabilities.
4 GDP in Rs. millions at a factor cost at constant prices (base year 2004–05) has been used for the analysis.

References

Agénor, Pierre-Richard and Moreno-Dodson, Blanca (2006) Public infrastructure and growth: New channels and policy implications. World Bank Policy Research Working Paper 4064. Retrieved from: http://documents.worldbank.org/curated/en/485431468141267544/pdf/wps4064.pdf Accessed on 12 July 2016.

Akpan, N. (2005) Government expenditure and economic growth in Nigeria: A disaggregated approach. *Central Bank of Nigeria Financial Review, 43*(1), pp. 1–10.

Aschauer, D.A. (1989) Is public expenditure productive? *Journal of Monetary Economics, 23*(2), pp. 177–200.

Aschauer, D.A. (1990) Why is infrastructure important? [Paper presentation]. Conference of the Federal Reserve Bank of Boston at Harwich Port, Massachusetts, June 28–29.

Baldwin, J.R. and Dixon, J. (2008) Infrastructure capital: What is it? Where is it? How much of it is there? *The Canadian Productivity Review, 16,* Statistics Canada, Ottawa. Retrieved from: http://citeseerx.ist.psu.edu/viewdoc/download?doi=10.1.1.607.5934&rep=rep1&type=pdf Accessed on 12 July 2016.

Barro, Robert J. (1990) Government spending in a simple model of endogenous growth. *Journal of Political Economy, 98*(S5), pp. 103–125.

Bristow, A.L. and Nellthorp, J. (2007) Transport project appraisal in the European Union. *Transport Policy, 7*(1), pp. 51–60.

Calderon, Cesar. (2009) Infrastructure and growth in Africa. Policy Research Working Paper Series 4914, The World Bank.

Calderón, C., Moral-Benito, E. and Servén, L. (2015) Is infrastructure capital productive? A panel heterogeneous approach. *Journal of Applied Econometrics, 30*(2), pp. 177–198.

Canning, D. and Bennathan, E. (2000) The social rate of return on infrastructure investments. World Bank Working Paper no. 2390. Retrieved from: http://documents.worldbank.org/curated/en/261281468766808543/pdf/multi-page.pdf Accessed on 19 October 2016.

Canning, David and Pedroni, Peter L. (2008) Infrastructure, long-run economic growth and causality tests for cointegrated panels. *The Manchester School, 76*(5), pp. 504–527, September, Special Issue. doi:10.1111/j.1467-9957.2008.01073.x.

Cobb, C.W. and Douglas, P.H. (1928) A theory of production. *The American Economic Review, 18*(1), pp. 139–165. [Papers presentation]. Supplement, Papers and Proceedings of the Fortieth Annual Meeting of the American Economic Association. Retrieved from: https://www.aeaweb.org/aer/top 20/18.1.139-165.pdf Accessed on 9 July 2016.

Dang, Giang and Low Sui Pheng. (2015) *Infrastructure Investments in Developing Economies: The Case of Vietnam*, p. 42, Springer, Singapore.

Demetriades, P.O. and Mamuneas, T.P. (2000) Intertemporal output and employment effects of public infrastructure capital: Evidence from 12 OECD economies. *The Economic Journal, 110*(465), pp. 687–712.

Devarajan, S., et al. (1996) The composition of public expenditure and economic growth. *Journal of Monetary Economics, 37*(2–3), pp. 313–344.

Easterly, W. and Rebelo, S. (1993, December) Fiscal policy and economic growth: An empirical investigation. *Journal of Monetary Economics, 32*(3), pp. 417–458.

Edison, J.C. (2016, January) An analysis of project scenario of India. *The Masterbuilder*, p. 66. Retrieved from: https://www.masterbuilder.co.in/data/edata/Articles/January2016/66.pdf Accessed on 19 October 2017.

Esfahani, H.S. and Ramírez, M.T. (2003) Institutions, infrastructure, and economic growth. *Journal of Development Economics, 70*(2), pp. 443–477.

Folster, S. and Henrekso, M. (2001) Growth effect of government expenditure and taxation in rich countries. *European Economic Review, 45*(8), pp. 1501–1520.

Futagami, K., Morita, Y. and Shibata, A. (1993) Dynamic analysis of an endogenous growth model with public capital. *The Scandinavian Journal of Economics, 95*(4), pp. 607–625.

Gibson, J. and Rozelle, S. (2003) Poverty and access to roads in Papua New Guinea. *Economic Development and Cultural Change, 52*(1), pp. 159–185.

GOI. (2007a) Report of the steering committee on urban development. *Eleventh Five-Year Plan*, Planning Commission, Government of India, p. 2. Retrieved from: https://niti.gov.in/planningco mmission.gov.in/docs/aboutus/committee/strgrp11/str11_hud1.pdf Accessed on 14 July 2016.

GOI. (2007b) Report of the steering committee on urban development. *Eleventh Five-Year Plan*, Planning Commission, Government of India, p. 26. Retrieved from: https://niti.gov.in/planningco mmission.gov.in/docs/aboutus/committee/strgrp11/str11_hud1.pdf Accessed on 14 July 2016.

GOI. (2007c) *Eleventh Five-Year Plan* (Chapter 8), Planning Commission, Government of India, pp. 239–240.

GOI. (2013) *Twelfth Five-Year Plan*, vol. 2, pp. 329–331.

GOI. (2013) *Twelfth Five-Year Plan*, vol. 2, Planning Commission, Government of India, p. 321. Retrieved from: https://mofpi.nic.in/sites/default/files/vol_2.pdf.pdf Accessed on 15 July 2016.

GOI. (2015) *Expenditure Budget*, vol. 1, 2015–2016, Government of India, p. 40.

GOI. (2016) *Expenditure Budget*, vol. 1, 2016–2017, Government of India, p. 43.

Government of India. (2002) *Tenth Five-Year Plan: 2002–07* (Chapter 8). Retrieved from: http://pla nningcommission.nic.in/plans/planrel/fiveyr/10th/volume2/v2_ch7_7.pdf Accessed on 19 October 2019.

Government of India. (2012) Economic survey 2016–17, Ministry of Finance, New Delhi, p. 241. Retrieved from: https://www.indiabudget.gov.in/budget2012-2013/es2011-12/echap-10.pdf.

Government of India. (2018) Industry and infrastructure (Chapter 8). In: *Economic Survey 2017–18, 2*, pp. 120–150.Economic Division, Department of Economic Affairs, Ministry of Finance, New Delhi.

Gu, W. and Macdonald, R. (2009) The impact of public infrastructure on Canadian multifactor productivity estimates. *The Canadian Productivity Review*. Research Paper (21). Catalogue No. 15-206-X, No. 021.

ILO. (2001) *The Construction Industry in the Twenty-First Century: Its Image, Employment Prospects and Skill Requirements*, p. 19. ILO, Geneva. Retrieved from: http://www.ilo.org/public/english/standards /relm/gb/docs/gb283/pdf/tmcitr.pdf Accessed on 9 July 2016.

India Brand Equity Foundation. (2008) Infrastructure sector in India. Retrieved from: https://www.ibe f.org/archives/detail/b3ZlcnZpZXcmMzc2NjMmMTA5 Accessed on 18 June 2017.

India Brand Equity Foundation. (2020) Retrieved from: https://www.ibef.org/archives/industry/indian -real-estate-industry-analysis-reports/indian-real-estate-industry-analysis-january-2020 Accessed on 07 June 2020.

Jacoby, Hanan G. (2000) Access to markets and the benefits of rural roads. *The Economic Journal, 110*(July), pp. 713–737.

Jalan, J. and Ravallion, M. (2003) Does piped water reduce diarrhoea for children in rural India? *Journal of Econometrics, 112*(1), pp. 153–173.

Jimenez, Emmanuel. (1995) Human and physical infrastructure: Public investment and pricing policies in developing countries. In: Hollis Chenery and T.N. Srinivasan (Eds), *Handbook of Development Economics*, 1st edition, Vol. 3, Chapter 43, pp. 2773–2843.Elsevier, Amsterdam.

Kaiser, Ronald W. (1997) The long cycle in real estate. *Journal of Real Estate Research, 14*(3), pp. 233–257. Retrieved from: https://core.ac.uk/download/pdf/7162449.pdf Accessed on 14 July 2016.

Lall, S.V. (1999) The role of public infrastructure investment in regional development: Experience of Indian state. *Economic and Political Weekly, 24*(12), pp. 717–725.

Laudau, D. (1983) Government expenditure and economic growth: A cross country study. *Southern Economic Journal, 49*(3), pp. 783–792.

Lipton, M. and Ravallion, M. (1995) Poverty and policy. In: Behrman, J., Srinivasan, T.N. (Eds), *Handbook of Development Economics*, Vol. 3, Chapter 41, pp. 2551–2657.North-Holland, Amsterdam.

Lokshin, M. and Yemtsov, R. (2005) Has rural infrastructure rehabilitation in Georgia helped the poor? *The World Bank Economic Review, 19*(2), pp. 311–333.

Macdonald, R. (2008) An examination of public capital's role in production. *Economic Analysis Research Paper Series. No. 50*, Statistics Canada, Ottawa.

Mamatzakis, E.C. (2008) Economic performance and public infrastructure: An application to Greek manufacturing. *Bulletin of Economic Research, 60*(3), pp. 307–326.

Martinkus, B. and Lukasevicius, K. (2008) Investment environment of Lithuanian resorts: Researching national and local factors in the Palanga case. *Transformations in Business and Economics, 7*(2), pp. 67–83.

Mattoon, R.H. (2004) Infrastructure and state economic development: A survey of the issues (I-G) Economic Conference. Ottawa, Canada.

Mehta, Rashi. (2007) A study on the Indian real estate market for investment: A qualitative approach (Doctoral dissertation), The University of Nottingham, Nottingham, England. Retrieved from: https://dokumen.tips/documents/07marashimehtapdf.html Accessed on 14 July 2016.

Murphy, Kevin M., Shleifer, Andrei and Vishny, Robert W. (1989) Industrialization and the big push. *Journal of Political Economy, 97*(5), pp. 1003–1026.

National Housing Bank. (2018). Retrieved from: https://nhb.org.in/wp-content/uploads/2019/03/ NHB-T&P-2018-Eng.pdf Accessed on 01 June 2020.

National Skill Development Corporation. (2009) Talent projections and skills gap analysis for the infrastructure sector (2022). Prepared by AON Hewitt for NSDC. New Delhi. p. 7.

Nijkamp, P. (1986) Infrastructure and Regional development: A multidimensional policy analysis. *Empirical Economics, 11*(1), pp. 1–21.

Ofori, G. (1990) *The Construction Industry: Aspects of Its Economics and Management*. Singapore University Press, Singapore.

Osei, Victor (2013) The construction industry and its linkages to the Ghanaian economy: Polices to improve the sector's performance. *International Journal of Development and Economic Sustainability, 1*(1), pp. 56–72. Retrieved from: https://www.eajournals.org/wp-content/uploads/THE-CONST RUCTION-INDUSTRY-AND-ITS-LINKAGES-TO-THE-GHANAIAN.pdf Accessed on 12 July 2016.

Palei, T. (2015) Assessing the impact of infrastructure on economic growth and global competitiveness. *Procedia Economics and Finance, 23*, pp. 168–175.

Park, S.H. (1989) Linkages between industry and services and their implications for urban employment generation in developing countries. *Journal of Development Economics*, *30*(2), pp. 359–379.

RBI Bulletin. (January 2018). Retrieved from: https://rbidocs.rbi.org.in/rdocs/Bulletin/PDFs/AFF ORDABLE609D506CB8C247DAB526C40DAF461881.PDF.

Ranjan, K.D. and Sharma, C. (2008) Government expenditure and economic growth: Evidence from India. *The ICFAI University Journal of Public Finance*, *6*(3), pp. 60–69.

Sanchez-Robles, B. (1998) The role of infrastructure investment in development: Some macroeconomic considerations. *International Journal of Transport Economics*, *25*(2), pp. 113–136.

Snieska, V. and Draksaite, A. (2007) The role of knowledge process outsourcing in creating national competitiveness in global economy. *Inzinerine Ekonomimka-Engineering Economics*, *3*(53), pp. 35–41.

Turin, D.A. (1973). The Construction Industry: Its Economic Significance and Its Role in Development. London: University College Environmental Research Group (UCERG).

Usman, A., Mobolaji, H.I., Kilishi, A.A., Yaru, M.A. and Yakubu, T.A. (2011) Public expenditure and economic growth in Nigeria. *Asian Economic and Financial Review*, *1*(3), pp. 104–113.

Van de Walle, D. (1996) Infrastructure and poverty in Vietnam. LSMS Working Paper No. 121, The World Bank, Washington, DC.

Vishwakarma, Vijay Kumar (2013, January/February) Is there a periodically collapsing bubble in the Indian real estate market? *The Journal of Applied Business Research*, *29*(1), pp. 167–172.

Wells, J. (1986) *The Construction Industry in Developing Countries: Alternative Strategies for Development.* Croom Helm, London.

World Bank. (1984) *The Construction Industry: Issues and Strategies in Developing Countries.* World Bank, Washington, DC.

World Bank. (2012) Transformation through infrastructure, World Bank, Washington, DC, p. 4. Retrieved from https://openknowledge.worldbank.org/handle/10986/26768.

Zugravu, Bogdan-Gabriel and Sava, Anca-Ştefania (2014) Patterns in the composition of public expenditures in CEE countries. *Procedia Economics and Finance*, *15*, pp. 1047–1054.

2 Construction industry in India

2.1 Development of physical infrastructure

The economic development of a country depends to a great extent on its physical infrastructure and development. Improvement in the physical infrastructure is crucial for broadbased and comprehensive growth of an economy. The quality, efficiency and productivity of the physical infrastructure influence the quality of life, health and happiness of society. The physical infrastructure comprises services such as electricity, railways, roads, ports, airports, irrigation, and urban and rural water supply and sanitation. It promotes the economic growth and development of an economy and is directly concerned with the needs of such production sectors as agriculture, industry, services, etc.

The development of a country's physical infrastructure and, consequently, its construction sector has been in focus from the beginning of the twenty-first century. It is well established that the influence of the construction industry spans several subsectors of the economy as well as infrastructure development, such as industrial and mining infrastructure, highways, roads, ports, railways, airports, power systems, irrigation and agriculture systems, telecommunication systems, hospitals, schools, townships, offices, houses and other buildings; urban infrastructure, including water supply, sewerage and drainage; and rural infrastructure. Consequently, it is an essential input in the socioeconomic development of a country.

The importance of construction activity in infrastructure, housing and other asset-building activities derives from the fact that the construction component comprises nearly 60 per cent to 80 per cent of the project cost of some infrastructure projects, for instance, roads and housing. Although the construction component of projects such as power plants and industrial plants is lower, it still remains critical. The construction industry also has linkages with the building and construction material manufacturing industry such as cement, steel, bricks, tiles, sand and aggregates, fixtures and fittings, paints, construction chemicals, construction equipment, petroleum products, timber, mineral products, aluminium, glass, plastics, tiles, plumbing and electrical materials. Construction materials generally account for nearly two-thirds of normal construction costs. On the basis of an analysis of the forward and backward linkages of construction, the multiplier effect of construction on the economy is estimated to be significant. The contribution of infrastructure development and the real estate sector to gross domestic product (GDP) is around 29.5 per cent. As per data of National Statistical Office, Govt. of India, Gross Value Added (GVA at basic prices) during 2015–16, 2016–17, 2017–18 and 2018–19 are Rs. 10491870 crores, Rs. 11318972 crores, Rs. 12104165 crores and Rs. 12906936 crores respectively. According to the Centre for Monitoring Indian Economy (CMIE) data, total income and sales of construction industry in India in 2017–18 was Rs. 5,553,587 million and Rs. 5,202,229 million respectively.

Table 2.1 Macro-economic aggregates (base year: 2011–12 at constant prices) (in Rs. crore)

Item/year	2012–13	2013–14	2014–15	2015–16	2016–17	2017–18	2018–19
GVA at basic prices	8,546,275	9,063,649	9,712,133	10,491,870	11,318,972	12,104,165	12,906,936
Net taxes on products	666,741	737,721	815,541	877,623	979,355	1,075,693	1,170,650
Gross domestic product	9,213,017	9,801,370	10,527,674	11,369,493	12,298,327	13,179,857	14,077,586
Consumption of fixed capital	1,010,661	1,100,610	1,178,644	1,270,890	1,380,954	1,502,962	1,602,641
Net domestic product	8,202,356	8,700,760	9,349,029	10,098,603	10,917,373	11,676,896	12,474,945
Primary income receivable from ROW (net)	–108,354	–122,343	–124,687	–134,922	–144,573	–145,736	–145,299
Gross national income	9,104,662	9,679,027	10,402,987	11,234,571	12,153,754	13,034,121	13,932,287
Net national income	8,094,001	8,578,417	9,224,343	9,963,681	10,772,800	11,531,159	12,329,646
Gross capital formation	3,639,296	3,448,236	3,659,763	3,917,358	4,146,020	4,679,689	–
Net capital formation	2,628,635	2,347,626	2,481,119	2,646,468	2,765,066	3,176,728	–
Per capita GDP (Rs.)	74,599	78,348	83,091	88,616	94,675	100,151	105,688
Per capita GNI (Rs.)	73,722	77,370	82,107	87,565	93,562	99,043	104,597
Per capita NNI (Rs.)	65,538	68,572	72,805	77,659	82,931	87,623	92,565
Per capita PFCE (Rs.)	41,936	44,423	46,667	49,738	53,149	56,364	60,185

Source: National Statistical Office (NSO). (2019, September) Handbook of statistics on Indian economy. Retrieved from: https://www.rbi.org.in/Scripts/AnnualPublications.aspx?head=Handbook%20of%20Statistics %20on%20Indian%20Economy Accessed on 7 October 2019.

Notes
[a] Data for 2015–16 are third revised estimates, for 2016–17 are second revised estimates and for 2017–18 are first revised estimates.
[b] Data for 2018–19 are provisional estimates.

The construction industry has the ability to register higher growth since India has sound fundamentals (refer to the macroeconomic aggregates data presented in Tables 2.1 and 2.2) and a physical infrastructure deficit.

2.2 Critical role of construction industry in economic development

The construction industry is an integral part of a country's infrastructure and economic development and an essential contributor to the development of an economy. Construction is the second largest economic activity in India after agriculture. It is one of the oldest industries and provides infrastructure to all other industries. It renders significant contributions

Table 2.2 Components of gross domestic product (base year: 2011–12) at constant prices (in Rs. crore)

Items/year	2012–13	2013–14	2014–15	2015–16	2016–17	2017–18	2018–19
Private final consumption expenditure	5,179,091	5,557,329	5,912,657	6,381,419	6,904,085	7,417,489	8,016,674
Government final consumption expenditure	974,263	979,825	1,054,151	1,132,802	1,199,041	1,378,563	1,506,035
Gross fixed capital formation	3,145,793	3,194,924	3,278,096	3,492,183	3,783,778	4,136,572	4,548,452
Changes in stocks	201,528	129,758	274,751	239,557	124,087	150,417	157,637
Valuables	259,949	148,879	187,957	185,986	150,784	192,120	174,780
Exports of goods and services	2,289,836	2,468,269	2,512,145	2,370,282	2,490,437	2,607,310	2,933,969
Import of goods and services	2,879,079	2,644,555	2,667,595	2,511,540	2,621,586	3,083,560	3,557,901
Discrepancies	41,635	(33,060)	(24,487)	78,804	267,700	380,947	297,939
Gross domestic product	9,213,017	9,801,370	10,527,674	11,369,493	12,298,327	13,179,857	14,077,586
(Base year: 2011–12) current prices							
Private final consumption expenditure	5,614,484	6,475,649	7,247,340	8,126,408	9,115,769	10,083,121	11,290,029
Government final consumption expenditure	1,062,404	1,156,509	1,301,762	1,436,171	1,583,312	1,885,613	2,134,615
Gross fixed capital formation	3,324,973	3,515,621	3,750,392	3,957,092	4,335,014	4,896,813	5,569,998
Changes in stocks	214,524	144,621	312,698	262,477	139,714	173,890	187,671
Valuables	273,775	161,761	209,407	203,506	166,559	218,706	193,992
Exports of goods and services	2,439,707	2,856,781	2,863,636	2,728,647	2,948,772	3,210,547	3,752,230
Import of goods and services	3,108,428	3,191,811	3,235,962	3,044,923	3,220,591	3,758,519	4,493,933
Discrepancies	122,574	114,389	18,687	102,495	293,838	384,835	375,562
Gross domestic product	9,944,013	11,233,522	12,467,959	13,771,874	15,362,386	17,095,005	19,010,164

Source: National Statistical Office (NSO). (2019, September) Handbook of statistics on Indian economy. Retrieved from: https://www.rbi.org.in/Scripts/AnnualPublications. aspx?head=Handbook%20of%20Statistics%20on%20Indian%20Economy Accessed on 7 October 2019.
As per data of National Statistical Office, Govt. of India, Gross Value Added (GVA at basic prices) during 2015–16, 2016–17, 2017–18 and 2018–19 are Rs. 10491870 crores, Rs. 11318972 crores, Rs. 12104165 crores and Rs. 12906936 crores respectively. According to the Centre for Monitoring Indian Economy (CMIE) data, total income and sales of construction industry in India in 2017–18 was Rs. 5,553,587 million and Rs. 5,202,229 million respectively.

Notes
[a] Data for 2015–16 are third revised estimates, for 2016–17 are second revised estimates and for 2017–18 are first revised estimates.
[b] Data for 2018–19 are provisional estimates.

to the economy by enhancing GDP, income and employment opportunities. It creates the physical foundations for economic development and improves the standard of living of a society.

Determining the demand for goods and services produced by the construction industry is a complicated process because it involves a number of factors such as political, economic, social and technological, legal/legislative based, procurement types and geographical location, cost, size, longevity and nature of investment, and partly because of the broad array of what constitutes construction activity. Similarly, factors such as interest rates, shocks to the economy, demand for goods, surplus manufacturing capacity, ability to remodel, monetary policy, fiscal policy, future expectation of continued increased demand for manufacturing goods, expectation of increased profits on the activities of those that demand construction, new technology, availability of credit, existing stock of constructed facilities, economic conditions/business cycle, exchange rates, disposable income of households, rates of interest, supply of money, people's tastes and preferences for housing, entertainment, planning regulations, political climate, size of population, structure and geographical distribution, prices (tender prices, property prices and import prices) output, inflation, taxation and weather conditions also determine the demand for goods and services produced by the construction industry as per research studies (Bickerton and Gruneberg, 2013; Hillebrandt, 2000; Briscoe, 1992). Some studies (Akintoye and Skitmore, 1994; Gruneberg, 1997) argue that the products of the construction industry are regarded as investment goods and state that construction investment can be seen as a derived demand which is growth dependent.

The output of the construction industry becomes the assets of a nation. Investment in the construction industry activates many other industries and it has the maximum linkage effect. In fact, construction has been ranked among the top 4 out of 20 economic sectors in terms of inter-sectoral linkages. With its backward and forward linkages, the construction industry has generated employment for 33 million people in the country. Around 16 per cent of the nation's working population depends on construction for its livelihood. Construction has accounted for about 35 per cent to 40 per cent of development investment in India during the last five decades. It has contributed 78 per cent of the gross capital formation. Today, India is one of the fastest-growing economies in the world and Asia's biggest infrastructure market after China.

2.2.1 Regulator of the economy

The construction industry is an essential contributor to the process of development. Roads, dams, irrigation works, schools, houses, hospitals, factories and other construction works are the physical foundations on which development efforts and improved living standards are established. Governments use investments in construction to regulate the economy as well as to introduce desired changes to it. Construction activity is perhaps the first activity to be given a boost by fiscal stimulus packages during a recession. Fiscal stimulus packages help an economy to overcome a liquidity crunch, a serious risk to the economic outlook during a recessionary period, and eventually the economy will be able to stimulate production, employment, income and demand.

2.2.2 Size

The size of the construction industry is important to any nation as change in its output affects the size of the national product, both directly and indirectly. As the provider of nearly half the country's fixed investment, if the output of the industry is down, gross investment in the economy goes down, and investment is of vital concern for the growth and development of the economy.

India is expected to become the third largest construction market in the world by 2022. According to India Brand Equity Foundation's Indian real estate industry report (April 2020), by 2040, Indian real estate market will to grow to Rs. 65,000 crore (USD 9.30 billion) from Rs. 12,000 crore (USD 1.72 billion) in 2019. Real estate sector in India is expected to reach a market size of USD 1 trillion by 2030 from USD 120 billion in 2017 and contribute 13 per cent of the country's GDP by 2025. Retail, hospitality and commercial real estate are also growing significantly, providing the much needed infrastructure for India's growing needs. Indian real estate increased by 19.5 per cent CAGR from 2017 to 2028.[1]

2.2.3 Construction sector and gross value added

A close examination of the dynamics of this industry clearly highlights its critical role in economic development. The construction sector contributed about Rs. 10.54 trillion to the country's gross value added in the financial year 2018–19, which is 8.17 per cent of the total GVA. Over the past 5 years, construction as a percentage of GVA has decreased from 8.6 per cent in 2014–15 to 8.17 per cent in 2018–19 (Table 2.3). The growth of GVA from construction has increased from 6.1 per cent in 2016–17 to 8.7 per cent in 2018–19 (Table 2.4). Over the past 5 years, the value of construction output has increased from Rs. 23.39 trillion in 2014–15 to Rs. 26.23 trillion in 2017–18 (Table 2.5). Over the past 5 years, gross domestic capital formation as a percentage of GDP at market price has increased from Rs. 32.78 trillion in 2003–04 to Rs. 45.48 trillion in 2018–19 (Table 2.6). The multiplier factor between the growth rates in construction and GDP appears to be about 1.5 to 1.62. Twenty per cent of gross global products, i.e. US$12 trillion per year, is spent on fixed capital projects worldwide.

2.3 Characteristics of the construction industry

The product of the construction industry is usually large, expensive and made to meet the requirements of each individual customer. A large number of the product's components are produced by other industries. Working conditions in the construction industry are dramatically different from those in most factories. In a factory, work activities are highly repetitive. The nature of the work and the layout of the workplace remain unchanged over long periods of time. The environment inside a factory is generally controlled and steady.

The products of the construction sector differ widely in terms of size, appearance, location and end use. They also show wide variations in terms of the materials and techniques used in their production, and their standards in terms of space, quality and durability as well as aesthetic considerations of the finished products. The heterogeneity of the products and particularly the great variety of end uses to which they are put are also frequently held to account for the fact that construction is different from other economic sectors. Therefore, there is no such thing as a standard construction product. Each project is a unique project. The completed products are generally not mobile but fixed for all time in the location where they are constructed. Construction products are site specific and cannot be produced in advance but rather have to be sold before they are produced or made to order. Generally, a vast majority of firms continue to operate in a product market where they have no control over demand, technology, materials, workplace, finance and labour supply. Consequently,

Table 2.3 Gross value added (GVA) by economic activity at constant basic prices (base year 2011–12) (Rs. trillion)

Year	Total GVA (in Rs. trillion)	Agriculture, forestry and fishing		Industry		Services		Construction	
		GVA	% share in total	GVA	% share in total	GVA	% share in total	GVA	% share in total
2001–02	42.42	11.3	26.64	12.32	29.05	19.31	45.52	2.91	6.85
2002–03	44.06	10.55	23.95	13.22	30	20.53	46.6	3.15	7.14
2003–04	47.57	11.51	24.19	14.15	29.75	22.04	46.34	3.54	7.44
2004–05	50.93	11.53	22.64	15.53	30.5	23.86	46.86	4.12	8.08
2005–06	55.14	12.08	21.91	17.02	30.86	26.04	47.22	4.65	8.43
2006–07	59.58	12.44	20.87	19.27	32.35	27.87	46.78	5.14	8.63
2007–08	63.98	13.12	20.51	20.82	32.54	30.04	46.95	5.74	8.98
2008–09	66.74	13.09	19.61	21.65	32.44	32	47.94	6.06	9.09
2009–10	71.32	12.98	18.19	23.57	33.04	34.78	48.76	6.48	9.08
2010–11	77.05	14.12	18.32	25.43	33	37.5	48.67	6.87	8.92
2011–12	81.07	15.02	18.53	26.35	32.5	39.7	48.97	7.77	9.59
2012–13	85.46	15.24	17.84	27.21	31.84	43.01	50.32	7.80	9.13
2013–14	90.64	16.09	17.75	28.24	31.16	46.3	51.09	8.01	8.83
2014–15	97.12	16.06	16.53	30.22	31.11	50.85	52.35	8.35	8.6
2015–16	104.92	16.16	15.4	33.11	31.56	55.64	53.04	8.65	8.25
2016–17	113.19	17.17	15.17	35.67	31.52	60.34	53.31	9.18	8.11
2017–18	121.04	18.03	14.9	37.8	31.22	65.22	53.88	9.69	8.01
2018–19	129.07	18.56	14.38	40.39	31.29	70.12	54.33	10.54	8.17

Source: Economic Outlook, CMIE. Retrieved from: https://economicoutlook.cmie.com/kommon/bin/sr.php?kall=wshreport&&tabcode=001001008005000000&repnum=75513&frequency=A&colno=2&parnum=75513&parfrq=A Accessed on 13 September 2019.

Table 2.4 Growth rates and composition of gross value added (GVA) at basic prices

Sector	Growth % share in GVA		
	2016–17	*2017–18*	*2018–19*
Agriculture, forestry and fishing	6.3	5	2.9
	15.2	14.9	14.4
Industry	8.3	6.1	6.2
	23.4	23.2	23.1
Services	8.1	7.8	7.7
	61.4	61.9	62.5
Construction	6.1	5.6	8.7
	8.1	8	8.2

Source: Reserve Bank of India, Annual Report 2018–19. Retrieved from: https://www.rbi.org.in/Scripts/AnnualReportPublications.aspx?Id=1270 Accessed on September 13, 2019.

their concern is to finish the job in hand and ignore other issues in the industry (Wells, 1986). Some of the major characteristics of the construction sector are listed below.

2.3.1 Output

The construction industry, excluding real estate developers, does not sell a tangible product; it sells a service. The services provided by the construction industry are determined

Table 2.5 Value of output from construction at constant prices (base year 2011–12) (Rs. trillion)

Year	Total	Dwellings, other buildings and structures						Plantation	Mineral exploration
		Total	Dwellings	Other buildings and structures					
				Total	Non-residential buildings	Roads and bridges	Other structures and land improvements		
2011–12	21.41	21.09	7.34	13.75	7.17	1.15	5.43	0.03	0.29
2012–13	21.58	21.23	6.18	15.05	8.66	1.26	5.13	0.02	0.34
2013–14	22.11	21.64	6.66	14.98	8.36	1.34	5.27	0.01	0.45
2014–15	23.39	23.1	6.22	16.88	9.54	1.44	5.9	0.01	0.28
2015–16	23.77	23.14	4.67	18.47	9.96	1.79	6.72	0.02	0.64
2016–17	25.05	23.9	5.03	18.87	9.61	1.9	7.37	0.02	0.4
2017–18	26.23	–	–	–	–	–	–	–	–

Source: Economic Outlook, CMIE. Retrieved from: https://economicoutlook.cmie.com/kommon/bin/sr.php ?kall=wshreport&&tabcode=001001008009000000&repnum=99699&frequency=A&colno=2&parnum=99699 &parfrq=A Accessed on 13 September 2019.

by clients and are performed by contractors at a time and place specified by the client. Contractors only have control over the demand for construction services.

2.3.2 Temporary nature of employment

Work is performed at the worksite. On a construction site, the various construction activities are of relatively short duration and, after a few days or weeks, a construction operation may be taking place in a different way at a different location under different climatic conditions. Furthermore, although construction work can be repetitive, cycle times are longer and days may elapse before any repetition occurs. Everyone moves on once the task is completed.

2.3.3 Employees work in groups

Due to the unskilled nature of the work, the piece-rate system of payment and the absence of protective legislation, workers encourage their womenfolk and children to help increase output. They consider their work as a family assignment.

2.3.4 Pricing

Contractors do not set rates for their service as rate setting is done by clients.

2.3.5 Financing

The financing of construction services is outside the control of contractors. It is done by the client who commissions the service and, until recently, the mode and periodicity of payments were decided unilaterally by him (negotiated contract terms are of recent origin).

2.3.6 Registration

To secure work, a company has to be on the approved list of contractors maintained by the government.

Table 2.6 Gross fixed capital formation (GFCF) at constant prices by economic activity (base year 2011–12) (Rs. trillion)

Year	Total GFCF	Industry		Mining and quarrying		Manufacturing		Electricity, gas, water supply and other utility services		Construction		Services		Real estate, ownership of dwelling and professional services	
		GFCF	%	GFCF	%	GFCF	%	GFCF	%	GFCF	%	GFCF	%	GFCF	%
2001–02	12.08	3.12	25.79	0.12	0.98	1.58	13.12	0.92	7.62	0.49	4.07	7.57	62.63	4.87	40.34
2002–03	11.99	3.19	26.58	0.11	0.92	1.69	14.09	0.87	7.29	0.51	4.29	7.54	62.9	4.78	39.82
2003–04	12.61	4	31.73	0.17	1.32	2.12	16.84	1.11	8.81	0.6	4.76	7.41	58.76	4.45	35.26
2004–05	14.05	5.03	35.83	0.32	2.26	2.51	17.84	1.41	10.06	0.8	5.67	7.76	55.24	3.8	27.02
2005–06	16.36	5.86	35.83	0.37	2.26	2.92	17.84	1.65	10.06	0.93	5.67	9.04	55.24	4.42	27.02
2006–07	18.63	6.68	35.83	0.42	2.26	3.32	17.84	1.88	10.06	1.06	5.67	10.29	55.24	5.03	27.02
2007–08	21.67	7.77	35.83	0.49	2.26	3.87	17.84	2.18	10.06	1.23	5.67	11.97	55.24	5.86	27.02
2008–09	22.37	8.01	35.83	0.51	2.26	3.99	17.84	2.25	10.06	1.27	5.67	12.35	55.24	6.04	27.02
2009–10	24.08	8.63	35.83	0.54	2.26	4.3	17.84	2.42	10.06	1.37	5.67	13.3	55.24	6.51	27.02
2010–11	26.74	9.58	35.83	0.6	2.26	4.77	17.84	2.69	10.06	1.52	5.67	14.77	55.24	7.23	27.02
2011–12	29.98	10.74	35.83	0.68	2.26	5.35	17.84	3.02	10.06	1.7	5.67	16.56	55.24	8.1	27.02
2012–13	31.46	11.3	35.91	0.73	2.33	5.47	17.38	2.98	9.47	2.12	6.73	17.72	56.34	7.55	24.01
2013–14	31.95	10.99	34.38	1.29	4.05	5.34	16.72	2.94	9.2	1.41	4.4	18.2	56.98	8.15	25.5
2014–15	32.78	10.61	32.36	0.57	1.74	5.75	17.53	2.98	9.08	1.32	4.02	19.53	59.58	9.2	28.07
2015–16	34.92	11.73	33.59	0.42	1.2	6.3	18.03	3.58	10.26	1.43	4.09	20.51	58.72	7.47	21.4
2016–17	37.84	13.19	34.85	0.54	1.42	7.3	19.3	3.52	9.3	1.83	4.83	22.23	58.74	8.33	22.03
2017–18	41.37	–	–	–	–	–	–	–	–	–	–	–	–	–	–
2018–19	45.48	–	–	–	–	–	–	–	–	–	–	–	–	–	–

Source: Economic outlook, CMIE. Retrieved from: https://economicoutlook.cmie.com/kommon/bin/sr.php?kall=wshreport&&tabcode=0010100800500000&repnum=98741 &frequency=A&colno=2&pamum=98741&patfrq=A Accessed on 13 September 2019.

2.3.7 Payment

After securing the 'contract', a company recruits labour and starts work. Usually, workers are paid advances and then wages at short intervals. Most workers are paid a piece-rate wage by measuring their output. Compensation for work is always in cash, and no one is entitled to perquisites or overhead benefits. But the firms in the construction industry get 'payment' for the work executed only at long intervals. This involves considerable financial outlay. As work is executed mostly either at the standard rates or at a small percentage variation, only a large-scale operation would yield sizeable profits.

2.4 Structure of construction industry

2.4.1 Fragmentation

The construction industry in India is highly fragmented. There are numerous unorganised players in the Indian construction industry that work on a subcontracting basis. The Indian construction industry comprises about 4000 listed firms. Of these, the percentage of firms earning more than Rs. 50 billion in revenue is less than half a per cent and more than 79 per cent of firms earn less than Rs. 100 million in revenue (Table 2.7). In addition to listed firms, lakh contractors are registered with various government construction bodies. A number of small firms compete for small jobs/work as subcontractors of prime or other contractors. To execute more critical projects, nowadays bids are increasingly placed by consortiums because different segments of such a project may require different types of expertise. Generally, there is a higher return on these types of projects. However, the profitability of projects in the construction sector varies across different segments. Complex technology-oriented projects can make larger profit margins for construction firms as compared to low technology projects, viz., road construction. A large number of small firms lead to lack of organisation in the industry and they also have a low level of management quality. Additionally, small firms dominate the construction industry, which is not helpful to skill formation and technical upgrading in the construction industry.

2.4.2 Skill gap

In construction, work is often seasonal, employment is casual and all employment relationships are contractual. Often, labour is migratory and labourers move from one site to

Table 2.7 Distribution of listed firms as per revenue (based on 2017–18 revenue)

Sr. no.	Rs. million	Companies	% of total
1.	50,000 and above	17	0.43
2.	20,000–49,999	33	0.83
3.	10,000–19,999	63	1.58
4.	5,000–9,999	64	1.6
5.	2,000–4,999	141	3.53
6.	1,000–1,999	137	3.43
	500–999	119	2.98
	100–499	248	6.21
	Less than 100	3,170	79.41
	Total	3,992	0.43

Source: Compiled from CMIE data.

another. As many construction activities are labour-intensive, construction firms have been focusing on mechanisation over the past decades. Consequently, growth in the number of labourers required declined from 1.6 per cent in the financial year 2004 to 0.9 per cent in the financial year 2008. Generally, construction projects are working capital–intensive in nature.[2] There are around 1.2 million job openings yearly in project-oriented businesses. Studies indicate that average yearly job openings are over 3.2 lakh in India.

There are over 30 million construction workers in the country classified as unskilled, semi-skilled and skilled, constituting masons, carpenters, bar benders, plumbers, electricians, tile layers, glass fitters, metal fabricators, concrete workforce, etc. In addition to those directly involved in the construction process, the industry also accounts for a large proportion of secondary employment created due to forward and backward linkages with ancillary industries such as the urban infrastructure sector, construction materials industries and real estate development. The distribution of different occupational categories in the construction sector is presented in Table 2.8. The data reveal that there are less technically qualified people in the Indian construction sector.

The World Bank study on 'India's Road Construction Industry: Capacity Issues, Constraints and Recommendations' states that the supply of skilled human resources required by the construction industry will fall short by 55 per cent to 64 per cent over the next 8 years. In order to meet this demand, the number of civil engineering graduates and diploma holders would have to increase by at least a factor of 3. The industry has not kept pace with this growth, as evidenced by the under-utilisation of funds allocated to road projects and perennial time and cost overruns on national and state highway projects. As per estimates, India today has around 110,000 highway engineers, in contrast to China which, in 1989–97, was supported by over 500,000 trained highway engineers. The paper by Laskar and Murty (2004) states that 'every Rs.1 crore investment on construction project generates employments of 22,000 unskilled man-days, 23,000 skilled or semiskilled man-days and 9,000 managerial and technical man-days approximately'. The Indian construction sector's requirements for construction manpower, materials and equipment are given in Table 2.11.

2.4.3 Employment size

The unorganised manufacturing sector constitutes up to 32 per cent of total unorganised sector units and approximately 37 per cent of total unorganised sector employment. It remains to be explored what the reform process has done for the informal non-manufacturing sector units and workers, particularly across the sectors of construction; trade and repair services;

Table 2.8 Skill categories

Occupation	Numbers (in thousands)	%	Numbers (in thousands)	%
	1995		2005	
Engineers	687	4.71	822	2.65
Technicians, foremen, etc.	359	2.46	573	1.85
Clerical	646	4.4	738	2.38
Skilled workers	2,241	15.34	3,267	10.57
Unskilled workers	10,670	73.08	25,600	82.45
Total	14,600	100	31,000	100

Source: Report of the working group on construction for the Eleventh Five-Year Plan, p. 35.

Table 2.9 Distribution of listed firms as per no. of employees (no. based on 2008–09)

Sr. no.	Rs. crores	No. of firms	%	No. of employees
1.	2,000 and above	14	1.08	75,863
2.	1,000–1,999	16	1.23	14,062
3.	500–999	27	2.08	11,122
4.	100–499	50	3.9	2,155
5.	50–99	32	2.49	1,244
6.	Less than 50	1,159	89.22	515
	Total	1,299	100	104,961

Source: Compiled from CMIE data.

transport, storage and communication services; and community and social services. These sectors constitute approximately 60 per cent of total informal enterprises and approximately 52 per cent of total informal employment (Karan and Selvaraj, 2008).

Labour productivity in the construction sector has declined over the years as reflected in its negative growth rate, but there has been a noticeable increase in real wages in construction, particularly for women workers.

Construction work is basically a contract job. Due to the nature of the industry, tenure of employment is purely temporary and no one may view his job beyond the contract period. There is constant change in the workplace, both of employers and of workers. Most workers are paid a piece-rate wage by measuring their output. Compensation for work is always in cash, and no one is entitled to perquisites or overhead benefits. Table 2.9 shows that two-thirds of employees work for 1 per cent of companies.

2.5 Nature of investment

The construction industry is greatly dependent on investments in infrastructure and the industrial and real estate sector. Massive investment in the construction industry is being driven by the growing requirements of sectors such as transportation, power, urban infrastructure, housing and irrigation. Construction is a capital-intensive activity and represents a very long-term investment. Broadly, the services of the sector can be classified into infrastructure development (54 per cent), industrial activities (36 per cent), residential activities (5 per cent) and commercial activities (5 per cent). Investment in construction is the fixed capital formation of the nation.

Construction is a working capital–intensive business activity. The working capital requirement for any company depends on the order mix of the firm. The construction industry functions on the basis of contractual agreements, a contract[3] between the employer and the contractor(s). The Constitution of India provides for union and state governments to enter into contracts.[4] A wide range of contracts have been developed depending on the size and nature of the project. The type of contract largely depends on the scale and nature of the work, any special design requirements, the availability of funds and the complexities of the job. Construction projects can generally be executed through different parts or packages or the types of obligations and tasks involved can be outsourced to other parties. Consequently, subcontracting is a universal phenomenon in the construction industry.

The main actors in the construction industry are clients, consultants, architects, construction contractors, equipment suppliers, material suppliers, labour suppliers and solution providers. India's construction equipment industry is growing at an average rate of 13 per cent per annum driven by the huge infrastructure development taking place in the country. Architects provide detailed design and preparation of (i) contract drawings, (ii) schedules and (iii) specifications to facilitate tendering; consultants support the client in choosing the contractors to be invited to tender for the construction project; and construction contractors execute the project.

2.6 Opportunities in the construction industry in India

This section attempts to forecast the sales and requirements of resources over the next 5-year period (up to 2025) in the construction industry in India.

Projected investment, monetary requirements for construction materials, equipment and manpower and detailed requirements for cement, steel and manpower (engineers, technicians, support staff, skilled workers and unskilled/semi-skilled workers) are estimated. A linear regression model was developed from the sales data from 1996–97 to 2017–18 of the construction and real estate industry, electricity generation industry, crude oil and natural gas industry and irrigation, collected from the Centre for Monitoring Indian Economy (CMIE Prowess database) to predict investment.[5] The data have been forecasted up to 2024–25. The predicted total sales in the construction industry are presented in Table 2.10. Since the construction industry's sales can give a real representation of investment in construction, the study considers sales as investment in construction.

The requirement of resources in the construction sector was found using ratios from the report of the working group on construction for the Eleventh Five-Year Plan for the period 2019–20 to 2024–25 (refer to Table 2.11).

2.6.1 Forecasted sales of construction industry

Table 2.10 presents the predicted sales of the construction industry in India.

2.7 Present scenario of the construction industry

It is well recognised that the influence of the construction industry spans numerous subsectors of the economy as well as infrastructure development. The present economic slowdown is impacted new order inflows for industrial and infrastructural construction. According to CMIE data, the new investment proposals have dropped very sharply, especially in the transport services sector. Overall the Indian economy witnessed a 56 per cent fall in new investment proposals in 2019-20. Two current initiatives of the Government of India, i.e. Housing for All and Smart Cities Mission along with the capital and credit infusion policies of the government, will, to some extent, reduce the problem of habitat and infrastructure bottlenecks. Opportunities in the sector have increased with the introduction of these programmes. This section provides an understanding of project investment, financial performance, project scenario, new investment in industrial, infrastructural and real estate construction.

2.7.1 Project investment

The Centre for Monitoring Indian Economy reports that the share of private proposals grew to an all-time high of 67.5 per cent in 2018–19. However, there is a somewhat

Table 2.10 Predicted total sales of construction industry and prediction limits, 1996–97 to 2024–25 (sales in Rs. crore)

Row	Year	Sales	Predicted sales	Standard error	Lower control limit at 95% confidence level	Upper control limit at 95% confidence level
1	1997	51,314	−109,286.83	87,033.88	−290,836.33	72,262.67
2	1998	60,940	−57,406.53	86,189.7	−237,195.08	122,382.02
3	1999	68,476	−5,526.23	85,422.79	−183,715.05	172,662.59
4	2000	82,274	46,354.07	84,735.27	−130,400.61	223,108.75
5	2001	88,909	98,234.37	84,129.09	−77,255.83	273,724.57
6	2002	99,725	150,114.67	83,606	−24,284.39	324,513.73
7	2003	132,753	201,994.97	83,167.58	28,510.44	375,479.5
8	2004	162,478	253,875.27	82,815.17	81,125.85	426,624.69
9	2005	200,347	305,755.57	82,549.88	133,559.54	477,951.6
10	2006	232,157	357,635.87	82,372.54	185,809.75	529,461.98
11	2007	294,597	409,516.17	82,283.73	237,875.31	581,157.03
12	2008	392,989	461,396.47	82,283.73	289,755.61	633,037.33
13	2009	465,104	513,276.77	82,372.54	341,450.65	685,102.88
14	2010	532,958	565,157.07	82,549.88	392,961.04	737,353.1
15	2011	613,335	617,037.37	82,815.17	444,287.95	789,786.79
16	2012	711,588	668,917.67	83,167.58	495,433.14	842,402.2
17	2013	747,781	720,797.97	83,606	546,398.91	895,197.02
18	2014	835,252	772,678.27	84,129.09	597,188.07	948,168.46
19	2015	885,519	824,558.57	84,735.27	647,803.88	1,001,313.25
20	2016	953,835	876,438.87	85,422.79	698,250.05	1,054,627.69
21	2017	981,529	928,319.17	86,189.7	748,530.61	1,108,107.72
22	2018	986,179	980,199.47	87,033.88	798,649.96	1,161,748.97
23	2019	–	1,032,079.77	87,953.14	848,612.74	1,215,546.79
24	2020	–	1,083,960.07	88,945.12	898,423.8	1,269,496.33
25	2021	–	1,135,840.37	90,007.43	948,088.15	1,323,592.58
26	2022	–	1,187,720.67	91,137.61	997,610.93	1,377,830.4
27	2023	–	1,239,600.97	92,333.17	1,046,997.34	1,432,204.59
28	2024	–	1,291,481.27	93,591.61	1,096,252.6	1,486,709.93
29	2025	–	1,343,361.57	94,910.41	1,145,381.92	1,541,341.21

Notes
[a] Data up to 2017–18 is calculated from total sales of construction and real estate industry, electricity generation industry, crude oil and natural gas industry and irrigation collected from CMIE (Prowess database).
[b] Data in parentheses from 2018–19 to 2024–25 is forecasted.
[c] The prediction interval estimates the predicted value of Y for a single individual with this value of X.

bizarre twist to this record. One-fifth of the private sector's investment proposals made in 2018–19 are already stalled. There was a fall in new investment proposals during the first quarter of 2019–20 across both public and private sectors, and the fall was greater in private sector investment proposals. New investment proposals by the public sector were down to Rs. 184 billion during the quarter. This was just 16.5 per cent of the proposals made a year ago or 20 per cent of the average proposals per quarter made during the preceding four quarters. See Figures 2.1 to 2.4 for a clearer picture of the new investment scenario.

2.7.1.1 New investment in industrial and infrastructural construction

The Government of India's recent announcement of infrastructure worth Rs. 100 trillion over the next 5 years instantly boosted the inflow of new orders, but they may remain

Table 2.11 Requirement of resources in the construction sector for the period 2019–20 to 2024–25

Projected investment (in Rs. crore)	61,32,755 (lower 95%)	84,31,175 (upper 95%)	72,81,965 (average)
Monetary requirements			
For construction materials	Rs. 20,93,596 crores	Rs. 28,78,229 crores	Rs. 24,85,912 crores
For construction equipment	Rs. 7,61,307 crores	Rs. 10,46,629 crores	Rs. 9,03,968 crores
Manpower	Rs. 4,56,784 crores	Rs. 6,27,877 crores	Rs. 5,42,381 crores
Detailed requirement of materials			
Cement	1,611 million tons	2,215 million tons	1,913 million tons
Steel	634 million tons	872 million tons	753 million tons
Manpower	389 million man years	535 million man years	462 million man years
Engineers	15.73 million man years	21.63 million man years	18.68 million man years
Technicians	18.27 million man years	25.12 million man years	21.70 million man years
Support staff	15.44 million man years	21.22 million man years	18.33 million man years
Skilled workers	98.76 million man years	135.77 million man years	117.26 million man years
Unskilled/semi-skilled workers	240.91 million man years	331.20 million man years	286.06 million man years

Source: Calculated from Table 2.10.

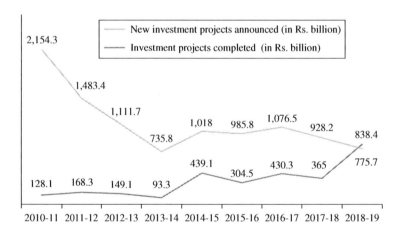

Figure 2.1 Projects announced and completed.

subdued in the coming quarters due to the existing economic slowdown. Recent industrial and infrastructural construction orders fell sharply in the June 2019 quarter. New order inflows stood at Rs. 139.1 billion in the June 2019 quarter as compared to Rs. 712.8 billion in the June 2018 quarter (Table 2.12). The combined orders won by industrial and infrastructural construction firms in 2017–18 stood at Rs. 2470.6 billion. Firms reported aggregate orders worth just Rs. 2180 billion in the subsequent year, i.e. 2018–19 (Table 2.13).

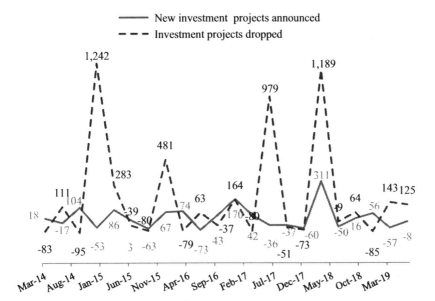

Figure 2.2 Investment projects in construction – Q-on-Q percentage change.

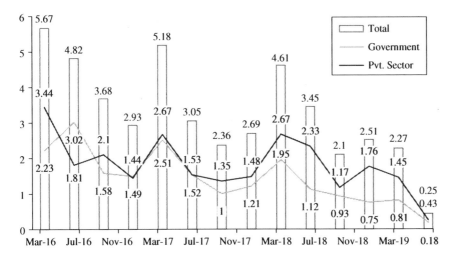

Figure 2.3 New investment announcements (Rs. trillion).

2.7.2 New investment in real estate industry

New investment proposals in the real estate industry caved in during the second quarter of 2019 (Table 2.14). Only 82 projects were announced in the June 2019 quarter, much lower than the 224 projects announced in the June 2018 quarter, which may be due to the liquidity crisis (paucity of non-bank financial companies [NBFC] funding), excess unsold inventory, the implementation of the real estate regulatory authority (RERA) and a fall in

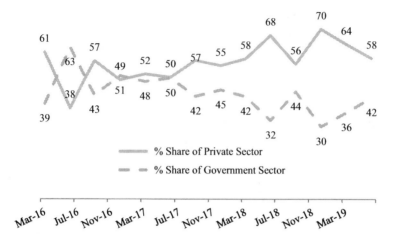

Figure 2.4 Investment announcements – percentage share of private sector and government.

Table 2.12 New quarterly order inflows

Quarter	New order inflows – quarterly (in Rs. billion)
March 2018	1,298.1
June 2018	712.8
September 2018	557.4
December 2018	521.3
March 2019	388.4
June 2019	139.1

Source: Industry Outlook, CMIE. Retrieved from: https://industryoutlook.cmie.com/kommon/bin/sr.php?kall=wshowov&nvdt=20190910162900140&nvpc=055000000000&nvtype=OVERVIEW&icode=0101060000000000 Accessed on 11 October 2019.

Table 2.13 New annual order inflows

Year	New order inflows – annual (in Rs. billion)
2013–14	1403.5
2014–15	1191.4
2015–16	1450.2
2016–17	1695.8
2017–18	2470.6
2018–19	2180

Source: Industry Outlook, CMIE. Retrieved from: https://industryoutlook.cmie.com/kommon/bin/sr.php?kall=wshowov&nvdt=20190910162900140&nvpc=055000000000&nvtype=OVERVIEW&icode=0101060000000000 Accessed on 11 October 2019.

Table 2.14 New quarterly order inflows

Quarter	New order inflows – quarterly (in Rs. billion)
March 2018	327,768.2
June 2018	164,729.6
September 2018	191,578.1
December 2018	298,837.2
March 2019	128,653.8
June 2019	118,727.4

Source: CMIE Industry Outlook. Retrieved from: https://industryoutlook.cmie.com/kommon/bin/sr.php?kall=wshreport&nvdt=20190819171833150&nvpc=055000000000&nvtype=ANALYSIS+%26+OUTLOOK&icode=0101060100000000 Accessed on 11 October 2019.

demand (CMIE). The subdued demand for housing is likely to continue in the real estate sector up to 2023.

2.7.3 *Quarterly financial performance of industrial and infrastructural construction*

As with any industry, financial factors are significantly related to the performance of firms in the construction industry too. The net sales of the industrial and infrastructural construction industry segment increased by 15.2 per cent during the quarter under review, i.e. June 2019 quarter (Table 2.15). According to CMIE, this was the fourth successive quarter wherein the industry witnessed double-digit growth in revenues. CMIE further states that this was due to the rapid execution of projects in hand by the construction firms. However, the operating expenses of the industry grew much slower than the increase in revenues, i.e. by 10 per cent, which resulted in a 16.5 per cent increase in the operating profits of the industry. On the other hand, as reported by CMIE, higher post-operating expenses together with a fall in other income resulted in a 3.1 per cent decline in the profits of the industry

Table 2.15 Quarterly financial performance: Industrial and infrastructural construction

Quarter	Total income	Net sales	Operating profit	Net profit	Operating profit margin (%)	Net profit margin (%)
	Year-on-year % change					
June 2018	0.55	0.15	−8.42	−47.24	9.98	1.19
September 2018	23.61	22.92	5.3	88.11	8.43	3.24
December 2018	14.97	15.76	−6.04	–	9.52	−0.35
March 2019	16.2	17.18	25.48	35.7	12.21	6.41
June 2019	14.07	15.15	16.53	−3.07	12.29	4.12

Source: CMIE Industry Outlook. Retrieved from: https://industryoutlook.cmie.com/kommon/bin/sr.php?kall=wshowov&nvdt=20190910162900140&nvpc=055000000000&nvtype=OVERVIEW&icode=0101060000000000 Accessed on 14 October 2019.

Note
[a] All income and profit figures are net of prior period and extraordinary transactions.

Table 2.16 Quarterly financials of listed non-financial firms

(% change)	Sep 18	Dec 18	Mar 19	Jun 19
Income	23.4	17.4	8.5	2.6
Expenses	26.7	17.3	8.2	2.1
Net profit	18.6	2.5	1.8	−9.7
PAT margin (%)	6.1	6	6.5	6.4

Source: Industry outlook, CMIE.

Table 2.17 Quarterly income and expenditure summary: Construction and real estate industry (Rs. million)

		Jun 2018	Sep 2018	Dec 2018	Mar 2019	Jun 2019
Total income	Rs. million	535,564.3	634,637.4	647,398.1	885,151.9	609,165.4
	% growth	–	18.5	2	36.7	−31.2
Sales	Rs. million	511,996.2	566,757.3	625,140.4	852,841.7	576,447.7
	% growth	–	10.7	10.3	36.4	−32.4
Total expenses	Rs. million	557,564.1	673,209	681,384.7	815,500.4	577,578
	% growth	–	20.7	1.2	19.7	−29.2
Operating expenses	Rs. million	480,466.3	523,582.2	580,223.6	723,734.8	511,773.4
	% growth	–	9	10.8	24.7	−29.3
Net profit	Rs. million	854.4	2,263.7	−18,999.6	46,792.4	37,567.5
	% growth	–	164.9	−939.3	−346.3	−19.7
Count	Sample size	221	219	222	211	207

Source: CMIE Industry Outlook. Retrieved from: https://industryoutlook.cmie.com/kommon/bin/sr.php ?kall=wshreport&&kall=wshreport&repcode=20500500500000000000000000000000000000000000000 &repnum=41532&frequency=Q&icode=0101060000000000 Accessed on 17 September 2019.

after tax. This declining trend is seen even in non-financial firms (Table 2.16) and the construction and real estate industry together (Table 2.17).

2.7.4 Profit and project scenario

However, the profit and project scenario this is unlikely to drive the inflow of new orders immediately as they are expected to stay muted in the coming quarters due to the current economic slowdown. Also, firms should focus more on the execution of their current orders due to a robust order book position rather than bidding for new orders.

Also, as per bank credit data released by the Reserve Bank of India (RBI), outstanding credit to the infrastructure segment fell by 2 per cent in July 2019 at Rs. 10.4 trillion as compared to Rs. 10.6 trillion at the end of March 2019.

However, when compared to July 2018, outstanding credit to the infrastructure segment grew by a robust 14 per cent. Government spending on infrastructure helped push the bank credit growth during the year. All the segments, namely power, telecommunications, roads and cement, witnessed an uptick in credit demand due to government spending on infrastructure.

On the financial front, the revenues of the industrial and infrastructure construction industry increased by 15.2 per cent in the June 2019 quarter. This was on account of the quicker execution of projects in hand by firms. This was the fourth successive quarter wherein the industry witnessed double-digit growth in revenues.

The operating expenses of the industry grew by 10 per cent, much slower than the increase in revenues. This resulted in a 16.5 per cent rise in the operating profits of the industry. However, higher post-operating expenses and a fall in other income resulted in a 3.1 per cent decline in the net profit of the industry.

The CMIE data reveal that the profitability of firms has been steadily declining. Figure 2.5 shows that the profitability of firms is diminishing and, at the same time, direct tax incidence is rising (year-on-year percentage change). The rise in taxation and the fall in profit margins have adversely affected the returns of the construction industry. Figure 2.1 reveals that new investment project announcements have declined considerably compared to new investment project announcements made in 2017–18. This is the fourth successive year of a decline in new investment announcements. Figure 2.2 also shows a sharp drop in projects. Figure 2.3 reveals that new investment announcements are declining, both private and government (refer to Figure 1.3 in Chapter 1 for data on total quarterly new investment announcements). The share of private sector investment announcements is also declining (Figure 2.4). Figure 2.5 and Table 2.18 show that the profitability of firms is diminishing and at the same time direct tax incidence is rising (year-on-year percentage change). The diminishing returns of the construction industry and the decline in new investment projects are leading to a deplorable situation.

2.8 Growing firms of the construction industry in India

A business is only successful if it is growing. Growth has different implications. It can be defined in terms of revenue generation and value addition, and expansion in terms of volume of business (Gupta, Guha and Krishnaswami, 2013). The growth of a firm is quantitatively

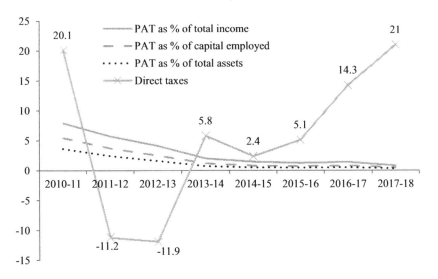

Figure 2.5 Profitability of the construction industry and percentage increase in direct taxes (Y-on-Y percentage change).

Table 2.18 Construction and real estate sector: Profitability ratios and direct taxes

	PAT as % of total income	PAT as % of capital employed	PAT as % of total assets	Direct taxes (Y-o-Y % change)
2010–11	7.9	5.45	3.61	20.1
2011–12	5.72	3.71	2.43	–11.2
2012–13	4.12	2.51	1.61	–11.9
2013–14	2.03	1.21	0.75	5.8
2014–15	1.48	0.81	0.51	2.4
2015–16	1.28	0.73	0.47	5.1
2016–17	1.43	0.82	0.52	14.3
2017–18	0.81	0.5	0.32	21

Source: Prowess, CMIE.

and qualitatively measured. 'Quantitative growth can be characterised by the company size (turnover, added value, volume), the profitability of the company and the value of the company (shareholder value)' (Carton, Hofer and Meeks, 1998). According to Carton, Hofer and Meeks (1998), growth can be measured in the form of qualitative features such as market position, the quality of the product and the goodwill of customers. Growth is measured in financial terms such as: (i) turnover; (ii) profit; (iii) total assets; (iv) net assets; (v) net worth; and (vi) increase in number of employees (Kruger, 2004).

As stated above, growth is an essential indicator of a growing endeavour. There are numerous factors similar to the features of an entrepreneur. Resources, for instance, finance and manpower affect the growth of a business enterprise and distinguish it from a non-growing endeavour. Gilbert, McDougall and Audretsch (2006) put forward 'how' and 'where' questions as important in the context of the growth of a firm. Growth is a function of the decisions an entrepreneur makes, i.e. how to grow (internally or externally) and where to grow (domestic market or overseas market). There are different studies on the main factors underlying the growth of a firm. Evans (1987), Heshmati (2001) and Morone and Testa (2008) emphasised the influence of a firm's size and age on its growth. Another group of studies by Fazzari, Hubbard and Petersen (1988), Lumpkin and Dess (1996) and Freel and Robson (2004) consider the influence of variables such as strategy, organisation and the characteristics of the firm's owners on the growth of a firm. Another study by Mateev and Anastasov (2010) found a relation between growth and size as well as other specific characteristics such as financial structure and productivity. Their study highlighted that total assets is one of the measures of an enterprise's size which has a direct impact on sales revenue.

The studies of Capon, Farley and Hoenig (1990), Chandler and Jansen (1992), Mendelson (2000), Cowling (2004), Serrasquerio, Nunes and Sequeira (2007), Asimakopoulous, Samitas and Papadogonas (2009), Serrasquerio (2009) and Jang and Park (2011) indicate the positive effect of growth on profitability. However, the studies of Reid (1995), Roper (1999) and Nakano and Kim (2011) could not find a positive effect of growth on profitability. The study of Asimakopoulous, Samitas and Papadogonas (2009) shows that the size of sales growth positively affects profitability but negatively affects leverage and current assets. Some of the latest studies show that firms that grow successfully first secure profitability and then grow (Davidsson, Achtenhagen and Naldi, 2005).

The current study ranks the fastest-growing firms in the construction industry in India by taking growth in revenue and growth in profit as the only two parameters. The analysis is based on the revenue, profits, assets and net worth of the firms in the construction and real estate sector of India.[6] The total assets and net worth of the firms in the construction and real estate sector of India are taken as qualifiers for the selection of firms.

2.8.1 Methodology

The data source for the study is the database of the Centre for Monitoring Indian Economy. It has data on 4015 listed firms in the construction and real estate industry. The study primarily ranked the construction real estate firms by constructing a growth index. Of these, 419 firms reported total revenue for 2018–19. From this, 124 reported losses, 7 firms reported zero profit and 288 had profits ranging from Rs. 0.10 million to Rs. 66777 million. Data have been collected for these firms from 2013–14 onwards as per the methodology adopted for the ranking of firms detailed below.

2.8.1.1 Ranking of firms

Even though there are 4015 listed firms in the sector, only 31 firms were qualified for inclusion in the study. Other firms were excluded after adopting the following methodology. Average revenue of the firms under study was Rs. 5828.2 crores, average profit after tax was Rs. 373.3 crores, average total assets were Rs. 8078.1 crores and average net worth was Rs. 2909.2 crores during 2018–19.

2.8.1.2 Definitions

   i. **Construction and real estate industry**: Taken to include firms in construction and infrastructure development activities, construction of commercial complexes, housing construction and contractual engineering works.
  ii. **Revenue**: Includes sales revenue, income on the sale of assets and other income from other sources, and revenue of subsidiaries. It is the sum of all kinds of income generated by an enterprise during an accounting period and includes income from continuing operations as well as income from discontinuing operations, income generated during the normal course of business as well as extraordinary or exceptional income, income generated from the sale of goods as well as services, income from investment activity, income accruing even without any sale of goods or rendering of services or as a result of an investment activity.
 iii. **Profit**: Taken as net profit, net of depreciation and taxes. It is the residual after all revenue expenses are deducted from the sum of the total income and the change in stocks.
  iv. **Assets**: Includes the total value of all assets, i.e. fixed assets, investments and net current assets, held by a firm on the last day of an accounting period. More precisely, it includes: net fixed assets; capital work in progress and net pre-operative expenses pending allocation, if any; investments; inventories; receivables; loans and advances; cash and bank balances; deferred tax assets and miscellaneous expenses not written off.
   v. **Equity**: Forms the sum of paid-up capital and reserves, excludes revaluation reserves and goodwill reserves and preference shares. Therefore, the net worth of a firm is what it owes its equity shareholders and consists of the monies put into the firm by the equity shareholders in the form of equity capital and the profits generated and retained as reserves by the firm.

2.8.1.3 Qualification for selection

i. **Company form**: Should be a legal entity created in any one of three modes:
   - registered as a public or private limited firm under the Indian Companies Act, 1956, or
   - registered as a partnership firm under the Indian Partnership Act, 1932, or
   - a public sector undertaking created under an Act of Parliament/State Legislature.

ii. **Audited accounts**: The annual statement of accounts, balance sheet and profit and loss accounts should be audited by an accredited firm of chartered accountants, and the same submitted annually to the body of share/stakeholders.

iii. **Size of firm**: Average revenue was above Rs. 100 crores, paid-up capital plus reserves were Rs. 50 crores or more and total assets were Rs. 300 crores or more during the 6-year period under study. Foreign firms working in India but not incorporated in India were excluded.

iv. **Age**: The firm should have been incorporated for at least 9 years. Thus, the statements of accounts should be available for the financial years 2013–14, 2014–15, 2015–16, 2016–17, 2017–18 and 2018–19.

v. **Nature of business**: The mainline business of the firm should primarily be real estate, civil works construction and contracting:
   a. For multi-product firms, the contribution of construction to their gross annual revenue should not be less than 40 per cent in each of the accounting year.
   b. Firms that serviced the in-house requirements of the group and did not cater to the open market were excluded from the scope of this selection.
   c. Loss-making firms were not considered in the ranking.

2.8.1.4 Data source

The data source of the study is Prowess. Prowess is a database of the financials of Indian firms prepared and maintained by the Centre for Monitoring Indian Economy, India.

2.8.1.5 Ranking

i. The growth in revenue and the growth in profit for each of the 6 years were the only two parameters used for ranking.

ii. Based on the above, the growth in revenues and the growth in profits of all firms were calculated and a 'growth index statement' was compiled.

iii. At the time of compiling these tables, data were not available from some of the firms.

iv. A weighted index of growth of revenue and of profit with profits as the weight was compiled using the following method:

Let X_{id} represent the value of the *i*th factor in the *d*th firm ($i = 1, 2, 3, \ldots, n; d = 1, 2, 3, \ldots, m$, say). Let us write:

$$Y_{id} = \frac{X_{id} - \operatorname*{Min}_{i} X_{id}}{\operatorname*{Max}_{i} X_{id} - \operatorname*{Min}_{i} X_{id}}$$

where $\operatorname*{Min}_{i} X_{id}$ and $\operatorname*{Max}_{i} X_{id}$, respectively, are the minimum and maximum of $(X_{1d}, X_{2d}, \ldots, X_{nd})$.

Obviously, the scaled value, Y_{id}, varies from zero to 1 from this matrix of scaled values. $Y = (Y_{id})$ as a measure for the various factors for different firms has been constructed as follows:

$$Y = W_1 Y_{i1}, W_2 Y_{i2}, \ldots, W_n Y_{in}$$

where W ($0 \geq W_i \leq 1$ and $W_1, W_2, \ldots, W_n$) is the weight reflecting the relative importance of the individual indicators (Table 2.19).

2.9 Constraints of Indian construction industry

A review of the literature reveals the following constraints of the Indian construction industry.

2.9.1 Investment climate

Weakening private investment has been of major concern to India over the last few years. New investment announcements are declining. The rate of project completion as well slowed down and an increasing proportion of the stock of projects is being abandoned (refer to Table 2.20). The prime constraints on the investment climate relate to skilled manpower and its availability, operational issues such as government approvals, clearances, etc., and taxation (World Bank, 2008). Subsequent constraints relate to the cost of materials, contract enforcement and dispute resolution, barriers to entry, subsidies and fiscal concessions.

Foreign contractors surveyed by the World Bank study cited (a) cultural bias in project management style, (b) poor governance, (c) bureaucracy and corruption, (d) risk allocation practices and contract conditions, (e) visa and travel document processing for expatriates, and (f) lack of information on the road construction industry as the most critical issues. In addition, foreign contractors observed some intangible limitations like preference given to domestic contractors in the bidding process (World Bank, 2008).

2.9.2 Pre-construction activities

Project formulation is the major activity in the pre-execution phase of a project. Project formulation is the culmination of a series of investigations of innumerable aspects that have a bearing on the structures involved. The pre-construction activities start with field investigations and the preparation of feasibility reports. The rest of the pre-construction activities, broadly, are detailed designs, preparation of construction drawings, tender document preparation, tendering and award of works (Figure 2.6). Contract management is a follow up to what is provided for in the contract. Therefore, all aspects that go into the formulation of proper contracts and their subsequent management during execution, have to be very carefully considered and provided for at the time of project planning. Consequently, project formulation must be done in a scientific manner without resorting to shortcuts or assumptions. A poorly planned project leads to innumerable complications owing to inherent uncertainties. A well-planned project helps timely completion in an orderly manner with satisfactory results for the client and contractor. Thus, a properly planned project based on adequate investigations is a must for the proper management of construction contracts.

Table 2.19 Fastest-growing firms in the construction industry in India

Rank	Company name	Weighted index	Total income (Rs. crore) 2018/2019	2017/2018	2016/2017	2015/2016	2014/2015	2013/2014	Total assets (Rs. crore) 2018/2019	2017/2018	2016/2017	2015/2016	2014/2015	2013/2014	Net worth (Rs. crore) 2018/2019	2017/2018	2016/2017	2015/2016	2014/2015	2013/2014	Profit after tax (Rs. crore) 2018/2019	2017/2018	2016/2017	2015/2016	2014/2015	2013/2014
1	Techno Electric & Eng. Co. Ltd.	0.394	1,048 / 1,331	163	124	112	122		2,268 / 2,064	993	965	1,004	1,053		1,404 / 1,223	662	574	580	571		182 / 200	56	36	35	35	
2	Capacit'e Infraprojects Ltd.	0.326	1,825 / 1,360	1,140	815	510	234		2,163 / 1,806	1,121	870	506	305		843 / 750	294	168	64	19		96 / 79	69	44	27	10	
3	H G Infra Engineering Ltd.	0.308	2,021 / 1,398	1,060	715	332	442		1,577 / 1,485	575	369	225	209		659 / 541	176	127	94	84		124 / 84	53	30	9	17	
4	Montecarlo Ltd.	0.104	2,470 / 1,975	2,005	1,648	1,070	790		2,437 / 1,675	1,549	1,268	949	679		737 / 593	434	331	238	175		146 / 159	118	93	63	40	
5	P N C Infratech Ltd.	0.076	3,208 / 1,929	1,761	2,078	1,575	1,163		3,727 / 2,829	2,443	1,936	1,662	1,351		2,115 / 1,807	1,572	1,378	716	629		325 / 251	210	235	100	70	
6	Atcons Infrastructure Ltd.	0.073	7,935 / 6,001	6,080	4,456	3,212	2,828		10,016 / 7,891	7,044	6,755	4,489	3,656		1,264 / 1,140	1,042	996	656	594		125 / 129	75	88	79	67	
7	K N R Constructions Ltd.	0.059	2,201 / 1,971	1,572	978	889	854		2,336 / 2,069	1,689	1,298	1,011	960		1,414 / 1,158	895	738	567	511		263 / 272	157	161	73	61	
8	Ashoka Buildcon Ltd.	0.056	3,936 / 2,546	2,085	1,647	2,004	1,665		5,391 / 3,975	3,497	3,099	2,670	2,218		2,212 / 1,926	1,723	1,696	1,054	933		286 / 237	176	139	142	103	
9	Apco Infratech Pvt. Ltd.	0.05	2,016 / 1,636	1,223	1,268	1,193	786		1,663 / 1,575	1,324	961	885	785		641 / 478	351	290	236	168		172 / 127	69	62	71	43	
10	R P P Infra Projects Ltd.	0.048	590 / 506	369	320	268	240		589 / 506	458	360	386	276		207 / 185	173	149	129	112		24 / 13	24	21	2	12	
11	J Kumar Infraprojects Ltd.	0.046	2,815 / 2,079	1,635	1,426	1,357	1,198		3,569 / 3,424	2,814	1,964	1,692	1,653		1,668 / 1,509	1,391	1,292	789	575		177 / 137	107	99	94	84	
12	Jamshedpur Utilities & Services Co.	0.035	1,129 / 928	788	779	669	577		1,013 / 857	716	596	521	468		137 / 89	64	12	31	14		45 / 26	52	40	14	4	
13	Rail Vikas Nigam Ltd.	0.034	10,424 / 7,889	6,245	4,781	3,293	2,597		11,553 / 7,815	8,461	21,785	15,817	13,011		3,739 / 3,351	3,083	3,012	2,679	2,538		607 / 470	382	304	186	157	
14	Gayatri Projects Ltd.	0.026	3,503 / 2,988	2,214	1,874	1,655	1,855		5,906 / 5,006	4,479	3,889	3,512	3,319		1,330 / 1,123	741	846	683	666		211 / 188	70	58	22	48	
15	Kalpataru Power Transmission Ltd.	0.022	7,188 / 5,827	5,063	4,480	4,547	4,225		8,323 / 7,346	6,165	5,316	4,837	4,613		3,151 / 2,769	2,478	2,215	2,070	1,954		401 / 322	269	192	166	146	
16	Power Mech Projects Ltd.	0.022	1,751 / 1,317	1,299	1,368	1,362	1,199		1,943 / 1,548	1,460	1,435	1,152	954		776 / 681	617	553	352	274		96 / 66	64	74	71	68	
17	N C C Ltd.	0.02	12,201 / 7,758	8,133	8,580	8,524	6,319		13,344 / 10,941	9,109	9,325	9,546	9,211		4,757 / 4,242	3,442	3,261	3,204	2,520		564 / 287	226	240	112	41	
18	I R B Infrastructure Developers Ltd.	0.017	3,623 / 3,326	3,645	3,167	2,201	2,528		10,026 / 10,648	9,247	7,970	7,147	5,916		2,620 / 2,497	2,494	2,372	2,150	1,727		329 / 444	203	292	138	288	
19	J M C Projects (India) Ltd.	0.016	3,327 / 2,815	2,366	2,436	2,432	2,678		4,081 / 3,532	2,909	2,717	2,258	1,859		923 / 789	690	638	429	404		142 / 106	58	43	30	23	

No.	Company																										
20	Larsen & Toubro Ltd.	0.015	94,635	80,678	70,735	67,146	60,418	60,033	129,125	118,725	105,000	101,939	89,597	80,964	52,398	48,966	45,794	41,906	36,992	33,570	6,678	5,387	5,454	5,000	5,056	5,493
21	Modern Road Makers Pvt. Ltd.	0.015	4,391	3,701	3,206	2,841	2,030	2,511	4,522	4,481	3,622	3,384	3,095	3,277	1,526	1,311	1,234	942	819	870	324	416	334	272	113	204
22	Ahluwalia Contracts (India) Ltd.	0.014	1,768	1,655	1,436	1,263	1,072	1,003	1,496	1,264	1,241	1,160	1,046	1,054	735	622	506	422	337	225	117	115	86	84	64	22
23	Sadbhav Engineering Ltd.	0.013	3,650	3,595	3,408	3,297	3,038	2,414	4,617	4,389	4,123	3,637	3,525	3,025	2,034	1,867	1,661	1,488	1,337	939	187	221	188	132	114	106
24	Ircon International Ltd.	0.013	5,068	4,255	3,470	2,964	3,349	4,453	13,291	13,019	9,472	8,167	7,478	6,901	3,950	3,751	3,828	3,667	3,354	2,993	445	388	369	395	579	907
25	G P T Infraprojects Ltd.	0.01	542	472	371	258	337		653	639	586	541	508	525	1,404	1,223	662	574	580	571	182	200	56	36	35	35
26	S P M L Infra Ltd.	0.008	1,493	1,412	1,744	1,459	1,446	1,268	2,930	2,914	2,483	2,216	2,163	1,958	843	750	294	168	64	19	96	79	69	44	27	10
27	Simplex Infrastructures Ltd.	0.004	6,276	5,948	5,805	6,021	5,639	5,559	10,014	9,477	8,756	8,372	7,956	7,364	659	541	176	127	94	84	124	84	53	30	9	17
28	Engineers India Ltd.	0.004	2,685	1,985	1,695	1,832	1,995	2,181	5,448	5,123	4,870	4,675	4,607	4,586	737	593	434	331	238	175	146	159	118	93	63	40
29	Reliance Industrial Infrastructure Ltd.	0.003	114	117	122	113	110	103	379	368	355	354	322	313	2,115	1,807	1,572	1,378	716	629	325	251	210	235	100	70
30	B G R Energy Systems Ltd.	0.001	3,279	3,340	3,468	3,253	3,453	3,306	6,806	6,682	6,405	6,622	6,244	6,343	1,264	1,140	1,042	996	656	594	125	129	75	88	79	67
31	Madhucon Projects Ltd.	0	685	686	712	739	898	1,165	2,967	2,910	2,762	3,280	3,382	3,170	1,414	1,158	895	738	567	511	263	272	157	161	73	61

Table 2.20 Stages of investment projects (Rs. trillions)

Year	Completed	Dropped	Outstanding	Under implementation
2011–12	4.3	8.9	167.6	83.3
2012–13	3.7	10.2	168.4	87.6
2013–14	3.6	10.8	169.9	88.5
2014–15	4.1	14.1	180.5	94.1
2015–16	5.9	13.3	189.6	99.5
2016–17	6.5	10.6	196.4	104
2017–18	5	17	195.1	107.8
2018–19	5.9	20.8	185.3	110.1

Source: CMIE Economic outlook. Retrieved from: https://economicoutlook.c
mie.com/kommon/bin/sr.php?kall=wshreport&nvdt=20190604182034666&nvpc
=035000000000&nvtype=INSIGHTS&oporder=1 Accessed on 9 July 2019.

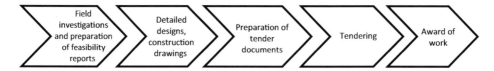

Figure 2.6 Pre-construction activities.

2.9.3 *Present form of contract and contractual procedures*

The current forms of contract often contain vague clauses that are open to interpretation, paving the way for disputes. The lack of training and inexperience of the client's staff and domestic supervision consultants in contract and contract conditions often lead to violation of the conditions of a contract, delayed decision-making, etc. These combined with inexperienced subcontractors make project execution complicated and cumbersome and contribute to project delays, claims and disputes.

According to the Planning Commission's working group report of the Eleventh Five-Year Plan[7]: (i) contract procedures for the procurement of contractors are highly cumbersome and costly, both for the project owner and the contractors; (ii) there are no standardised contract procedures or evaluation criteria; (iii) there are no restrictions on entry to this sector by unqualified players; (iv) contract conditions are not equitable – conditions such as damages to contractors due to delays by project owner, resource mobilisation through advances and cost escalation, which can lead to disputes and time overruns, are not effectively laid out; (v) the selection of the lowest cost bidder hampers the process of adopting superior technology, best practices and quality; (vi) delayed dispute resolution leads to business losses and capital blockage; (vii) lack of a clear definition of the status of construction as an industry, and the absence of support systems, mean that the construction industry is not able to easily access institutional sources of finance.

> As per prevailing laws, an organization engaged in construction activity requires registration under five different legislations and is subject to inspection by officers appointed under twelve enacted laws having prosecution powers. Further, they are required to obtain licenses under three enactments. It is pertinent to note that all the applicable legislation requires periodic returns and dealing with the notices issued by different authorities.[8]

Various statutes pertaining to construction labour are:

1. Children (Pledging of Labour) Act, 1938
2. Employment of Children Act, 1938
3. Factories Act, 1948
4. Mines Act, 1952
5. Employment Exchange (Compulsory Notification of Vacancies) Act, 1959
6. Industrial Employment (Standing Orders) Act, 1946
7. Industrial Disputes Act, 1947
8. Workmen's Compensation Act, 1923
9. Indian Trade Unions Act, 1926
10. Employer's Liability Act, 1938
11. Employer's Sate Insurance Act, 1948
12. Employees Provident Funds Act, 1952
13. Maternity Benefits Acts, 1961
14. Payment of Wages Act, 1936
15. Motor Transport Workers, Act, 1951
16. Contract Labour (Regulation and Abolition) Act, 1970
17. Payment of Gratuity Act, 1972
18. Apprentices Act, 1961
19. Minimum Wages Act, 1948
20. Equal Remuneration Act, 1976
21. Payment of Bonus Act, 1965
22. Weekly Holidays Act, 1942
23. Collection of Statistics Act, 1953
24. The Inter-State Migrant Labour (Regulation of Employment and Conditions of Service) Act, 1973
25. The Building and Construction Workers (Regulation of Employment and Conditions of Service) Act, 1996
26. The Building and Other Construction Workers Welfare Cess Act, 1996
27. The Employees Provident Fund and Misc. Provisions (Amendment) Act, 1996

2.9.4 Institutional structure

The prevailing institutional arrangements do not ensure a clear separation of the functions of the client and the provider. 'Construction firms are regulated under multiple laws and there is no unified regulatory framework' (Government of India, 2013). Lack of a single construction law is another barrier to entry of firms. The overlap in the functions of regulatory agencies often creates problems. The lack of an independent supervisory body to safeguard the interests of the different stakeholders is also a factor in the construction sector's slow development.

2.9.5 Entry barriers

Contracting firms from other sectors face entry barriers, for instance, rigorous qualification requirements requiring high managerial capabilities (Male, 1991) related to prior technical experience in the sector. Uncontrolled cartelisation and collusion among contracting firms in many states also inhibit non-regional bidders from even submitting their bids.

Furthermore, it is impossible for small- and medium-contracting firms to obtain a rating that would allow them easier access to credit to expand their business (World Bank, 2007).

2.9.6 Supply-side constraints

2.9.6.1 Number of contracts and contractors

Table 2.21 shows that the present capacity of firms in the construction sector is limited. The predicted total sales of the construction industry and prediction limits (1996–97 to 2024–25) given elsewhere in this chapter indicate that contractor capacity would have to be greatly augmented (Figure 2.7).

Table 2.21 Distribution of firms in the construction and real estate industry as per sales and percentage share in sales

Sales[a] (in Rs. million)	Number of companies	% share in total
Below 100	322	27.83
100–1999	89	7.69
200–499	171	14.78
500–999	116	10.03
1000–1499	88	7.61
1500–1999	42	3.63
2000–2999	82	7.09
3000–3999	46	3.98
4000–4999	22	1.9
5000–9999	77	6.66
10000–19999	58	5.01
20000–70000	44	3.8
Total	1,157	100

Source: Industry outlook, CMIE. Retrieved from: https://industryoutlook.cmie.com/ Accessed on 10 August 2019.

Note
[a] Sales of firms as in the construction and real estate industry, 2016–17.

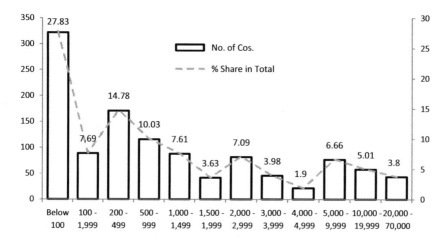

Figure 2.7 Number of companies – sales wise.

2.9.6.2 Inadequacy of skilled human resources

The inadequacy of skilled human resources is a major constraint across all subsectors of the construction industry in India. The estimated requirement for resources in the construction sector for the period 2019–20 to 2024–25 reveals that India needs 389–535 million man years. The non-availability of engineers with knowledge of construction management and the paucity of skilled technicians, building supervisors, skilled and semi-skilled workers, masons, carpenters, etc., will not keep up with these massive requirements.

2.9.6.3 Construction equipment

Modern construction is highly mechanical. Productivity in construction can be improved by the introduction of modern construction equipment. India requires construction equipment worth Rs. 9,03,968 crores over the next 6 years under a medium-growth scenario, thus the supply needs to increase. Under a high-growth scenario, India needs construction equipment worth Rs. 10,46,629 crores. This demands significant enhancement of the production capacity of the equipment manufacturers and the streamlining of import procedures. Alternatively, the possibility of equipment rental and equipment bank and leasing facilities should be investigated.

2.9.6.4 Construction materials

Construction consists of a sequence of activities that transforms materials from a given to a desired form using the most appropriate techniques. India requires construction materials worth Rs. 24,85,912 crores over the next 6 years under a medium-growth scenario, thus the supply needs to increase. Under a high-growth scenario, India needs construction materials worth Rs. 28,78,229 crores. The requirement for cement under medium- and high-growth scenarios is 1913 million tons and 2215 million tons, respectively. The requirement for steel under medium- and high-growth scenarios is 753 million tons and 872 million tons, respectively (Table 2.11). Arrangements for such huge requirements may be planned in advance along with other construction materials.

In September 2019, the government announced plans to build infrastructure to the tune of Rs. 100 trillion over the next 5 years after the sharp decline in the June 2019 quarter GDP growth at 5 per cent.

2.10 Project delays and disputes

Delays in activities are a frequent problem across all construction projects. The study of Gebrehiweta, Tsegay and HanbinLuo (2017) identified and ranked the top ten most pivotal causes of delay, in the pre-construction stage, as corruption, ineffective project planning and scheduling, inflation/price increases in materials, unavailability of utilities at site, improper project feasibility study, late design and design documents, unclear and inadequate details and specification of design, slow delivery of materials, and design mistakes and errors. An analysis of the Ministry of Statistics and Programme Implementation data over the last 18 years reveals that time overruns had reduced from 44.99 per cent in March 2000 to 23.52 per cent in September 2017. Similarly, cost overruns had declined from 36 per cent in March 2000 to 12.60 per cent in September 2017 (Figure 2.8).

The data of the Ministry of Statistics and Programme Implementation show that they monitored 1453 projects with an investment cost of Rs. 150 crore and above. The total

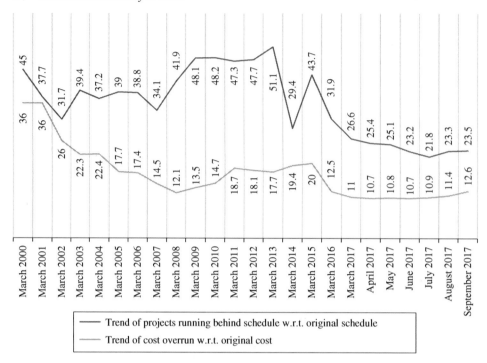

Figure 2.8 Project delay.

cost of these projects at the time of sanctioning was of the order of Rs. 18,32,579.17 crore. The total cost of these projects was subsequently revised to Rs. 2161313.18 crore implying a cost overrun of 17.9 per cent (Table 2.22) because the time overrun of 345 projects ranged from 1 month up to 220 months in some cases (Table 2.23). Railways have the highest cost overrun (Figure 2.9). India requires Rs. 100 lakh crore (approximately US$1.45 trillion) for infrastructure development in the next 5 years (Sharma, 2019, July 14).

Project delays are often due to cumbersome procedures required to get the necessary clearances for projects, rework, vague laws and regulations, variations, lack of coordination between the various government departments, erroneous design, imperfect documentation, delayed authority approvals, etc. Many of these reasons often lead to disputes on construction projects. Enormous amounts of money are locked up because of disputes between contractors and clients, leading to cost and time overruns. The reasons for time overruns reported by the Ministry of Statistics and Programme Implementation in the projects they are monitoring are delay in land acquisition; delay in obtaining forest/environment clearances; lack of infrastructure support and linkages; delay in tie-up of project financing; delay in finalisation of detailed engineering; changes in scope; delay in tendering, ordering and equipment supply; law and order problems; geological surprises; pre-commissioning teething troubles; and contractual issues. The reasons for cost overruns reported are under-estimation of original cost; changes in rates of foreign exchange and statutory duties; high cost of environmental safeguards and rehabilitation measures; spiralling land acquisition costs; changes in the scope of projects; monopolistic pricing by vendors of equipment services; general price rise/inflation; disturbed conditions; and time overrun.

Table 2.22 Extent of cost overrun in projects with respect to original cost (Rs. 150 crore and above) (Rs. crore)

Sl. No.	Sector	No. of projects	Original cost	Anticipated cost	Cost overrun (%)	Projects with cost overrun[a]			
						No.	Original cost	Anticipated cost	Cost overrun (%)
1	Atomic energy	4	67,120	75,589	12.62	2	14,951	23,420	56.65
2	Civil aviation	1	314.61	441.33	40.28	1	314.61	441.33	40.28
3	Coal	98	99,419.73	99,873.07	0.46	11	25,725.94	28,037.55	8.99
4	Fertilisers	5	1,299.07	1,310.72	0.9	1	197.79	209.44	5.89
5	Mines	5	7,078.62	7,078.62	0	0	0	0	0
6	Steel	14	32,793.01	39,597.41	20.75	2	15,868	23,030.4	45.14
7	Petroleum	140	2,29,497.78	2,34,857.89	2.34	19	27,989.65	36,251.99	29.52
8	Power	77	2,87,073.06	3,45,343.53	20.3	33	1,61,106.73	2,19,894.25	36.49
9	Heavy industry	2	3,272	5,381.3	64.47	1	1,718	3,827.3	122.78
10	Health and family welfare	20	8,129.23	8,246.52	1.44	3	1,076.25	1,193.54	10.9
11	Railways	364	4,84,549.3	7,01,059.04	44.68	205	1,72,046.28	3,98,507.24	131.63
12	Road transport and highways	652	4,28,477.95	4,41,091.58	2.94	49	28,447.55	41,837.22	47.07
13	Shipping and ports	7	6,557.29	6,999.74	6.75	3	760.89	1,545.44	103.11
14	Telecommunications	5	16,310.53	27,553.13	68.93	3	13,781.1	25,123.7	82.31
15	Urban development	57	1,60,233.35	1,66,436.66	3.87	12	28,109.46	34,312.77	22.07
16	Defence production	2	453.64	453.64	0	0	0	0	0
	Total	1,453	18,32,579.17	21,61,313.18	17.94	345	4,92,093.25	8,37,632.17	70.22

Source: Government of India. (April 2019). Flash report on central sector projects (Rs. 150 crore and above), Infrastructure and Project Monitoring Division, Ministry of Statistics and Programme Implementation, Government of India, New Delhi, p. 5. Retrieved from: http://www.cspm.gov.in/english/flr/Fr_apr_Report_2019.pdf Accessed on 11 July 2019.

Note
[a] Cost overrun = anticipated cost minus original cost.

Table 2.23 Extent of time and cost overruns on projects with respect to original schedule with latest approved cost (Rs. 150 crore and above)

SI. no.	Sector	Projects with time[a] and cost[b] overruns			
		No.	Latest approved cost	Anticipated cost	Time overrun in months
1	Atomic energy	1	3,492	6,840	133–133
2	Civil aviation	1	314.61	441.33	21–21
3	Coal	1	152.43	459.49	60–60
4	Fertilisers	0	0	0	0–0
5	Mines	0	0	0	0–0
6	Steel	1	343	488.4	41–41
7	Petroleum	2	3,382.56	3,488.31	33–110
8	Power	8	38,106.88	56,808.87	12–147
9	Heavy industry	0	0	0	0–0
10	Health and family welfare	1	414.11	513.89	83–83
11	Railways	47	45,002.06	1,09,195.4	3–220
12	Road transport and highways	15	13,212.66	16,239.91	1–140
13	Shipping and ports	0	0	0	0–0
14	Telecommunications	1	13,334	24,664	58–58
15	Urban development	4	7,446.31	8,338.27	5–40
16	Defence production	0	0	0	0–0
	Total	82	1,25,200.62	2,27,477.87	–

Source: Extracted from Government of India. (April 2019). Flash report on central sector projects (Rs.150 crore and above), Infrastructure and Project Monitoring Division, Ministry of Statistics and Programme Implementation, Government of India, New Delhi, p. 11. Retrieved from: http://www.cspm.gov.in/english/flr/Fr_apr_Report_2019 .pdf Accessed on 11 July 2019.

Notes
[a] Time overrun = anticipated date of commissioning minus original date of commissioning.
[b] Cost overrun = anticipated cost minus original cost.

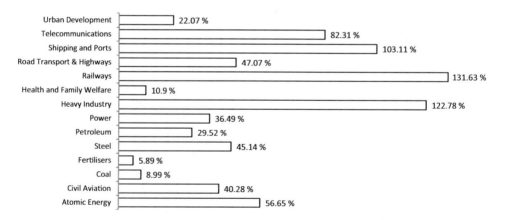

Figure 2.9 Extent of cost overrun on projects with respect to original schedule.

2.10.1 *Reasons for delays in execution of construction projects*

Delays in the execution of construction projects are a common phenomenon in India as is evident in the reports of the Government of India's Ministry of Statistics and Programme Implementation. Delays in the execution of construction projects lead to time and cost overruns. These delays need to be minimised, considering that construction, or infrastructure capital, is an essential component of the growth and development of an economy. Expenditure in the construction sector through its backward and forward linkages, as well as the multiplier effect, generates further investment in other sectors of the economy and generates further employment opportunities. Therefore, minimising delays in project execution is crucial for a developing country like India. Edison's (2017, July 1–2) study on project delays examines the causes of delays and their frequency in construction projects in India. The nine sources (groups) of delay considered are: (i) owner related; (ii) contractor related; (iii) consultant related; (iv) project related; (v) design team related; (vi) materials related; (vii) plant/equipment related; (viii) labour related; and (ix) external factors. From a sample of 186 executives in the construction sector in India, the study found that the major causes of delays in project execution in India were delays in revising and approving design documents, delays in subcontractors work, type of project bidding and award, mistakes and discrepancies in design documents, delay in material delivery and delay in obtaining permits from local administration. Category-wise delay factors considered in the study are given below:

2.10.1.1 *Owner-related delays*

To the client, delay denotes loss of revenue through the delayed availability of production facilities. There are several causes for delay on the part of the owner which have negative impacts on the interests of the owner himself, for instance, late approval of drawings and specifications, frequent change orders by the owner during construction and incorrect site information generate claims from both the main contractors and subcontractors which often involve disputes and litigation with enormous financial consequences. Other probable delays caused by the owner, reported by different studies, relate to conflicts between joint ownership of the project; delays in progressing payments; delays in furnishing and delivering the site to the contractor; delays in revising and approving design documents; poor communication and coordination with other parties; slowness in the decision-making process; suspension of work by the owner; unavailability of incentives for contractor to finish ahead of schedule, etc. The contractor is entitled to an extension of time for delays caused by the owner's actions.

2.10.1.2 *Contractor-related delays*

Often, contractor-related delays are attributed to lack of adequate managerial skills; difficulties in financing a project; inefficient subcontractors; poor aptitude of the contractor's technical staff; lack of planning and poor understanding of the project; poor site management and supervision; delays in site mobilization; rework due to errors during construction; poor communication and coordination with other parties; improper construction methods implemented by the contractor; and conflicts between the contractor and the other two parties (consultant and owner) and stakeholders. Generally, the client is compensated for contractor-related delays.

2.10.1.3 Consultant-related delays

Generally, construction consultants are independent professional engineers who carry out detailed engineering services for their clients. Their services begin even before the feasibility reports and include engineering design, detailed design and preparation of contract documents and construction supervision (general and residential supervision), etc. Construction consultants guide the owner in almost all construction activities. They interpret plans, drawings and specifications; check the accuracy of the drawings and the data supplied; process, estimate and certify progress payments; and inspect and validate the project upon completion. Consulting firms send resident staff/engineers on a detailed inspection of the structure during construction to verify that the construction is as per specifications and plans. There are several causes for delay on the part of the consultant such as late in reviewing and approving design documents; delays in approving major changes in the scope of the work; delays in performing inspection and testing; inflexibility (rigidity); poor communication/coordination with other parties; conflicts with design engineers; inadequate experience; and inaccurate site investigations.

2.10.1.4 Design-related delays

Design-related delays include mistakes and discrepancies in design documents; delays in producing design documents; unclear and inadequate details in drawings; the complexity of the project design; insufficient data collection and survey before design; misunderstanding of owner's requirements by design engineer; advanced engineering design software not employed; inadequate design-team experience, etc.

2.10.1.5 Project-related delays

Project-related delays include inadequate definition of substantial completion; ineffective delay penalties; legal disputes between project participants; very short original contract duration; and unfavourable contract clauses.

2.10.1.6 Labour, equipment, materials and external delay–related factors

The majority of the studies indicate labour, equipment, materials and external delay–related factors are responsible for project delays. Labour productivity is affected for a variety of reasons. Frequently, a shortage of skilled labour leads to the deployment of semi-skilled or unskilled labour. This not only delays the project but also affects the quality of the project. Labour, equipment, materials and external delay–related factors are given in Figure 2.10.

The reasons for time overruns given in the report of the Ministry of Statistics and Programme Implementation, Government of India (Government of India, 2018, July–September) are as follows:

(A) *Issues with Union Ministries*: (i) environment, forest and wildlife clearances; (ii) eco sensitive zone clearance; (iii) tree cutting permission; (iv) grant of working permission; (v) approval for private railways siding construction; (vi) industrial license permission; (vii) road crossing of pipelines/transmission lines; (viii) grant of right of way; and (ix) shifting of utilities.

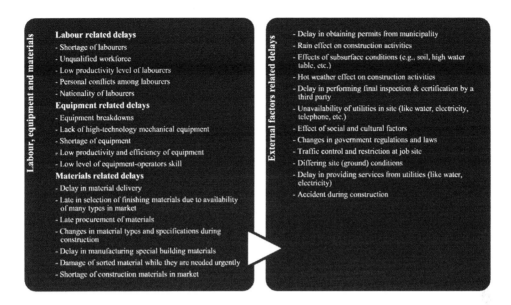

Figure 2.10 Labour, equipment, materials and external delay–related factors in construction projects.

(B) *Issues with State Governments*: (i) land acquisition issues; (ii) removal of encroachments; (iii) relief and rehabilitation plan; (iv) No objection certificate under forest rights act; (v) power and water supply; (vi) consent to establish and operate from state pollution control board; (vii) transfer of government land; (viii) law and order issues; (ix) right of way permission; and (x) diversion of forest land.

The Economic Survey (2019)[9] lists the major constraints facing road sector development in India: (i) availability of funds for financing large projects; (ii) lengthy processes in the acquisition of land and the payment of compensation to the beneficiaries; (iii) environmental concerns, and time and cost overruns due to delays in project implementation; (iv) procedural delays; (v) less traffic growth than expected increasing the riskiness of the projects resulting in stalled or languishing projects; and (vi) shortfall in funds for maintenance. The Economic Survey further reveals that massive investments have been made in the sector in 2018–19. The total investment was Rs. 158,839 crore in 2018–19 from only Rs. 51,914 crore in 2014–15, a three-fold increase (Economic Survey, 2019). The share of private investment was only 14 per cent in 2018–19.

2.11 Factors influencing construction productivity

According to the International Labour Organisation, productivity refers to how efficiently resources are used; it can be measured in terms of all factors of production combined (total factor productivity) or in terms of labour productivity, which is defined as output or value added divided by the amount of labour used to generate that output.[10] It measures how efficiently inputs are used to produce a given level of output. Therefore, it is a ratio of the output produced and the inputs used for production (labour, capital, raw materials, etc.) or, in simple terms, how much output is obtained from a given set of inputs. From a macro-economic point of view, the productivity concept refers to the relationship between the economic resources used in production and the resultant output of goods and services. An increase in the productivity of the construction sector not only raises the earnings and profits of those working in the construction sector but it also contributes to an improvement in productivity in other sectors, thereby improving general standards of living. However, the productivity of the construction process in India is made complex by the type of environment in which it exists.

Productivity can be measured at different levels. The three main measures of productivity are: (i) industry or sector level; (ii) project level; and (iii) activity or process level. The factors influencing productivity in construction are generally categorised into: (i) management; (ii) technological; (iii) labour; and (iv) external. Several factors significantly influence construction productivity (Box 2.1).

BOX 2.1 GENERAL FACTORS INFLUENCING CONSTRUCTION PRODUCTIVITY

- Accidents at worksites
- Availability of materials
- Change orders
- Changing of foremen
- Coordination among design disciplines
- Delay in responding to requests for information
- Design changes/change orders
- Errors/incomplete design drawings
- Inclement weather
- Incompetent supervisors
- Incomplete drawings
- Inspection delays
- Instruction time

- Interference from other trades or other crew members
- Labour skills
- Lack of incentive scheme
- Lack of labour supervision
- Lack of material
- Lack of tools and equipment
- Level of skills and experience of the workforce
- Method of construction
- Occasional working overtime
- Overcrowding
- Payment delay
- Poor communication
- Poor estimation

- Poor site conditions
- Poor site layout
- Project management
- Rework
- Scheduled overtime work
- Shift work
- Specification and standardisation
- Stringent inspection by the engineer
- Tools/equipment breakdown
- Wastage of materials on site
- Work delay caused by inspection delays by local authority
- Workforce absenteeism and changing crew members
- Working overtime

The study by Enshassi, Mohamed and Mustafa (2009) identified and analysed 61 factors (refer to Box 2.2) in 10 categories affecting labour productivity in building construction works in the Gaza Strip. The paper recommends that: 1) project owners must work collaboratively with contractors and facilitate regular payments in order to overcome delays, disputes and claims; 2) project participants should actively have their input in the process of decision-making; and 3) continuous coordination and relationship between project participants are required through the project life cycle in order to solve problems and develop project performance (Enshassi, Mohamed and Mustafa, 2009).

BOX 2.2 FACTORS AFFECTING THE PERFORMANCE OF CONSTRUCTION PROJECTS

Cost factors
- Cash flow of project
- Cost control system
- Cost of rework
- Cost of variation orders
- Differentiation of currency prices
- Escalation of material prices
- Liquidity of organisation
- Market share of organisation
- Material and equipment cost
- Motivation cost
- Overhead percentage of project
- Profit rate of project
- Project design cost
- Project labour cost
- Project overtime cost
- Regular project budget update
- Waste rate of materials

Time factors
- Average delay because of closures leading to materials shortage
- Average delay in claim approval
- Average delay in regular payments
- Percentage of orders delivered late
- Planned time for construction
- Site preparation time
- Time needed to implement variation orders
- Time needed to rectify defects
- Unavailability of resources

Quality factors
- Conformance to specification
- Quality assessment system in organisation
- Quality of equipment and raw materials
- Quality training/meeting
- Unavailability of competent staff
- *Productivity factors*
- Absenteeism rate through project
- Management–labour relationship
- Number of new projects/year
- Project complexity
- Sequencing of work according to schedule

Client satisfaction factors
- Information coordination between owner and project parties
- Leadership skills for project manager
- Number of disputes between owner and project parties
- Number of rework incidents
- Speed and reliability of service to owner

Regular and community satisfaction factors
- Cost of compliance to regulators requirements
- Number of non-compliance events

- Quality and availability of regulator documentation
- Site condition problems

People factors
- Belonging to work
- Employees attitude
- Employees motivation
- Recruitment and competence development

Health and safety factors
- Application of health and safety factors in organisation
- Assurance rate of project
- Project location is safe to reach
- Reportable accidents rate in project

Innovation and learning factors
- Learning from best practice and experience of others
- Learning from own experience and past history
- Review of failures and solving them
- Work group

Environmental factors
- Air quality
- Climate condition
- Noise level
- Waste around the site

Source: Enshassi, A., Mohamed, S. and Mustafa, Z.A. (2009) Factors affecting labour productivity in building projects in the Gaza strip projects in the Gaza strip, *Journal of Civil Engineering and Management*, 16(3), pp. 269–280. Retrieved from: https://journals.vgtu.lt/index.php/JCEM/article/download/6428/5569 Accessed on 14 October 2019.

2.12 Conclusion

Physical infrastructure development is essential for economic development. Infrastructure development is not possible without the construction industry. Therefore, the critical role of the construction industry in economic development, the characteristics of the construction industry, the structure of the construction industry, the nature of investments, the present scenario of the construction industry, the constraints of the Indian construction industry, reasons for delays in the execution of construction projects and factors influencing construction productivity have been discussed at length in this chapter. In order to find the fastest-growing firms in the construction industry in India, the study primarily ranked construction real estate firms by constructing a growth index with a weighted index of growth of revenue and of profit with profit as the weight. One of the major findings of the study is that the turnover leaders were not the growth leaders.

Notes

1 *See* https://www.ibef.org/industry/real-estate-india.aspx.
2 *See*   http://www.prlog.org/10279740-new-report-indian-construction-industry-available-through -aarkstore-enterprise.html.
3 Section 2(h) of the Indian Contract Act, 1872 defines a contract as an agreement enforceable by law. Section 2(e) defines an agreement as 'every promise and every set of promises forming consideration for each other'. Section 2(b) defines a promise in these words: 'When the person to whom the proposal is made signifies his assent thereto, the proposal is said to be accepted. A proposal when accepted becomes a promise'.
4 Article 299, Constitution of India provides the manner of execution of contracts with the union and state governments.
5 Estimated model = (−103714245.591756) + (51880.2998306011) × (Time, i.e. Year).
   The equation of the straight line relating Sales and Time (Year) is estimated as: Sales = (−10371 4245.59) + (51880.30) Year using the 22 observations in this dataset. The y-intercept, the estimated value of Sales when Year is zero, is −103714245.59 with a standard error of 5428329.65. The slope, the estimated change in Sales per unit change in Year, is 51880.30 with a standard error of 2704.01. The value of R^2, the proportion of the variation in Sales that can be accounted for by variation in Year, is 0.9485. The correlation between Sales and Year is 0.9739.
      A significance test that the slope is zero resulted in a t-value of 19.19. The significance level of this t-test is 0.00. Since 0.00 < 0.05, the hypothesis that the slope is zero is rejected. The estimated slope is 51880.30. The lower limit of the 95 per cent confidence interval for the slope is 46239.83 and the upper limit is 57520.77. The estimated intercept is −103714245.59. The lower limit of the 95 per cent confidence interval for the intercept is −115037542.82 and the upper limit is −92390948.36.
6 A similar study was published in Edison, J.C. (2013) An analysis of the fastest growing construction firms in the real estate sector of India. *International Journal of Business and Management Studies*, 2(2), pp. 163–187. Retrieved from: http://www.universitypublications.net/ijbms/0202/html/H 3V206.xml. The same methodology was used for ranking the firms here.
7 *See* http://planningcommission.nic.in/aboutus/committee/wrkgrp11/wg11_constrn.doc Accessed on 23 July 2019.
8 Ibid., p. 53.
9 *See* Economic Survey (2019) Industry and infrastructure (Chapter 8), pp. 210–11. Retrieved from https://www.indiabudget.gov.in/economicsurvey/doc/vol2chapter/echap08_vol2.pdf Accessed on 11 October 2019.
10 *See* https://www.ilo.org/global/topics/dw4sd/themes/productivity/lang--en/index.htm Accessed on 12 October 2019.

References

Akintoye, A.S. and Skitmore, M. (1994) Models of UK private sector quarterly construction demand. *Construction Management and Economics*, 12(1), pp. 3–13.

Asimakopoulous, I., Samitas, A. and Papadogonas, T. (2009) Firm-specific and economy wide determinants of firm profitability: Greek evidence using panel data. *Managerial Finance, 35*(11), pp. 930–939.

Bickerton, M. and Gruneberg, S.L. (2013) London interbank offered rate and UK construction industry output 1990–2008. *Journal of Financial Management of Property and Construction, 18*(3), pp. 268–281.

Briscoe, G. (1992) *The Economics of Construction Industry*, Batsford Ltd., London B.T.

Capon, N., Farley, J.U. and Hoenig, S. (1990) Determinants of financial performance: A meta-analysis. *Management Science, 36*(10), pp. 1143–1159.

Carton, R.B., Hofer, C.W. and Meeks, M.D. (1998) The entrepreneur and the entrepreneurship: Operational definitions of their role in society. In: Kruger, M.E. (Ed) (2004) *Entrepreneurial Theory and Creativity* (Chapter 2). University of Pretoria, South Africa. Retrieved from: https://repository.up.ac.za/bitstream/handle/2263/27491/02chapter2.pdf Accessed on 20 November 2019.

Chandler, G.N. and Jansen, E. (1992) The founder's self-assessed competence and venture performance. *Journal of Business Venturing, 7*(3), pp. 223–236.

Cowling, M. (2004) The growth-profit nexus. *Small Business Economics, 22*(1), pp. 1–9.

Davidsson, P., Achtenhagen, L. and Naldi, L. (2005) Research on small firm growth: A review. In: *Proceedings European Institute of Small Business*, 61. Retrieved from: http://eprints.qut.edu.au/2072/1/EISB_version_Research_on_small_firm_growth.pdf Accessed on 25 November 2019.

Economic Survey. (2019). Industry and infrastructure (Chapter 8), pp. 210–211. Retrieved from: https://www.indiabudget.gov.in/economicsurvey/doc/vol2chapter/echap08_vol2.pdf Accessed on 11 October 2019.

Edison, J. C. (2013) An analysis of the fastest growing construction firms in the real estate sector of India. *International Journal of Business and Management Studies. 2*(2), pp. 163–187. http://www.universitypublications.net/ijbms/0202/html/H3V206.xml

Edison, J.C. (2017, July 1–2) *An Analysis of Causes of Delay in Construction Projects in India.* [Paper presentation], Oxford Business & Economics Conference (OBEC), Saïd Business School, Oxford University, Oxford, United Kingdom.

Enshassi, A., Mohamed, S. and Mustafa, Z.A. (2009) Factors affecting labour productivity in building projects in the Gaza Strip projects in the Gaza Strip. *Journal of Civil Engineering and Management, 16*(3), pp. 269–280. Retrieved from: https://journals.vgtu.lt/index.php/JCEM/article/download/6428/5569 Accessed on 28 September 2019.

Evans, D.S. (1987) The relationship between firm growth, size and age: Estimates for 100 manufacturing industries. *The Journal of Industrial Economics, 35*(4), pp. 567–581.

Fazzari, S.M., Hubbard, R.G. and Petersen, B.C. (1988) Financing constraints and corporate investment. *Brookings Papers on Economic Activity, 1*(1), pp. 141–195.

Freel, M.S. and Robson, P.J.A. (2004) Small firm innovation, growth and performance. *International Small Business Journal: Researching Entrepreneurship, 22*(6), pp. 561–575.

Gebrehiweta, Tsegay and Luo, Hanbin (2017) Analysis of delay impact on construction project based on RII and correlation coefficient: Empirical study. *Procedia Engineering, 196*, pp. 366–374. Retrieved from: https://reader.elsevier.com/reader/sd/pii/S1877705817330825?token=98772700A5893D53EE38B3B20BDB8C062C448E0CC62D39E6EB4B37DD8E3ED9511DB2BEC27100584E96CF9DF2BF87DDF9 Accessed on 10 July 2019.

Gilbert, B.A., McDougall, P.P. and Audretsch, D.B. (2006) New venture growth: A review and extension. *Journal of Management, 32*(6), p. 926.

Government of India. (2013) *Twelfth Five-Year Plan (2012–2017) Economic Sectors* (volume II), Planning Commission, p. 364. Retrieved from: http://planningcommission.gov.in/plans/planrel/12thplan/pdf/12fyp_vol2.pdf Accessed on 27 July 2019.

Government of India. (2018) *Annual Report 2017–18.* Ministry of Statistics and Programme Implementation, pp. 91–92.

Government of India. (2018, July–September) Project implementation status report of central sector projects costing Rs. 150 crore & above. Ministry of Statistics and Programme Implementation,

Infrastructure and Project Monitoring Division, Government of India. Retrieved from: http://www.cspm.gov.in/english/qr/jul-sep_2018.pdf Accessed on 27 July 2019.

Gruneberg, S.L. (1997) *Construction Economics: An Introduction*. Macmillan Press, London.

Gupta, P.D., Guha, S. and Krishnaswami, S.S. (2013) Firm growth and its determinants. *Journal of Innovation and Entrepreneurship*, 2(15), pp. 1–14.

Heshmati, A. (2001) On the growth of micro and small firms: Evidence from Sweden. *Small Business Economics*, 17(3), pp. 213–228.

Hillebrandt, P.M. (2000) *Economic Theory and Construction Industry*, 3rd edition. The Macmillan Press Ltd., London. Retrieved from: https://reader.elsevier.com/reader/sd/pii/S1877705817330825?token=98772700A5893D53EE38B3B20BDB8C062C448E0CC62D39E6EB4B37DD8E3ED9511DB2BEC27100584E96CF9DF2BF87DDF9 Accessed on 10 July 2019.

Jang, S. and Park, K. (2011) Inter-relationship between firm growth and profitability. *International Journal of Hospitality Management*, 30(4), pp. 1027–1035.

Karan, Anup K. and Selvaraj, Sakthivel. (2008, May) Trends in wages and earnings in India: Increasing wage differentials in a segmented labour market. *International Labour Organization*, ILO, New Delhi, India. Retrieved from: https://pdfs.semanticscholar.org/82c6/bf47350fc214b682c3b70e43ce6ced4d2039.pdf Accessed on 10 November 2019.

Kruger, Maria Elizabeth. (2004) Creativity in the entrepreneurship domain. (Doctoral dissertation), University of Pretoria, Pretoria, South Africa. Retrieved from: https://repository.up.ac.za/bitstream/handle/2263/27491/Complete.pdf?sequence=11 Accessed on 20 November 2019.

Laskar, Arghadeep and Murty, C.V.R. (2004) Challenges before construction industry in India. Retrieved from: https://www.iitk.ac.in/nicee/RP/2004_Challenges_Construction_Industry_Proceedings.pdf Accessed on 10 November 2019.

Lumpkin, G.T. and Dess, G.G. (1996) Clarifying the entrepreneurial orientation construct and linking it to performance. *Academy of Management Review*, 21(1), pp. 135–172.

Male, S. (1991) Strategic management in construction: Conceptual foundations. In: Male, S. and Stocks, R. (Eds), *Competitive Advantage in Construction*, pp. 5–44. Butterworth-Heinemann, Oxford.

Mateev, M. and Anastasov, Y. (2010) Determinants of small and medium sized fast growing enterprises in central and eastern Europe: A panel data analysis. *Financial Theory and Practice*, 34(3), pp. 269–295.

Mendelson, H. (2000) Organizational architecture and success in the information technology industry. *Management Science*, 46(4), pp. 513–529.

Morone, P. and Testa, G. (2008) Firms' growth, size and innovation – An investigation into the Italian manufacturing sector. *Economics of Innovation and New Technology*, 17(4), pp. 311–329.

Nakano, A. and Kim, D. (2011) Dynamics of growth and profitability: The case of Japanese manufacturing firms. *Global Economic Review: Perspectives on East Asian Economies and Industries*, 40(1), pp. 67–81. Retrieved from: https://www.tandfonline.com/doi/full/10.1080/1226508X.2011.559329 Accessed on 25 November 2019.

Reid, G.C. (1995) Early life-cycle behaviour of micro-firms in Scotland. *Small Business Economics*, 7(2), pp. 89–95.

Roper, S. (1999) Modelling small business growth and profitability. *Small Business Economics*, 13(3), pp. 235–252. https://link.springer.com/article/10.1023/A:1008104624560#citeas.

Serrasquerio, Z.S. (2009) Growth and profitability in Portuguese companies: A dynamic panel data approach. *Economic Interferences*, 11(26), pp. 565–573. https://core.ac.uk/download/pdf/6283826.pdf.

Serrasquerio, Z.S., Nunes, P.M. and Sequeira, S.T.N. (2007) Firms' growth opportunities and profitability: A nonlinear relationship. *Applied Financial Economics Letters*, 3(6), pp. 373–379.

Sharma, S.N. (2019, July 14) Inside Modi's game plan to bring in investment for infrastructure. *Economic Times Daily*. Retrieved from: https://economictimes.indiatimes.com/news/economy/infrastructure/inside-modis-game-plan-to-bring-in-investment-for-infrastructure/articleshow/70208190.cms Accessed on 22 July 2019.

Wells, Jill. (1986) *The Construction Industry in Developing Countries: Alternative Strategies for Development*, Croom Helm Ltd., London.

World Bank. (2007) *Indian Road Construction Industry: Capacity Issues, Constraints and Recommendations*, p. 4. Retrieved from: http://siteresources.worldbank.org/INTSARREGTOPTRANSPORT/Res ources/Main_Report_Oct30.pdf Accessed on 27 July 2019.

World Bank. (2008) *India – Indian Road Construction Industry: Capacity Issues, Constraints and Recommendations*, pp. 3–5, World Bank, Washington, DC. Retrieved from: http://documents.wor ldbank.org/curated/en/974171468267350704/India-Indian-road-construction-industry-capacity-issues-constraints-and-recommendations Accessed on 25 November 2019.

3 Real estate industry in India

A financial analysis[1]

3.1 Introduction

The real estate sector in India is expected to grow to US$1 trillion by 2030[2] from its current market value of Rs. 12,000 crore.[3] It is forecasted to grow to Rs. 65,000 crore (US$9.30 billion) by 2040.[4] The Tenth Five-Year Plan (2002–7) of the Government of India defines 'real estate' as 'land, including: (i) the air above it; (ii) the ground below it; and (iii) any buildings or structures on it'.[5] It further states that it covers residential housing; commercial complexes and offices; trading and commercial spaces such as theatres, hotels and restaurants; retail outlets and industrial structures such as factories and government administrative buildings. Real estate comprises the (i) purchase, (ii) sale and (iii) development of land and residential and non-residential buildings. According to the Real Estate Bill (2011), a real estate project includes (i) the development of immovable property including construction thereon or alteration thereof and its management; and (ii) the sale, transfer and management of immovable properties.[6] The Economic Survey (2010–11) includes the development of commercial and residential real estate with the participation and involvement of both government agencies and private developers in the real estate sector. The global economic crisis significantly impacted the Indian real estate industry which has now started to recover. As stated by the survey, the real estate sector accounted for 9.3 per cent of gross domestic product (GDP) in 2009–10. Real estate investors play an important role in the development of the Indian economy (Gill, Sharma, Mand and Mathur, 2012). Gill, Mand and Tibrewala (2012) studied the factors that influence the decision of Indian investors to invest in the real estate market and found that investment behaviour and decision-making differ based on the age of the investor. In view of the fact that the Indian real estate market is one of the major contributors to the GDP of the economy and that real estate investors play a significant role in the development of the economy, it is important to examine this industry. The present study examines the relationship between real estate sector investment and GDP and analyses the financial parameters of the Indian real estate industry.

3.2 Literature review

Although studies have been made of industries around the world, literature related to the study of the fastest-growing construction and real estate firms is not available. Real estate is often considered synonymous with real property, in contrast to personal property (Singh, Vandna and Komal, 2009). Considering that it is not a product or service per se, but a whole sector comprising distinct businesses, studying the nature of real estate is complex (Porter, 1989). The Indian real estate market is one of the emerging markets in less developed countries (Sanford, 2006). Investment in the real estate market is a popular investment because everyone needs a place to live.[7] The real estate sector is the second largest employer in India. Estimates show that for every rupee invested in housing and construction, Rs. 0.78 is added to the GDP. Housing ranks fourth in terms of the multiplier effect on the economy and third among 14 major industries in terms of total linkage effect. In 2010–11, the GDP share of the real estate sector (including ownership of dwellings) along with business services was 10.6 per cent.[8] Studies show that there is a high degree of positive correlation between real estate prices and GDP (Singh, Vandna and Komal, 2009).

Real estate in India has strong demand-based growth due to India's population growth, a higher allocation of savings to real estate, the actualisation of mortgaging by consumers and the necessity to bridge the gap in the current housing deficit (Mehta, 2007). The study of Vishwakarma (2013) has found signs of a positive periodically collapsing bubble in the Indian real estate market. Even though 'real estate is long cycle prone business' (Kaiser, 1997), the size of the Indian real estate market is expected to reach US$180 billion by 2020.[9] At the start of the Twelfth Five-Year Plan (2012–17), it was estimated that the total housing deficiency in India would be 18.78 million housing units.[10] Tahsin Ahmad and Joy Sen's (2014, June) paper reveals that the real estate market in India is growing rapidly. On the contrary, in his financial analysis of real estate companies, Knight (2014) points out that the high interest rate regime and higher prices coupled with uncertain job prospects have discouraged real estate investment in India.

The study by Grant Thornton India LLP and CII (2012) shows that the Indian real estate sector has emerged as an expanding base for developers, investors and global stakeholders, buoyed by the growing construction industry in India, which has been undergoing corporatisation and professionalisation and is now recognised as a vital sector for the economic growth and development of the country. The sector witnessed a slight correction due to dwindling demand because of the global economic downturn, a slowdown in the domestic economy, an increase in input costs and controversies over land acquisition. However, in the long run, urbanisation is unavoidable and this will contribute significantly to the demand for real estate. The study states that 'finance has unequivocally been the major challenge for Indian real estate sector'. It has affected the sector both by increased borrowing costs for developers and by demand for real estate which is largely driven by bank finances. Developers are focusing on the latest and most advanced technology to optimise the value of their businesses in the marketplace. This has resulted in a reduction in time wastage, from design to project execution and even in marketing and customer service.

The Deutsche Bank Research (2006) report states that about one in every six people in the world lives in India. The rate of population growth has moderated to just 1.5 per cent per annum. The high birth rate and the reduction in infant mortality during the past few decades suggest that the population of India is very young. One in every three Indians is under the age of 15, and only one in three is older than 35. By 2030, India may become the most populous country on earth. Further, by 2050, the Indian subcontinent's population may reach nearly 1.6 billion, 200 million more than in China. The United Nations Population Division (UNDP) expects the degree of urbanisation to grow by over 40 per cent by 2030, implying that the urban population will grow by 2.5 per cent per annum over the next 25 years. Hence, by 2030, while the rural population will increase only marginally, the urban population will double to approximately 600 million (Deutsche Bank Research, 2006). Therefore, there is enormous scope for developing the real estate industry.

The analysis of Kearney (2012a) ranked India as the fifth most attractive country for retail investment and pronounced that India remains a high-potential market with an accelerated retail growth of 15–20 per cent expected over the next 5 years. The retail sector employs approximately 8 per cent of India's population, with demand for skilled workers expected to increase (Kearney, 2012b). The Indian office market has benefited from offshoring activities. A report by NCAER (2011) revealed that by 2015–16, India would have 53.3 million middle-class households, translating into 267 million people falling into this category.[11] Currently, India has 31.4 million middle-class households (160 million individuals). By 2025–26, the number of middle-class households in India is expected to reach 113.8 million, or 547 million people, an almost three-fold growth from current levels. As per the study, households with an annual income of between Rs. 3.4 lakh and Rs. 17 lakh (at 2009–10 price levels) fall into the middle-class category. According to the report, a typical Indian middle-class household spends about 50 per cent of their total income on day-to-day expenses with the remainder saved. All this proves the sustainability of the economy with the high numbers of middle-class households with sizable disposable incomes and savings, and reveals that there is enormous potential for India's real estate market.

3.3 Real estate industry in India

All firms within the real estate sector of India constitute the real estate industry of India. The industry has moved from single buildings to large layouts to integrated townships. Commercial buildings are contemporary in design and malls are springing up on a national scale like developed countries. The arrival of multinational firms is catalysing the transformation of the industry through the infusion of much required capital, contemporary designs, best practices, etc. Large developers, particularly those who are listed, have already begun to professionalise their organisations, and are developing unique brand identities.[12]

The real estate industry has significant linkages (both direct and indirect) with nearly 250 sectors such as cement, steel, paints and building hardware, which not only contribute to capital formation and the generation of employment and income opportunities, but also catalyse and stimulate economic growth. Therefore, investment in housing and real estate activities can be considered a barometer of growth for the entire economy.[13]

Real estate sales declined by Rs. 275.6 billion in 2014–15 from Rs. 302.8 billion in 2013–14. While the demand for real estate remained low for 2 years (2008–10), i.e. after the beginning of the economic slowdown, the real estate sector expected to complete Rs. 2.11 trillion worth of projects by September 2016.[14]

3.3.1 Financial aggregates of the Indian real estate industry

Firms in the real estate sector generated a total income of Rs. 329 billion and sales of Rs. 276 billion in 2014–15. Net worth and profit after tax (PAT) were Rs. 679 billion and Rs. 22 billion, respectively, during the same period (Table 3.1).

Tables 3.1 and and Figures 3.1 and 3.2 show that all the financial aggregates of the Indian real estate industry, such as total income, sales, net worth and profit after tax, are decreasing. Currently, the real estate industry in India is facing a slowdown in sales. This brings about a moderate decline in demand for the input industries of the real estate industry. A probable revival in the construction and real estate sector in the coming years may drive the demand for input industries upward. The real estate demand has been weak largely due to the slowdown in the economy, weak macro-economic indicators and weakness in the global market, increased property prices and high interest rates due to the tight monetary policy.[15] On the other hand, the return on the assets of the real estate industry in India is

Table 3.1 Financial aggregates of the real estate industry in India (in Rs. billion)

Particulars	2005–06	2006–07	2007–08	2008–09	2009–10	2010–11	2011–12	2012–13	2013–14	2014–15
Total income	111	193	396	359	360	432	384	350	373	329
Sales	100	169	354	297	319	370	328	299	303	276
Net worth	104	163	506	616	753	821	779	787	816	679
Profit after tax	15	46	93	60	61	84	45	34	27	22
Total assets	482	703	1,731	2,185	2,162	2,324	2,198	2,199	2,367	2,003
Total expenses	108	195	397	363	356	403	374	342	385	330
Operating expenses	90	163	309	265	268	303	282	255	277	227
Investments	39	49	214	165	279	285	289	292	290	220
Return on assets	0.024	0.009	0.03	0.016	0.027	0.033	0.024	0.023	0.013	0.027
Sample size	346	502	699	929	1,083	1,079	896	757	624	568

Source: Compiled from industry outlook, Centre for Monitoring Indian Economy, India.

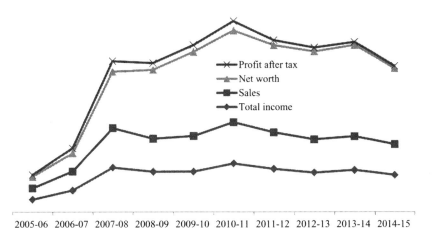

2005-06 2006-07 2007-08 2008-09 2009-10 2010-11 2011-12 2012-13 2013-14 2014-15

Figure 3.1 Financial aggregates of the Indian real estate industry.

growing (Figure 3.3). This indicates an increase in the efficiency level of the real estate industry.

Buyers expected a correction in prices in 2011 owing to an increase in the cost of borrowing. The Reserve Bank of India is also expected to cut rates in the near term. However, as a result of the high cost of construction and cost overruns due to execution delays, builders are not expected to reduce their prices considerably (Centre for Monitoring Indian Economy, 2013, January). It seems the decade beginning 2010–11 was a lost decade for housing sector because all over the world witnessed a slump in demand for housing during this period. Impact of current COVID-19 and the resultant lockdown and the stagnant job market is likely to negatively affect the housing demand further. It appears that the housing demand situation may begin to rise from mid of 2022.

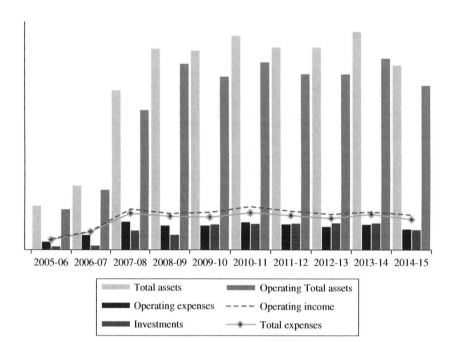

Figure 3.2 Total assets, investments, operating income and expenses.

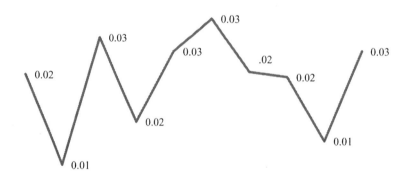

Figure 3.3 Return on assets of the Indian real estate industry.

3.4 The study

A financial analysis of the real estate industry in India was carried out by taking financial aggregates, financial ratios, ratios pertaining to margins on income, returns on investments and efficiency ratios. A macro-level analysis of the real estate industry was also carried out to examine the relationship between real estate sector investments (taken as sales) and GDP.

3.5 Methodology

The data source for the study is the database of the Centre for Monitoring Indian Economy, which has data on over 1000 listed firms in the real estate industry. The financial data of the firms used in the study range from 346 to 1,083 (Table 3.1). In order to validate this, industry sales have been projected up to 2014–15 using a linear regression analysis. The analytical methods used in this study are detailed below.

3.5.1 Data source

The data source for the study is Prowess, Economic Outlook and Industry Outlook database on the financials of Indian companies prepared and maintained by the Centre for Monitoring Indian Economy.

3.5.2 Real estate industry analysis

In order to carry out a ratio analysis of the firms under study, current ratio, quick ratio, debt to equity ratio and return on assets are applied. The analytical tools used for an industry analysis are the financial ratios of the real estate industry pertaining to margins on income, returns on investments and efficiency ratios. The following financial ratios are applied for a real estate industry analysis.

Margins on income on total income

i. Profits before interest, taxes, depreciation and amortisation (PBDITA) as a percentage of total income
ii. Profits before tax (PBT) as a percentage of total income
iii. PAT as a percentage of total income
iv. Cash profit as a percentage of total income

Margins on income on total income net of prior period income and extraordinary income (P&E)

v. PAT net of P&E as a percentage of total income net of P&E
vi. Cash profit net of P&E as a percentage of total income net of P&E

Margins on income on sales

vii. PBDITA net of P&E & OI (other income) & FI (income from financial services) as a percentage of sales

PBDITA, net of FI, is profits before depreciation, interest, tax and amortisation, net of prior period and extraordinary transactions and excluding other income. It also excludes income from financial services. Essentially, these are interest and dividends. This is a close approximation to what is usually called the operating profits of a firm.

Returns on investments on net worth

viii. PAT net of P&E as a percentage of net worth
ix. PAT as a percentage of net worth
x. Cash profit as a percentage of net worth

Returns on investments on capital employed

xi.   PAT net of P&E as a percentage of capital employed
xii. PAT as a percentage of capital employed

Returns on investments on total assets

xiii. PAT net of P&E as a percentage of total assets excluding revaluation
xiv. PAT as a percentage of total assets excluding revaluation

Returns on investments on gross fixed assets (GFA)

xv.   PAT net of P&E as a percentage of GFA excluding revaluation
xvi. PAT as a percentage of GFA excluding revaluation

Asset utilisation (times)

xvii. Total income/total assets
xviii. Total income/compensation to employees

3.6 The real estate industry and the growth of the Indian economy

The general belief is that there is a strong correlation between the real estate industry and economic growth. In order to substantiate this, a model was fitted, using a simple linear regression analysis, by taking a log of the sales of the real estate industry (real estate sector investment) and GDP (at constant prices) (Table 3.2).

A linear regression model is fitted in order to establish the relationship between the real estate industry and the Indian economy (GDP). The results reveal that real estate industry sales are an important determinant of GDP.

The following is a model summary of a regression analysis. R^2 indicates that the model explains around 93 per cent.

Results

Model		Unstandardised coefficients		Sig
		B	Std. Error	
1	(Constant)	14.436	0.191	0
	Sales	0.255	0.017	0*
F			222.455	0.000.

a Dependent variable: GDP.
R^2 = 92.5%.
* Significant at 1%.

A summary statement of the linear regression report is as follows:

The equation of the straight line relating GDP and sales of the real estate industry is estimated as: GDP = (14.43) + (0.25) sales of the real estate industry using the 19 observations in this dataset. The y-intercept, the estimated value of GDP when sales of the real estate industry are zero, is 14.43 with a standard error of 0.19. The slope, the estimated change

Table 3.2 GDP and sales of real estate industry

Year	GDP (in Rs. million)	Growth of GDP* (in percentage)	Sales of real estate industry (in Rs. million)	Growth of sales* (in percentage)
1995–96	17,377,410	–	7,213	–
1996–97	18,763,190	7.97	14,154	96.24
1997–98	19,570,320	4.3	10,115	–28.54
1998–99	20,878,280	6.68	12,473	23.32
1999–00	22,462,760	7.59	12,734	2.09
2000–01	23,427,740	4.3	16,201	27.22
2001–02	24,720,520	5.52	22,554	39.21
2002–03	25,706,900	3.99	25,977	15.18
2003–04	27,778,130	8.06	47,198	81.69
2004–05	29,714,640	6.97	58,694	24.36
2005–06	32,530,720	9.48	100,358	70.99
2006–07	35,643,630	9.57	169,049	68.45
2007–08	38,966,360	9.32	353,803	109.29
2008–09	41,586,750	6.72	296,902	–16.08
2009–10	45,160,710	8.59	319,067	7.47
2010–11	49,185,330	8.91	370,286	16.05
2011–12	52,475,290	6.69	327,640	–11.52
2012–13	54,821,110	4.47	298,696	–8.83
2013–14	57,417,910	4.74	302,781	1.37
2014–15	–	–	275,602	–8.98

Source: Compiled from Economic Outlook and Industry Outlook, Centre for Monitoring Indian Economy, India.
Note: * Calculated.

in GDP per unit change in sales of the real estate industry, is 0.25 with a standard error of 0.02. The value of R^2, the proportion of the variation in GDP that can be accounted for by variation in sales of the real estate industry, is 0.929. Adjusted R^2 is 0.925. The correlation between GDP and sales of the real estate industry is 0.964.

A significance test that the slope is zero results in a t-value of 14.87. The significance level of this t-test is 0.00. Since 0.00 < 0.05, the hypothesis that the slope is zero is rejected.

The estimated slope is 0.25. The lower limit of the 95 per cent confidence interval for the slope is 0.22 and the upper limit is 0.29. The estimated intercept is 14.43. The lower limit of the 95 per cent confidence interval for the intercept is 14.03 and the upper limit is 14.84.

3.7 Real estate industry ratio: Analysis, results and discussions

3.7.1 *Margins on income*

In order to evaluate the profitability of the real estate industry, the parameters of margins on income, viz., (i) return on total assets, (ii) return on income, (iii) return on total income net of P&E and (iv) return on sales, are analysed. The results of the analysis are as follows:

3.7.1.1 *On the basis of return on total income*

To facilitate the evaluation of margins on the income of the real estate industry, parameters of margins on total income, viz., (i) PBDITA as a percentage of total income, (ii) PBT as a percentage of total income, (iii) PAT as a percentage of total income and (iv) cash profit as a percentage of total income, are analysed.

PBDITA is a reasonable measure of the operating profit. Normally, a firm/industry should make sufficient profits at the PBDITA level that it can account for depreciation and amortisation, pay its debts and then, if there is still a surplus, pay direct taxes. PBT as a percentage of total income measures the profit before tax as a percentage of the total income. This is among the most comparable measures of profitability when it comes to comparing companies, or even industries. PAT as a percentage of total income is the final net profit that is made over the total income generated by the firm. Cash profit measures the firm's/industry's ability to generate cash from the business it does in a year.

The study examined profitability ratios, viz., PBDITA as a percentage of total income, PBT as a percentage of total income, PAT as a percentage of total income and cash profit as a percentage of total income, for analysing the return on total income. The return on total income had a steep rise in 2006–08. It shows that the income of firms reduced after 2007–08. The PBDITA as a percentage of total income figures reveals that profitability as a ratio of gross income to net income reduced from its 2008–09 level and started gaining momentum in 2014–15 (Figure 3.4). The other profitability ratios also followed a similar trend except PBT as a percentage of total income. Therefore, it appears that margins on income on the basis of the return on total income of the real estate industry are yet to recover from recession.

3.7.1.2 On the basis of margins on total income net of P&E

A supplementary ratio considered for assisting the evaluation of margins on the income of the real estate industry, parameters of margins on total income net of P&E, viz., (i) PAT net of P&E as a percentage of total income net of P&E and (ii) cash profit net of P&E as a percentage of total income net of P&E, are analysed.

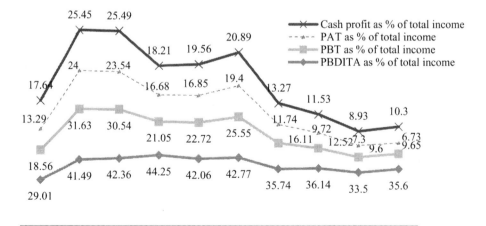

Figure 3.4 Margins on income on the basis of the return on total income.

To derive a more accurate estimate of the profits generated during an accounting period, it is useful to remove the impact of transactions that pertain to prior periods (P) or are extraordinary (E) in nature. PAT net of P&E is such a measure. Cash profit net of P&E as a percentage of total income net of P&E compares the cash generated during an accounting period against the total income generated during the same period after having netted out the prior period and extraordinary transactions from both the numerator and the denominator.

PAT net of P&E as a percentage of total income net of P&E shows that profitability as a ratio of gross income to net income is on an upward trend; however, it was negative in the first 3 years under study. Cash profit net of P&E as a percentage of total income net of P&E reached 25 in 2006–08, reduced during the subsequent 2 years, increased to around 20 in 2010–11 and then reduced continuously during 2012–15 (Figure 3.5). The above ratios indicate that the profit generated during an accounting period against the total income generated has continuously declining during the recent years.

3.7.1.3 On the basis of margins on income on sales

An additional ratio considered for supporting an evaluation of margins on the income of the real estate industry on the basis of PBDITA net of the prior period and extraordinary transactions and other income to sales is analysed (Figure 3.3). PBDITA net of PE&OI is a reasonably close measure of operating profits.

It indicates a decrease after 2009–10. Moreover, the average during the first half of the period under study (2005–10) was over 32 per cent, while during the second half it (average) was only around 27 per cent. Consequently, the trend of PBDITA net of P&E&OI&FI as a percentage of sales indicates that the operating profit is decreasing (Figure 3.6).

3.7.2 Returns on investments

In order to evaluate the profitability of the real estate industry, parameters of margins on income, viz., (i) return on net worth, (ii) return on capital employed, (iii) return on total assets and (iv) return on gross fixed assets, are investigated. The results are as follows:

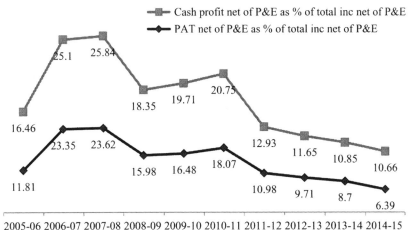

Figure 3.5 Margins on income on the basis of margins on total income net of P&E.

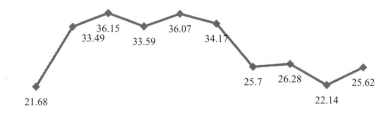

2005-06 2006-07 2007-08 2008-09 2009-10 2010-11 2011-12 2012-13 2013-14 2014-15

Figure 3.6 Margins on income on the basis of margins on income on sales.

3.7.2.1 On the basis of return on net worth

Profit after tax net of P&E as a percentage of net worth is one of the measures of the returns that a business generates on funds provided by its equity shareholders. The equity shareholders fund, or net worth, is the sum of the funds provided by the equity shareholders and the accumulated reserves of the firm. Net worth is always net of revaluation reserves, if any. PAT net of P&E is a better measure of returns on net worth than PAT alone. PAT as a percentage of net worth is the ratio of PAT generated by the firm during a year (an accounting period, to be more precise) and the average of the net worth of the firm at the beginning of the year and at the end of the year. Cash profit is the profit after tax adjusted for the effect of non-cash transactions. Principally, these non-cash transactions are depreciation, amortisation and write-offs. These and other similar non-cash charges are added back to the PAT. Correspondingly, non-cash incomes are deducted from the PAT to derive the cash profit generated by a business during a year.

PAT net of P&E as a percentage of net worth was very high during 2006–07 but began to deteriorate during the 2007–15 period. It declined to 3 per cent (2014–15) from 38 per cent during 2006–07. A similar trend can be observed in the case of cash profit as a percentage of net worth and PAT as a percentage of net worth (Figure 3.7). This proves that the returns that the Indian real estate industry generates on funds provided by its equity shareholders have declined considerably.

3.7.2.2 On the basis of return on capital employed

This is one of the measures of the returns that an enterprise generates on funds provided by its shareholders and lenders. PAT net of P&E is a measure of profits that is net of prior period and extraordinary transactions. Prior period and extraordinary incomes are removed and similar expenses are added back to derive a measure of PAT that better corresponds to the current year's activities. It removes the impact of transactions that are not directly related to the current year's operations. PAT as a percentage of capital employed is a ratio of PAT generated by a firm during a year and the average of the capital employed by a firm as of the beginning of the year and end of the year.

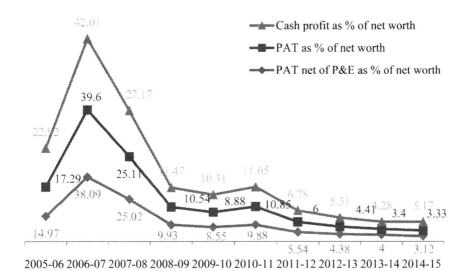

Figure 3.7 Returns on investments on the basis of the return on net worth.

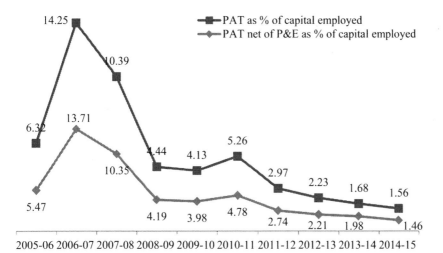

Figure 3.8 Returns on investments on the basis of the return on capital employed.

Both PAT net of P&E as a percentage of capital employed and PAT as a percentage of capital employed indicate that profitability as a ratio of capital employed is currently on a downward trend (Figure 3.8).

3.7.2.3 On the basis of return on total assets

This is one of the measures of the returns that an enterprise generates on the total funds deployed by it in the business. It is a measure of profits that is net of prior period

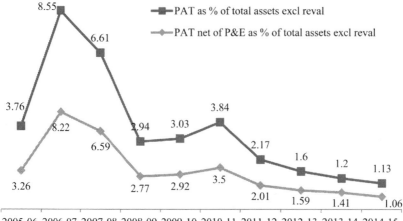

Figure 3.9 Returns on investments on the basis of the return on total assets.

and extraordinary transactions. In order to facilitate an assessment of the returns on investments of the real estate industry, PAT net of P&E as a percentage of total assets excluding revaluation and PAT as a percentage of total assets excluding revaluation are analysed.

Both the ratios reveal that returns generated on the total funds deployed by the real estate industry in the business was negative in the initial year under study, it was maximum in 2006–07 but fell drastically in subsequent years. During 2010–11, it started improving but subsequent years show a drastic decline (Figure 3.9). This establishes that there is a drastic slump in the returns that the Indian real estate industry generates on the total funds deployed by it in the business.

3.7.2.4 On the basis of return on gross fixed assets

This is one of the measures of the returns that an enterprise generates on the fixed assets created by it. Two ratios are studied on returns on investments on the basis of the return on gross fixed assets: (i) PAT net of P&E as a percentage of gross fixed assets excluding revaluation and (ii) PAT as a percentage of GFA excluding revaluation. Since fixed assets are usually maintained at prime productivity levels, the undepreciated gross value of all fixed assets is the denominator in the ratio. Prior period and extraordinary incomes are removed and similar expenses are added back to derive a measure of PAT that better corresponds to the current year's activities. The numerator of this ratio is the PAT net of P&E generated by a firm during a year. The data on both ratios indicate the same trend of the return on total assets. The return on the gross fixed assets of the real estate industry was highest during the boom period but started declining significantly. Both the ratios halved in 2008–09 from the previous year. Recovery started in the last year, i.e. 2010–11 but subsequent years showed a severe decline (Figure 3.10). These ratios reduced to little over six per cent from around 100 per cent (2005–06). This establishes that there is a drastic fall in returns that the Indian real estate industry generates on the fixed assets created by it.

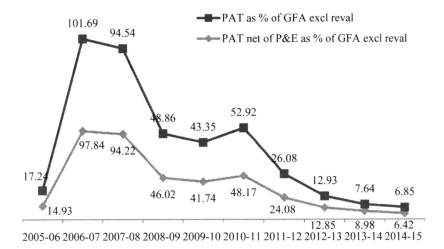

Figure 3.10 Returns on investments on the basis of the return on gross fixed assets.

3.7.3 Asset utilisation (times)

Low ratios of total income and total assets (Figure 3.11) show that the profitability of the industry is falling. This indicates that the real estate industry's efficiency in using its assets to generate revenue has reduced. The ratios of total income and compensation to employees were 29.18 in 2006–07 reaching 20.26 in 2007–08 and increasing to 21.59 in 2014–15 from 19.34 in 2012–13 (Figure 3.11). This indicates a trend of increasing expenditure.

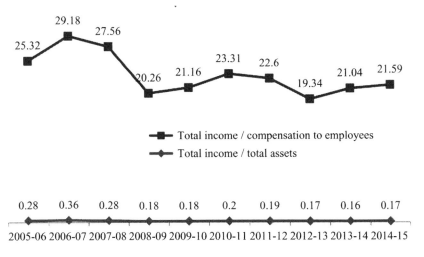

Figure 3.11 Asset utilisation ratios.

3.7.4 Liquidity ratios

The current ratio of the Indian real estate industry began to show a slight upward trend in 2014–15. This indicates that the real estate industry may be able to meet its short-term obligations and regain financial health in the coming years. This is supported by an improvement in the quick ratio and debt to equity ratio of the Indian real estate industry (Figure 3.12). The return on assets indicates the inefficient performance of the industry because it is able to give only a small amount of returns.

The Centre for Monitoring Indian Economy's (CMIE) data on total expenses on raw material, power, fuel and water charges, compensation to employees, etc., indicate that the firms in the real estate industry are reducing their raw material stock as a result of the fall in sales in 2008–09. The situation started improving from 2009–10. The figures for current liabilities confirm an increase in liabilities due to low sales in recent years. A reduction in stock to –48.88 per cent in 2008–09 from 63.23 per cent in the preceding year indicates that firms were reducing their raw material stock. Another notable point is that gross fixed assets reduced considerably in 2007–08; growth in gross fixed assets was 0.31 per cent in 2007–08 from 59.10 per cent in 2006–07. This indicates that firms purchased land and sold it. Similarly, a three-fold growth from the previous year is recorded in 2008–09 in depreciation allocations. This means firms in the real estate industry were buying large quantities of equipment in 2007–08. All this indicates the fact that the firms expected a steady growth in subsequent years. Therefore, it appears that the industry did not have an awareness of business cycles and could not foresee a downturn in the economy.

The demand for real estate is determined by population growth, personal incomes, employment rates, interest rates and access to capital. The profitability of individual firms depends on property values and demand, which are both impacted by general economic conditions.

The decadal growth of the population from 2000–01 to 2010–11 was 16.84. Employment for the five-year period from 2005–06 to 2009–10 was 2.70 crore, 2.73 crore, 2.75 crore, 2.82 crore and 2.87 crore, respectively[16]. This indicates that there was a steady growth in employment in India.

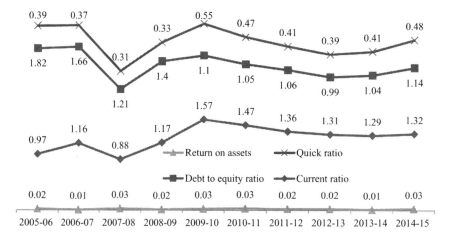

Figure 3.12 Liquidity ratios.

The per capita income at current prices during 2011–12 was estimated to be Rs. 60,972 compared to Rs. 53,331 during 2010–11, showing an increase of 14.3 per cent[17] from Rs. 46,492 in 2009–10.[18]

3.8 Conclusions

The study has considered sales of the real estate industry as real estate sector investment and carried out a linear regression analysis by taking GDP as a dependent variable and sales of the real estate industry as a dependable variable. The results indicate a positive relationship between both variables with 92.5 per cent R^2 and 96.4 per cent correlation. The data shown in Table 3.2 reveal that the real estate industry's sales are in decline. Therefore, it is essential to augment the real estate industry's sales so that growth of the economy can be improved through the real estate industry's backward and forward linkages.

A financial analysis of the Indian real estate industry reveals that: (i) the real estate industry is yet to recover from recession; (ii) the profit generated during an accounting period against the total income generated has continuously declined over the recent years; (iii) operating profit is decreasing; (iv) returns on funds provided by its equity shareholders have declined considerably; (v) profitability as a ratio of capital employed is currently on a downward trend; there is a drastic slump in returns generated on the total funds deployed by it in the business; (vi) there is a drastic fall in the returns that it generates on the fixed assets created by it; (vii) asset utilisation points to a trend of increased expenditure; and (viii) it is able to give only a minimum amount of returns.

The data reveal that firms in the real estate industry were accumulating fixed assets and raw materials before the economic slowdown (2007–08) due to an expectation of steady growth in demand in the subsequent years. They have since considerably reduced their fixed assets and raw material stock as a result of the fall in sales in 2008–09. It appears that the industry's lack awareness of business cycles meant that they did not foresee a downturn in the economy.

Subsequent to the economic slowdown, the scale of operations and sales of the real estate industry had declined. The situation was further aggravated by an increase in the exchange rate. The burden of interest payment on the firms increased due to the increase in the exchange rate and the expensive monetary policy of the Reserve Bank of India. Therefore, the firms have reduced fixed assets and raw-material stock considerably as a result of fall in sales in 2008–09. It appears that the industry lack awareness on the business cycles and could not foresee a downturn in the economy.

In view of the fact that the real estate industry reflects all the macro-economic factors and decisions of the economy, firms should plan their scale of operations by taking into account the macro-economic factors of the economy as well the global economic scenario.

Notes

1 An earlier version of this paper was published in the *Journal of Business and Economic Development*. Edison J.C. (2017, January) Financial analysis of the real estate industry in India. *Journal of Business and Economic Development*, 2(1), pp. 44–56, doi: 10.11648/j.jbed.20170201.16

2 *See* India Brand Equity Foundation. (2019, August) Retrieved from: https://www.ibef.org/industry/indian-real-estate-industry-analysis-presentation Accessed on 17 October 2019.

3 *See Times of India.* (2019, February 22) Size of real estate market to grow to Rs. 65,000 crore by 2040 from Rs. 12,000 crore now: NITI Aayog. Retrieved from: https://timesofindia.indiatimes.com/business/india-business/size-of-real-estate-market-to-grow-to-rs-65000-crore-by-2040-from-rs-12000-crore-now-niti-aayog/articleshow/68114577.cms Accessed on 17 October 2019.

4   *See* India Brand Equity Foundation. (2019, December) Retrieved from: https://www.ibef.org/industry/real-estate-india.aspx Accessed on 20 January 2020.
5   *See* Government of India. (2002) *Tenth Five Year Plan Document* (2002–07). Planning Commission, Government of India, p. 829.
6   *See* Government of India. (2011) *The Real Estate (Regulation & Development) Bill.*
7   Gill, Amarjit S., Harvinder S. Mand and Rajen Tibrewala. (2012), p. 113.
8   Economic Survey 2010–12, p. 241.
9   *See* http://www.ibef.orgartdispview.aspx?in=60andart_id=32963andcat_id=381andpage=1 Accessed on 20 August 2019.
10 *See* http://pib.nic.in/newsite/erelease.aspx?relid=87915
11 *See Economic Times.* (2011, February 6) Retrieved from: http://articles.economictimes.indiatimes.com/2011-02-06/news/28424975_1_middle-class-households-applied-economic-research Accessed on 18 February 2013.
12 *See Times of India.* (2011, August 27) Times Property, Pune.
13 *See* Economic Survey 2011–12, Govt. of India, p. 241.
14 *See* Centre for Monitoring Indian Economy.
15 *See* Industry Outlook, Centre for Monitoring Indian Economy.
16 *See* CMIE. (2013) Economic Outlook. Data retrieved on 12 February 2013.
17 Press Note, Advance Estimates of National Income, 2011–12, Press Information Bureau, Government of India, 7 February 2012, p. 3.
18 *See Economic Times.* (2012, 7 February) Retrieved from: http://articles.economictimes.indiatimes.com/2011-02-07/news/28433672_1_capita-income-data-on-national-income-economy-at-current-prices Accessed on 12 February 2013.

References

Ahmad, Tahsin and Sen, Joy. (2014, June) Evaluating the growth potentials of real estate market in Noida. *International Journal of Advanced Research in Management and Social Sciences*, 3(6), pp. 220–232.

ATKearney. (2012a) Global retail expansion: Keeps on moving. *Global Retail Development Index 2012*. Retrieved from: http://www.atkearney.com/documents/10192/302703/Global+Retail+Expansion+Keeps+On+Moving.pdf/4799f4e6-b20b-4605-9aa8-3ef451098f8a Accessed on 18 February 2013.

ATKearney. (2012b) *Global Retail Index 2012*. Retrieved from: http://www.atkearney.in/consumer-products-retail/global-retail-development-index/past-report/-/asset_publisher/r888rybcQxoK/content/2012-global-retail-development-index/10192 Accessed on 28 December 2016.

Centre for Monitoring Indian Economy. (2013, January) *Indian Industry: A Monthly Review*. Mumbai, India.

Deutsche Bank Research. (2006) *Building up India: Outlook for India's Real Estate Markets*. DB Research, Frankfurt, Germany.

Frank, Knight. (2014, March) *Financial Analysis of the Real Estate Companies*. E&R@Glance. Retrieved from: http://content.knightfrank.com/research/319/documents/en/financial-analysis-of-re-companies-2478.pdf Accessed on 19 October 2016.

Gill, A., Sharma, S.P., Mand, H.S. and Mathur, N. (2012) Factors that influence Indian propensity to invest in the real estate market. *Journal of Finance and Investment Analysis*, 1(2), pp. 137–156.

Gill, A. S., Mand, H.S. and Rajen, T. (2012) Factors that influence the decision of Indian investors to invest in the real estate market. *International Research Journal of Finance and Economics*, 100, pp. 112–121.

Grant Thornton India LLP and CII. (2012) Emerging trends in real estate-India 2012. [Paper presentation]. In: *8th International Conference on Real Estate*, New Delhi.

Kaiser, Ronald W. (1997) The long cycle in real estate. [Paper presentation]. In: *American Real Estate Society Conference Sarasota*, Florida.

Mehta, Rashi. (2007) A study on the Indian real estate market for investment: A qualitative approach (Doctoral dissertation), The University of Nottingham, Nottingham, England. Retrieved from: https://dokumen.tips/documents/07marashimehtapdf.html Accessed on 14 July 2016.

Porter, Michael E. (1989) Competitive strategy and real estate development. [Paper presentation]. In: *Harvard Business School-Remarks to the 1989 Harvard Business School Real Estate Symposium.* Cambridge, MA.

Sanford, J. (2006) Foreign attraction. *Canadian Business, 79*(10), pp. 163–165.

Singh, Vandna and Komal. (2009) Prospects and problems of real estate in India. *International Research Journal of Finance and Economics, 24,* pp. 242–254.

Times of India. (2019, February 22) Size of real estate market to grow to Rs. 65,000 crore by 2040 from Rs. 12,000 crore now: NITI Aayog. *The Times of India.* Retrieved from: https://timesofindia.indiatimes.com/business/india-business/size-of-real-estate-market-to-grow-to-rs-65000-crore-by-2040-from-rs-12000-crore-now-niti-aayog/articleshow/68114577.cms Accessed on 17 October 2019.

Vishwakarma, Vijay Kumar. (2013, January/February) Is there a periodically collapsing bubble in the Indian real estate market? *The Journal of Applied Business Research, 29*(1), pp. 167–172.

4   Economic analysis of projects

4.1 Introduction

The basic economic problem facing all economies is that of allocating scarce resources to a variety of different uses in such a way that the net benefit to society is as great as possible. The tools of an economic analysis facilitate decision makers in examining the impact of a project on a society, an economy, an entity undertaking the project, on stakeholders and the risks and sustainability of the project on the welfare of a country. Given the scarcity of resources, choices must be made among competing uses of resources, and the feasibility and economic analysis of projects are some of the methods used to evaluate alternatives. An economic analysis assesses the benefits and costs of a project, thereby reducing the investment risk. The most viable alternatives with benefits that outweigh costs and with technical, market, financial and economic feasibility and with the least environmental impact are considered as investible projects. Investment projects in the infrastructure development sector are assessed more in relation to their net contribution to the economic growth and development of a country. 'Economic analysis is most useful when used early in the project cycle to identify poor projects and poor project components'(Belli, Anderson, Barnum, Dixon and Tan, 2001). A feasibility study generally contains a description of the business, market feasibility, technical feasibility, financial feasibility, organisational feasibility, etc.(Figure 4.1).[1]

4.2 Economic analysis

An economic analysis helps to understand the economic value of a project. 'Economic analysis aims at identifying economic, social and environmental benefits from the perspective of the national economy' (Konstantin and Konstantin, 2018).The overall economic impact of a construction project will be a multiple of the initial impact because of its linkages with other sectors of the economy (Davis, 1990). Furthermore, information on a project's economic feasibility is used as the most basic and indispensable data in a policy analysis. As such, an economic analysis is the most essential part of preliminary feasibility studies. An appropriate economic analysis can reveal that the selected project is the most efficient or least-cost alternative among all the feasible options for achieving the intended project benefits. Such projects will generate 'a positive economic net present value' (expected net present value[ENOVA]) using the minimum required economic internal rate of return (EIRR) as the discount rate. The project will have an EIRR higher than the discount rate. Even though the process of an economic analysis as per the World Bank is given in Box 4.1, this chapter refers to the methods of the Asian Development Bank.

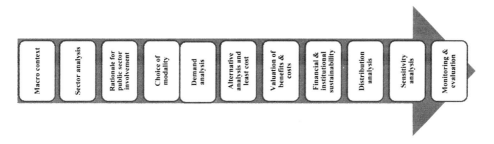

Figure 4.1 Select steps in project economic analysis.

BOX 4.1 THE PROCESS OF ECONOMIC ANALYSIS

After identifying with-project and without-project situations, selecting the best of the alternatives considered and dropping bad project components, the analyst prepares a financial analysis of the project. This step, which examines the net benefits to the project-implementing agency, conveys important information about incentives. It helps assess whether the project would be of interest to the private sector. Once the financial analysis is complete, the analyst needs to adjust the flows and prices to reflect the net benefits to society. As discussed in Chapter 4 of the handbook *Economic Analysis of Investment Operations: Analytical Tools and Practical Applications*, the analyst must get the flows right by removing all subsidies and taxes from the adjusted financial flows, taking into account the project's externalities, especially the environmental externalities. To assess the project's fiscal and financial sustainability, it is important to keep track of who receives or pays for the benefits and costs of the environmental externalities and for the implicit and explicit transfers (typically, income taxes, direct subsidies and property taxes).

After correctly identifying the streams of costs and benefits, the analyst needs to accuratelyprice them. Market prices seldom reflect the economic values of inputs and outputs, and adjustments need to be made. Chapter 5 explains that the main price adjustments include using 'border' prices for all tradeable goods and services and a 'shadow' exchange rate to convert foreign to domestic currency. Information about the sources of divergence between border and market prices and between shadow and market exchange rates will help identify the groups that benefit from and pay for the differences.

The final price adjustments affect nontradeables. If nontradeables are a sizable part of a project's costs, their prices need to be adjusted to reflect opportunity costs to society. As Chapter 5 discusses, labour is one of the most important nontradeables; thehandbook suggests that analysts use a sensitivity analysis to determine whether the project's net present value (NPV) turns negative when using an upper bound for the shadow price of labour (usually the market price). If it does not, then there is no need for further analysis. In many cases, especially health and education projects, volunteer labour is an important component. If project costs and sustainability are to be assessed correctly, such contributions need to be priced at their opportunity costs.

Next, the analyst needs to put this information together and identify sources of divergence between the financial and the economic analysis of the project. The sources of divergence convey very useful information that enables the analyst to answer a number of important questions.

First, by identifying the groups that enjoy the benefits and pay for the costs of the project, this comparison helps identify the impact of the project on the main stakeholders and assess its sustainability. In particular, since taxes and subsidies are usually important sources of difference, this step is essential to assess the project's fiscal impact.

Second, by identifying the causes of the differences between the financial and the economic evaluations, the analyst can tell whether the differences are marketinduced or policyinduced. If they are policyinduced, the analyst needs to consider whether any types of policy changes would bring the economic and financial assessments closer to each other; in short, is the project timely, or might it be preferable to convince the authorities that what is needed is policy reform?

Finally, the comparison also sheds light on the size and incidence of the environmental externalities that can be evaluated in monetary terms.

Source: Adapted from Belli, Pedro, Jock R. Anderson, Howard N. Barnum, John A. Dixon and Jee-Peng Tan. (1998) *Economic Analysis of Investment Operations: Analytical Tools and Practical Applications.* World Bank Institute, The World Bank, Washington, DC. Retrieved from: https://www.adaptation-undp.org/sites/default/files/downloads/ handbookea.pdf Accessed on 18 April 2018.

4.2.1 Benefits vary with the type of project

The details of an economic analysis in preliminary feasibility studies differ depending on the type of project. The benefits of a highway construction project can be estimated by travel time savings, vehicle operating cost savings, such as cuts in fuel expenses, and accident cost savings (Figure 4.2). The benefits of a cultural facility construction project can be estimated by non-use value such as the value of its existence and use value.

4.2.2 The process

Afinancial analysis of a project examines the profit in monetary terms accruing to the authority owning and operating the project, while an economic analysis evaluates the significance of the project on the fundamental objectives of the economy as a whole.An economic analysis starts byestimatingthe demand for a project. Demand estimation requires various statistical methods. Estimated demand should be validated with past experience and similar domestic and foreign projects. It is also used to estimate benefits. To estimate total project costs, it should be possible to estimate initial investment costs such asconstruction and land acquisition costs as well as operation and maintenance (O&M) costs.

An economic analysis is a process of ascertaining a project's economic and financial feasibility through calculating the benefit–cost ratio (BCR), net present value, internal return

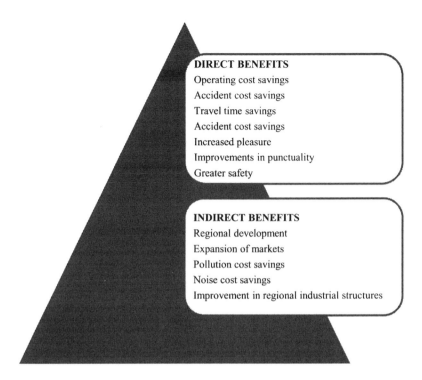

DIRECT BENEFITS
Operating cost savings
Accident cost savings
Travel time savings
Accident cost savings
Increased pleasure
Improvements in punctuality
Greater safety

INDIRECT BENEFITS
Regional development
Expansion of markets
Pollution cost savings
Noise cost savings
Improvement in regional industrial structures

Figure 4.2 The benefits of a highway construction project.

rate (IRR), etc. When it is necessary to address the errors of various estimates used in an economic analysis, a sensitivity analysis is conducted with respect to changes in the primary variables, such as the demand, unit price and discount rate, to determine their impact on the economic feasibility.

4.2.3 Sector analysis

A sector analysis reveals binding constraints (related to policy, legal and regulatory framework, physical infrastructure) onthe efficient functioning of the concerned sector, the reasons for the proposal of aparticular project and how it will facilitate alleviation of the sector constraints.

A sector analysis examines the demand (current and future), the costs, existing sources of supply and any planned investment that may compete with the project. Similarly, it studies the contribution of the planned project to sector demand, technological innovation and cost reduction. Asector analysis also examines the extent of direct government participation in the sector either as a producer or a financier and any government subsidiesor taxes. Another crucial factor to be considered is whether additional physical investment embodied in the concerned project is the optimum solution to the problem at hand (Figure 4.3).

4.2.4 Demand analysis

A demand analysis is of crucial importance to any business. It provides a base for demand forecasting and reducing the business risk. Ademand analysis focuses on the future expectation

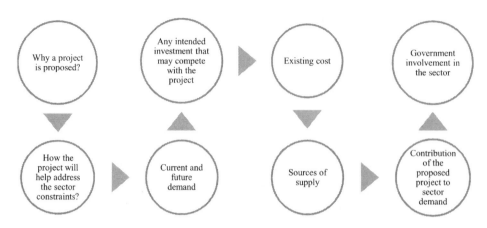

Figure 4.3 Sector analysis.

perspective of the need in question. It identifies the need for an investment by assessing the current demand using models and actual data and then forecasting the future demand. It permits the government or investors to propose a project that is better able to accommodate the present and future reasonable needs of a community. The financial viability and the success of the project will largely depend on robust economic forecasts.

Ademand analysis of an infrastructure project depends essentially on the services that need to be delivered, the size of the project and alternatives for cost recovery. This is especially true of a physical infrastructure that has large capital investment requirements, such as seaports, airports, roads and water and sanitation systems. Anassessment of the future demand for a municipal corporation's waste management project will consider demographic growth and migratory flows. In the case of an industrial waste management project, the prime criterion would be the projected industrial growth in the pertinent economic sectors. Ademand analysis for road projects requires a traffic study, which estimates future traffic growth on the basis of the socioeconomic profile, past traffic and vehicle growth rates and indicative transport elasticity for the various vehicle categories.

4.2.5 *Alternative analysis*

Cost-effectiveness and economic efficiency relate to evaluating alternative means of attaining the same goals and the proposed project should represent the most competent and efficient option among available feasible alternatives for addressing the identified problem. Often, this implies that the chosen project should have the lowest discounted cost per unit of output. Conversely, when project alternatives have very different benefit flows, for instance, because of quality differences, an alternative analysis cannot be supported on the basis of the most promising option based on a cost comparison alone, and the most efficient project option is the one with the highest economic net present value, provided the investment is within budget constraints. In some complex cases, an alternative analysis may be supplemented by a multi-criteria analysis, which concurrently considers a variety of objectives in relation to the evaluated intervention, to determine optimum trade-offs among alternatives, depending on the data available. An alternative analysis should be carried out as part of project preparation.

4.3 Benefit–cost analysis

Cost–benefit analysis entails transforming all benefits and costs into monetary terms, including non-market, environmental, social and other impacts. It compares the estimated costs and benefits arising from a project decision to find whether an investment makes sense, i.e. the benefits outweigh the costs, from a business angle (Keating and Keating, 2017). It plays a critical role in informing investment decision-making. It incorporates all costs to be invested, and subsequently identifies all benefits and converts them into monetary form. The benefits of a project are compared with its costs, as well as the opportunity costs, within a common analytical framework. The direct benefits are generally measured physically in differing units. Other benefits are intangible and difficult to estimate in physical or monetary terms (Figure 4.4). For instance, transportation projects can have a variety of impacts on a society's economic development objectives, for instance productivity, employment, business activity, property values, investment andtax revenues.[2] A society/community can transform from a small town to a city due to the increase in economic activities brought about by improved transportation facilities. The detailed process of a cost–benefit analysis is presented in Figure 4.5.

Generally, transport projects improve businesses' ability and increase economic productivity and development. Transport projects improve (i) the speed of movement of goods and services and (ii) people's ability to access education, employment and services; and theyreduce transportation costs including travel time, vehicle operating costs, road and parking facility costs, accidents and pollution. Many economic impacts are economic transfers (one person, group or area benefits at another's expense) while others are true resource changes (overall economic productivity increases or declines).[3]

Benefits and costs of each option should thus be converted into monetary values in a given time period and compared with the common scenario that would prevail if no action was taken. The net benefit of each alternative option is given by the difference between the costs and benefits. The most economically efficient option is that with the highest present value of net benefits (net present value). Economic efficiency requires selecting the option with maximum net present value, assuming that various options involve equal investments. Options are economically viable only where the net present value that they generate is positive or the present value of total benefits equals or exceeds the present values of total cost (B/C => 1).[4]

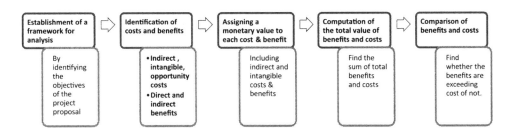

Figure 4.4 The general basic process of cost–benefit analysis.

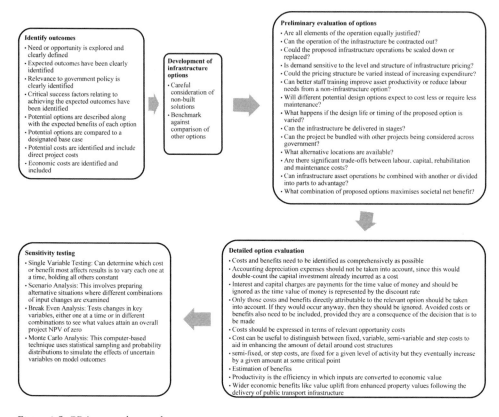

Identify outcomes
- Need or opportunity is explored and clearly defined
- Expected outcomes have been clearly identified
- Relevance to government policy is clearly identified
- Critical success factors relating to achieving the expected outcomes have been identified
- Potential options are described along with the expected benefits of each option
- Potential options are compared to a designated base case
- Potential costs are identified and include direct project costs
- Economic costs are identified and included

Development of infrastructure options
- Careful consideration of non-built solutions
- Benchmark against comparison of other options

Preliminary evaluation of options
- Are all elements of the operation equally justified?
- Can the operation of the infrastructure be contracted out?
- Could the proposed infrastructure operations be scaled down or replaced?
- Is demand sensitive to the level and structure of infrastructure pricing?
- Could the pricing structure be varied instead of increasing expenditure?
- Can better staff training improve asset productivity or reduce labour needs from a non-infrastructure option?
- Will different potential design options expect to cost less or require less maintenance?
- What happens if the design life or timing of the proposed option is varied?
- Can the infrastructure be delivered in stages?
- Can the project be bundled with other projects being considered across government?
- What alternative locations are available?
- Are there significant trade-offs between labour, capital, rehabilitation and maintenance costs?
- Can infrastructure asset operations be combined with another or divided into parts to advantage?
- What combination of proposed options maximises societal net benefit?

Sensitivity testing
- Single Variable Testing: Can determine which cost or benefit most affects results is to vary each one at a time, holding all others constant
- Scenario Analysis: This involves preparing alternative situations where different combinations of input changes are examined
- Break Even Analysis: Tests changes in key variables, either one at a time or in different combinations to see what values attain an overall project NPV of zero
- Monte Carlo Analysis: This computer-based technique uses statistical sampling and probability distributions to simulate the effects of uncertain variables on model outcomes

Detailed option evaluation
- Costs and benefits need to be identified as comprehensively as possible
- Accounting depreciation expenses should not be taken into account, since this would double-count the capital investment already incurred as a cost
- Interest and capital charges are payments for the time value of money and should be ignored as the time value of money is represented by the discount rate
- Only those costs and benefits directly attributable to the relevant option should be taken into account. If they would occur anyway, then they should be ignored. Avoided costs or benefits also need to be included, provided they are a consequence of the decision that is to be made
- Costs should be expressed in terms of relevant opportunity costs
- Cost can be useful to distinguish between fixed, variable, semi-variable and step costs to aid in enhancing the amount of detail around cost structures
- semi-fixed, or step costs, are fixed for a given level of activity but they eventually increase by a given amount at some critical point
- Estimation of benefits
- Productivity is the efficiency in which inputs are converted to economic value
- Wider economic benefits like value uplift from enhanced property values following the delivery of public transport infrastructure

Figure 4.5 CBA general procedure.

4.4 Investment decision criteria

'After identifying and valuing project benefit and cost flows accrued in different years of a project's life that normally spans over 20–30 years, the future flows should be converted to their present value (or the value of a base year) by discounting at a required economic discount rate. The discounting allows calculating aggregated indicators of economic viability of a project for making investment decisions'.[5] Once the economic values of the benefits and costs have been derived, the project proposal's economic viability can be presented in: (i) netpresent value; (ii) benefit–cost ratio (BCR); and (iii) economic internal rate of return.

4.4.1 Benefit–cost ratio

The benefit–cost ratio of a project measures economic efficiency as a ratio of the benefits to costs. It measures a stream of benefits and costs over time resulting from a project.

$$\text{BCR} = \frac{\sum_{t=0}^{n} \dfrac{B_t}{(1+i)^t}}{\sum_{t=0}^{n} \dfrac{C_t}{(1+i)^t}}$$

When the project costs C_t include only fixed investment cost and exclude operation and maintenance costs, this is called the net BCR.

4.4.2 The ENPV and the EIRR

The outcome of a cost–benefit analysis (CBA) is encapsulated in two complementary figures: the EIRR (also referred to as ERR) and the ENPV. The ENPV and the EIRR should be calculated for all projects in which benefits can be valued. Costs and benefits occurring at different time must be discounted. The discount rate in the economic analysis of investment projects, the social discount rate, attempts to reflect the social view on how future benefits and costs should be valued against present ones.

The EIRR is the interest rate at which aproject's discounted benefits equalsitsdiscounted costs, both valued from the whole of society's point of view. Therefore, it will reveal the actual average annual return to society on the capital invested over the lifetime of aproject. A project is accepted if the EIRR is equal to or exceeds a certain threshold (the social discount rate) (EIB, 2013). The ENPV of a project is the difference between the discounted benefits and the costs at a given discount rate. The general criterion for accepting a project proposal is achieving a positive economic net present value discounted at the minimum required economic internal rate of return, or achieving the minimum required EIRR. Every project with an EIRR less than 5 per cent or a negative ENPV after actualisation and with a discount rate of 5 per cent should be carefully appraised or even rejected.[6,7] Where a project's benefits cannot be adequately quantified in monetary terms, its cost-effectiveness must be demonstrated as part of an alternative analysis. In some exceptional cases for social sector projects and projects that generate environmental benefits, a negative ENPV could be accepted if there are important non-monetised benefits.

4.4.3 Net present value

The netpresent value is a measure of the absolute welfare gain over the lifetime of aproject. Aproject should be rejected if the discounted value of the benefits is less than the discounted value of the costs of the project, i.e. if the net present value of the project is negative.

Future benefits and costs are discounted at a compound rate, r, typically 12 per cent per annum.[8] There are two principal rationales for discounting: (i) to reflect the opportunity cost of capital; or (ii) to reflect social time preference. According to current evidence, these rationales lead to discount rates in the region of 10–12 per cent or 5 per cent, respectively. The latter might be appropriate for a project thatmade no draw on capital. Benefits, net of costs, are then summed across all years.[9]

$$NPV = \sum_{t=0}^{n} \frac{B_t - C_t}{(1+i)^t}$$

$(1+r)^t$ is called the discount factor in year t.

Note that if the present value of benefits (PVB) is the sum of the discounted benefit stream,

$$\sum_{t=0}^{n} \frac{B_t}{(1+r)^t}$$

andthe present value of costs (PVC) is the sum of the discounted cost stream,

$$\sum_{t=0}^{n} \frac{C_t}{(1+r)^t}$$

then NPV = PVB − PVC (Table 4.1).

4.4.4 Economic internal rate of return

The economic internal rate of return tells us whether or not the project can improve the overall economic well-being of society.

$$\text{EIRR} = \sum_{t=0}^{n} \frac{B_t}{(1+r)^t} = \sum_{t=0}^{n} \frac{C_t}{(1+r)^t}$$

where:

B_t = benefit at time t;
C_t = cost at time t;
r = EIRR;
n = number of years.

The proposed project is economically feasible if the benefits of the project exceed its costs. This will be indicated by an NPV greater than zero (NPV > 0). One cannot rank a number of alternative investment projects by calculating the NPV of a proposed project because the NPV of a project is predominantly positively linked to the project's investment cost or scale. The benefit–cost ratio is usefulin such a situation since it can evaluate the project in terms of benefits per one monetary unit of cost. A project is worth investing in only if it meets the criterion where the B/C ratio is greater than 1.

> Cost benefit analysis is an applied economic technique that attempts to assess a government program or project by determining whether societal welfare has or will increase (in the aggregate more people are better off) because of the program or project. At its greatest degree of usefulness, Cost-Benefit Analysis can provide information on the full costs of a program or project and weigh those costs against the dollar value of the benefits. The analyst can then calculate the net benefits (or costs) of the program or project, examine the ratio of benefits to costs, determine the rate of return on the government's original investment, and compare the program's benefits and costs with those of other programs or proposed alternatives.
>
> (Kee, 2005)

The IRR is the rate at which benefits are realised following an initial transport investment. It can be thought of as the constant compound rate of return which is equivalent to the actual – fluctuating – rate of return over the project's lifetime. The IRR is also closely related to the NPV: the IRR is the rate of discount at which the NPV of the project is reduced to zero.[10] The economic internal rate of return is the highest interest rate that a project owner

Table 4.1 Example calculations: Highway rehabilitation project

Benefits (B_t)	Costs (C_t)	Discount factor @12% (DF_t = 1/1.12^t)	Discounted benefits (B_t*DF_t)	Discounted costs (C_t*DF_t)	Discounted net benefits ((B_t – C_t)*DF_t)	Net benefit (B_t – C_t)
	325	1		325.0	–325	–325
	285	0.89		254.5	–254.5	–285
110	–	0.8	87.7	–	87.7	110
113.6	–	0.71	80.9	–	80.9	113.6
117.3	–	0.64	74.6	–	74.6	117.3
124.6	–	0.57	70.7	–	70.7	124.6
128.3	–	0.51	65	–	65	128.3
135.6	–	0.45	61.4	–	61.4	135.6
143	–	0.4	57.7	–	57.7	143
150.3	–	0.36	54.2	–	54.2	150.3
157.6	–	0.32	50.8	–	50.8	157.6
165	–	0.29	47.4	–	47.4	165
176	–	0.26	45.2	–	45.2	176
187	–	0.23	42.8	–	42.8	187
198	–	0.2	40.5	–	40.5	198
205.3	–	0.18	37.5	–	37.5	205.3
212.6	–	0.16	34.7	–	34.7	212.6
223.6	–	0.15	32.6	–	32.6	223.6
230.9	–	0.13	30	–	30	230.9
238.3	–	0.12	27.7	–	27.7	238.3
241.9	–	0.1	25.1	–	25.1	241.9
245.6	–	0.09	22.7	–	22.7	245.6
249.3	–	0.08	20.6	–	20.6	249.3
252.9	–	0.07	18.7	–	18.7	252.9
256.6	–	0.07	16.9	–	16.9	256.6
260.3	–	0.06	15.3	–	15.3	260.3
267.6	–	0.05	14.1	–	14.1	267.6
274.9	–	0.05	12.9	–	12.9	274.9
278.6	–	0.04	11.7	–	11.7	278.6
274.1	–	0.04	10.2	–	10.2	274.1
Present value of benefits (PVB)=			1,109.4			
Present value of costs (PVC)=				579.5		
Net present value (NPV)=					530	
Internal rate of return (IRR)=						20%

Source: Retrieved and adapted from: http://siteresources.worldbank.org/INTTRANSPORT/Resources/336291-1227561426235/5611053-1231943010251/trn-6EENote2.pdfAccessed on 15 November 2017.

should economically pay. In other words, the EIRR is the discount rate at which the NPV is just equal to zero. The criterion for project selection is that the EIRR must be greater than the social discount rate (Tangvitoontham and Chaiwat, 2012).

4.4.5 Cost-effectiveness analysis

A cost-effectiveness analysis is used in lieu of a benefit–cost ratio when a valuation of benefits is difficult; it also measures efficiency. It starts from the premise that the productor service concerned must be supplied. Acost-effectiveness analysis looks at a single quantified effectiveness measure of the cost per unit. Thus, there is no room for a 'do nothing' scenario, requiring as acounterfactual at least a 'do minimum' scenario. The appraisal then focuses on whether the chosen technology meets the minimum required cost performance criteria. Should there be room for selecting among alternative options, the result of the analysis may evaluate alternative 'do something' options to help identify the most efficient option, effectively comparing a 'do something' against a 'do something (else)'.[11]

The ratio of the present value of a project's investment and operating costs to the present value of the project's output or outcome is known as the cost-effectiveness ratio (CER). It is primarily used for:

1. choosing the optimum project alternative when project benefits cannot be sufficiently valued and economic viability involves deciding the option with the least cost per unit of output or outcome; and
2. circumstances where project benefits can be priced and project alternatives have similar benefit flows so that investment decisions contain two steps:
    a. choosing the project option with the lowest CER; and
    b. examining whether the economic net present value at the minimum required discount rate is positive or the EIRR is more than the minimum required discount rate.

The CER can be calculated as follows:

$$CER = \frac{\sum_{t=0}^{n} \frac{C_t}{(1+i)^t}}{\sum_{t=0}^{n} \frac{O_t}{(1+i)^t}}$$

where O_t is the output or outcome in year t, which is not in monetary terms.

4.5 Shadow prices

Shadow prices are the value of the contribution to an economy's vital socioeconomic objectives made by any marginal distortion in the availability of factors of production. One of the essential elements in shadow pricing is consumer surplus, the difference between what a consumer is willing to pay for a product and what he actually pays. The consumer surplus is considered part of the benefits arising from a project. General changes in the price level (inflation) should be taken into consideration when calculating shadow (relative) prices.

Shadow prices can be a useful construct in assessing the value of relaxing a resource constraint for the economy.[12] The methodology of shadow prices was developed in the 1970s primarily for evaluating projects in developing countries, where the market conditions were far from efficient and a correction of the market prices was unavoidable. As institutions such as the United Nations Industrial Development Organisation (UNIDO) and the World Bank were active in financing many projects in developing countries around the world, the demand for methodological progress in CBA application came into being and the shadow price methodology was worked out as a reaction to the demand. In CBA, shadow prices are employed to adjust the values of the benefits and costs, including externalities.[13]

In certain cases, the market price may not accurately reflect the economic value of the goods or services, for instance, when: (i) a buyer or seller has undue influence on the price; or (ii) where there are other limitations to efficient pricing. Under these conditions, it is common to apply a shadow price. A shadow price is a non-market-determined price that has been calculated to approximate the economic value of the resources involved in the provision of goods or services.[14] It is used in project appraisal when:

1. there is strong evidence of non-performing markets; or
2. administrated prices are far from matching supply and demand.

4.5.1 Common reasons for distortion of market price

1. Market prices generally include taxes and subsidies.
   These distortions are to be eliminated as they are classified as a transfer within society and not a use of resources.
2. Many impacts such as noise and other forms of pollution do not have any market value as no market currently exists.

Thus, in CBA, shadow prices are used to ensure that these distortionary impacts do not skew the results of an analysis.

The shadow price ratio required in CBA for multiplying flows originally estimated in actual (market) prices to get 'economic flows from the society perspective' instead of 'financial flows from the investor or project provider perspective' would then be the rate between the shadow price of the item and its appropriate market price.[15]

$$SPR = \frac{SP_i}{MP_i}$$

where:

SPR is the shadow price ratio of good i;
SP is the shadow price of good i;
MP is the market price of good i.

Shadow pricing involves converting financial values into economic values. The use of labour in a project prevents its use somewhere else. The forgone output of this labour in its best alternative use is a main constituent of the social cost of using that labour, since productive efficiency is presumably a basic objective. As per Asian Development Bank (ADB) guidelines, due to the scarcity of skilled labour, the actual wage rate paid by the project, together with benefits, can be considered its economic cost. The economic price of workers engaged by a project, who were openly unemployed, will be determined by the monetary compensation they will require to take on work. For unskilled or semi-skilled labour,[16] the shadow wage will be determined by the output forgone in informal sector activities in either rural or urban areas as a result of workers' employment on a project plus, where migration from rural to urban areas is involved, any additional costs of social infrastructure provision (such as housing, health and education services) not borne by the project itself (ADB, 2017).

4.6 Risk and sensitivity analysis

A risk analysis consists of studying the probability that a project will achieve a satisfactory performance (in terms of IRR or NPV), as well as the variability of the result compared to the best estimate previously made. It reveals the project risks related to the value of key project variables, and therefore the risk associated with the overall project result. A quantitative risk analysis examines the range of possible values for the key variables, and the probability with which they may occur. While deciding on a particular project, decisionmakers might take into account not only the expected scale of the project's net benefits but also the risk that they will not be achieved. A sensitivity analysis facilitates identifying (i) the key variables with major influence on the cost and benefits of the project; (ii) the consequences of adverse amendments to these key variables; and (iii) whether these changes will affect the project andthe decisions made; and discovering and implementing the actions that will help mitigate these adverse effects.[17]

Actual values may deviate from the expected values estimated for the analysis. Therefore, it is imperative to investigate the impact of such divergences on the NPV of the project. A generally accepted technique is to change the magnitude of the more important variables, individually or in combinations, by a certain proportion and then determine how sensitive the NPV is to such changes; or instead, to determine the extent to which a variable must change for the NPV to be reduced to zero. Such a sensitivity analysis allows a superior understandingof the crucial elements on which the outcome of the project depends and can also help management by indicating critical areas of the project that demand close supervision. Larger and more complex projects often require risk analysis too.

A sensitivity analysis is undertaken to identify the key variables that can influence project cost and benefit streams. Itis a calculating procedure often used in investment decision-making for predicting the effect of changes of input data (income, costs, value of investments, etc.) on output.

> A methodological approach to Sensitivity Analysis of the criteria for investment project evaluation can be presented, in quite general terms, in the following way. First, we define a set of quantitative criteria which will serve as the basis for the investment project evaluation. Defined after that is a set of input values observed in calculation of criteria, and we select the values whose influence will be analysed, e.g. income from the investment project (P), investment value (I), discount rate (i), etc. Then we determine the range over which these values can move (I_m-1 to I_m-+1), to be used for calculation of individual criteria values. Calculated after that are the values of individual criteria in order to define the values of certain input variables, to determine the maximum and minimum values that certain variables can take with the investment project still remaining profitable, as well as to present the obtained results. Finally, we analyse and interpret the results, and determine the measures and actions that would help us to possibly prevent or remove adverse impacts and make certain improvements.
>
> (Jovanovic, 1999)

An important step in a sensitivity analysis is to identify areasonable range of variability for the main parameters (ADB, 2004, June). The sensitivity analysis undertaken to identify the key variables that can influence project cost and benefit streams involves recalculating the EIRR or the ENPV with varying values of key variables, where the variations can be independent or a combination. A sensitivity analysis aims to assess the effect of adverse changes in key variables on the project's ENPV and EIRR and the implications of these changes for the project investment decision.[18] A sensitivity analysis usually entails four steps, as shown in Figure 4.6.

Guidelines for the Economic Analysis of Projects by the Asian Development Bank propose that 'ex post evaluation studies' and project experience may indicate both the type of

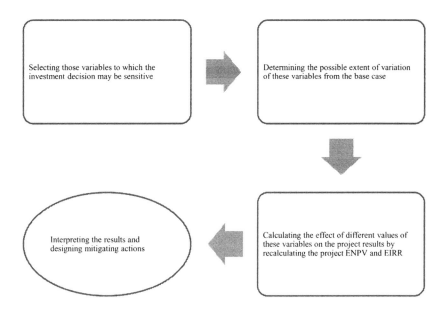

Figure 4.6 Steps in a sensitivity analysis.

variables that are uncertain and the possible extent of deviation from the base case. Ex post evaluation studies comprises (i) output demand; (ii) output prices; (iii) capital cost; and, in some projects, (iv) timing and delays as the key variables to include. Further, it is vital that a sensitivity analysis is not applied mechanically, such as a 10 per cent to 20 per cent reduction in benefits or a 10 per cent to 20 per cent increasein costs. 'Sensitivity analysis should focus on the specific parameters that lie behind the aggregate benefit or cost estimates, so that the true impact of a specific change can be assessed' (ADB, 2017).

Asensitivity analysis aids in: (i) properly assessing the risk and comprehending the benefits and drawbacks of a decision model; (ii) decision-making by predicting the outcome of a decision; and (iii) formulating more realistic, comprehensible and persuasiverecommendations. The outputs of asensitivity analysis generally include a table showing changes in the EIRR and the ENPV, a switching value for each key variable, etc. (Figure 4.7).

The percentage difference between the base case value and the switching value highlights the risk level of the project in relation to key variables. The results of this analysis will then need to be interpreted in terms of the likelihood of switching values occurring and the measures that could be taken to mitigate or reduce the sensitivity and risk analysis likelihood of such variations from the base case. Such measures can include long-term supply contracts for key inputs, better training for project personnel, technical assistance programmes to impart operational management skills and initiatives for institutional and policy reform.[19]

The basic purpose of a sensitivity analysis is to:

1. get an insight into the impact of changes indifferent parameters on changes incertain criteria values;
2. understand the impact of such changes on the total evaluation of a certain investment project's validity;

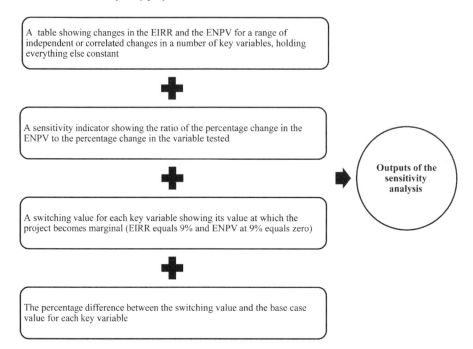

Figure 4.7 Outputs of asensitivity analysis.

3. define steps and actions for purposeful influence to be exerted on certain factors in order to avoid possible unwanted changes of some input values and of investment project evaluation.

To illustrate further, aproject would not be profitable if an increase in the realised total value of an investment exceeded 50 per cent of the starting value. If transgression over the limit is possible, in accordance with predictions, and higher than the maximum allowed (due to the effect of different actual and subjective factors), certain changes in the investment project itself are needed in order to increase its efficiency and reduce its sensitivity to anincrease inthe investment value.[20]

4.7 Sustainability analysis

The economic viability of a project is achieved when a project is designed in such a way that the project's net economic benefits are sustained during its economic life. A sustainability study will establish therelevance, acceptability, political expediency, viability and adaptability of aproject.[21] A financial analysis should consider the executing agency's capacity to implement the project and the project itself.[22] Therefore, two types of evaluation studies can facilitatethe assessment of a project's financial sustainability: (i) financial evaluation of the project; and (ii) financial analysis of the project-executing unit.

Afinancial evaluation will disclose the capability of the project to create sufficient incremental cash flows to cover its (i) financial costs, (ii) capital and (iii) recurrent costs. In this respect, an investigation of the cost recovery goals and mechanisms of the project is significant because a financial evaluation is worthless without full cost recovery. The weighted average cost of capital (WACC) is used as the benchmark to assess the financial viability of aproject. The WACC is determined by ascertaining the actual lending (or on-lending) rates, together with the cost of equity contributed as a result of the project. To obtain the WACC in real terms, the inflation factor is deducted from the estimated cost of borrowing and equity capital (ADB, 1999b). In the case of entire cost recovery, the financial NPV discounted at the project's WACC must be more than zero, and the financial internal rate of return (FIRR) must be more than the WACC. Afinancial evaluation helps in deciding whether or not to proceed with a project. Projects with low returns are riskier to implement and strain the financial sustainability of the corporate entity (public or private) charged with its operation and maintenance. Consequently, it is important to keep these issues in mind when comparing the FIRR of a project against a benchmark such as the WACC. These issues become particularly important as the role of government in the supply and O&M of infrastructure services changes and private sector participation becomes more prevalent (ADB, 1999b).

A financial analysis of the project-executing agency's capacity to implement the project aims to evaluate whether the concerned entity is financially healthy enough to carry out the project and operate and maintain the project's assets. This involves an evaluation of the recent historical, current and expected future performance of the executing agency, calculating the ability of the entity to (i) function as a going concern; (ii) operate and maintain the entire network of assets including the project; and (iii) fund recurrent costs. The institutional capacity of the project-operating entity to implement the project should also be assessed.

4.7.1 Environmental sustainability

An environmental sustainability assessment is an approach to exploring the environmental impacts of a project. The assessment should disclose that a project's outputs can be produced without permanent and unacceptable changestothe natural environment on which it and other economic activities depend, over the life of the project. Anenvironmental sustainability assessment consists of describing the system under study, the selection of sustainable development indicators and quantification. The quantification consists of the monetisation of the results and finding in monetary terms the environmental, social and economic benefits. If a project causes long-term environmental damage, it should compensate expenditure for mitigation. Likewise, the environmental benefits produced by a project shouldbe valued and incorporated into the project's economic analysis.[23] Generally, a sustainability analysis is carried out as part of the economic analysis of the project to assess possible threats to the sustainability of the project.

4.8 Conclusion

In sum, an economic analysis measures the effect of aproject on the national economy as a whole. An economic analysisof a project involves an assessment and comparison of the costs and benefits of a project. It attempts to assess the overall impact of a project on improving the economic welfare of the citizens of the country concerned. This chapter discussed economic analysis, benefit–cost analysis, investment decision criteria, shadow prices, risk and sensitivity analysis and sustainability analysis.

Appendix 1

Financial and economic analysis of a regional power transmission enhancement project[24,25]

(adapted from https://www.adb.org/sites/default/files/linked-documents/44183-013-geo-efa.pdf)

1. The financial and economic analyses of the proposed project were carried out in accordance with the *Financial Management and Analysis of Projects* (2005), *Guidelines for the Economic Analysis of Projects* (1997) and *Economic Analysis of Sub-Regional Projects* (1999) of the Asian Development Bank. All financial and economic costs and benefits are expressed in constant 2012 prices. Cost streams used to determine the financial internal rate of return and economic internal rate of return – capital investment and operation and maintenance – reflect the cost of delivering the estimated benefits and are projected for 35 years after project implementation. The assumed tariffs are based on existing tariffs used by Georgian State Electro-system (GSE).

A. **Methodology and major assumptions**
   i. **Project components**

2. The financial and economic analyses cover the following project components:
   i. **Construction of a new 220/110 kilovolt (kV) substation in Khorga.** A new substation equipped with two units of 220/110 kV transformers with a total capacity of 400 megavolt-amperes (MVA) will be fully commissioned by 2016. The substation will improve the reliability and security of power supply to the region, meeting the increased power demand of the Poti industrial zone and the region, and facilitating power export to Turkey.
   ii. **Substation rehabilitation and improvement.** Up to 11 existing substations will be rehabilitated and improved by replacing and installing a modern digital control and relay protection system to enable remote operation and automatic energy management for efficient power dispatching operation. The existing supervisory control and data acquisition (SCADA) system will be expanded to link major substations and control centres through optical fibres, enabling the system operator to dispatch more efficiently based on real-time data. The rehabilitation and improvement will reduce technical losses by avoiding overload, and enable faster detection and repair of faults, thereby improving the overall efficiency and reliability of power transmission and supply.

3. The third component of the project, namely the feasibility study for a future hydro-power plant is not considered in this analysis.
   ii. **Overall approach and methodology**

4. The financial and economic analyses including the determination of the EIRR, FIRR and weighted average cost of capital are based on streams of benefits and costs resulting from the construction, installation and operation of the project components over their economic lives. The benefits and costs and the EIRR and FIRR were determined separately for all components.

5. The potential project benefits for each component are projected over the project's economic life and are presented in quantitative terms.

6. The financial benefits, and therefore computation of the FIRR, are restricted to those accruing to GSE, even though the ADB guidelines permit the inclusion of financial

benefits that accrue on a regional context (to a country other than Georgia) where the project has contributed. The project's economic benefits are calculated as those that accrue to the Georgian economy and do not include those that can be attributed to other countries (for example, Turkey). The economic benefits are further adjusted to reflect the true economic benefits, costs, inflows and outflows based on the ADB guidelines.

B. **Least-cost solution**

7. The project components are selected in such a way that investment in the new Khorga substation and substation rehabilitation equipment will be at the minimum efficient cost, which increases the load-servicing capacity in the country, most notably to the industrial park at Poti and its surrounding area, resulting in the potential for a more efficient flow of electricity in the region, notably to and from Turkey. Least technically accepted costs are adopted for all project components as no major relevant quality, environmental or social considerations were identified.

C. **Project benefits**

8. The two components, construction of a new substation and rehabilitation and improvement of existing substations, are complementary as due to the rapid development of the network on the Turkish side and their grid code requirements for reliable imports, it is assumed that without adequate protection equipment exports cannot develop much beyond forecasts at the time the project is being constructed. As both the Khorga substation and the substation rehabilitation and improvement components are considered preconditions for the development of exports, the export-related economic and financial benefits of the rehabilitation project are considered together with the export-related benefits of the Khorga substation.

a. **Khorga substation**

9. The major benefits of this project component can be grouped under:
    (i)   enhancement of regional trade, due to the substation facilitating electricity exports to neighbouring countries; and
    (ii) facilitation of increased domestic demand.

10. The project will facilitate enhanced regional trade, as it will provide GSE with greater capacity to supply surplus electricity to cross-border interconnectors to be sold to export markets.[26]

11. On the domestic side, the development of the substation will enable GSE to supply significant increases in load forecast to the Poti industrial park.[27] GSE forecasts that demand in Poti and the surrounding area will increase by 300 MW over 5 years. Without the project, the alternative solution would be to build additional lines from other substations at a higher cost and with greater environmental impacts.[28]

12. Additional benefits not quantified in this analysis include the following:
    (i)   The Khorga substation will enhance the security of supply by increasing redundancy on the 500 kV line that supplies Tbilisi. The Khorga substation will become a secondary substation used to transmit electricity to eastern cities in the event of a failure in the 500 kV line, thereby freeing up capacity on other lines that would have been servicing the increase in load in the developing west.
    (ii) The quality of supply to customers in Georgia will be improved by creating additional redundancies to supply to major load centres such as Tbilisi.

b. **Substation rehabilitation and improvement**

13. The project's installation of protection equipment at a number of substations will provide benefits in the following areas:
    (i)   enhanced regional trade by increasing the capacity and stability of the electricity supply to meet the requirements of neighbouring countries;
    (ii) improved quality of supply to meet domestic demand in Georgia; and
    (iii) improved operating efficiencies at substations.

14. The substation rehabilitation project and the development of the Khorga substation will create economic and financial benefits by facilitating export trade. The substation rehabilitation project will facilitate enhanced regional trade. It will enable GSE to provide a higher-quality supply, thereby meeting the stability requirements for power supply imposed by neighbouring countries, which is a prerequisite for Georgia to become a major exporter of electricity. The major benefits, which are quantified in the analysis, are as follows:
    (i)   The benefit of increasing the transmission quality within the country will induce the private sector to construct generating facilities to export power to neighbouring countries. This will eventually displace existing government-owned and other generating facilities that are currently exporting, and can be used to meet the expected increase in domestic demand. The project is conservatively assumed to enable 10 per cent of exported electricity from existing generating facilities to be displaced by newly developed independent power producers.[29]
    (ii) The incremental benefit of transmission revenues to GSE will result from the project providing commercial and contractual comfort to the importing body in Turkey and other countries.[30] The increase in exports facilitated by the substation will provide economic benefits to customers in Georgia, reducing electricity transmission tariff rates by offsetting revenues generated by GSE by transmitting the export sales from Georgia against the counterfactual of no project. The financial benefits to GSE[31] will arise from increased export sales.
    (iii) Enhancing protection equipment will have non-incremental domestic benefits by improving the quality of supply and reducing the number of incidents and/or short circuits at substations, by increasing the stability of the system and by reducing the number of power outages. Based on data provided by GSE, the amount of undelivered electricity to customers in Georgia is around 5,000 megawatt-hours/year[32]. Over the course of a year, the project will reduce undelivered electricity to customers in Georgia by 60 per cent or 3,000 megawatt-hours. The economic benefit of the reduced outages is conservatively estimated assuming a willingness to pay $0.25/kWh based on the cost of alternative diesel generation sources.[33]

15. The project will also result in operating efficiencies, permitting greater automation of functions at substations, thereby reducing staffing requirements and lowering overall maintenance costs. As eliminating redundancies in staffing is politically sensitive (staffing, currently at 20–30 persons per substation, and may be lowered by up to 90 per cent due to automation), a gradual reduction is expected over the next 10–20 years. This benefit, therefore, has not been quantified under this analysis.

D. Projectcosts

16. The project costs comprise the following:
    (i)   Capital costs incurred during the 3-year installation and construction period covering all the components discussed previously (2013–2015). The base year is assumed as 2012, which means all costs are based on 2012 prices.

(ii)   ADB's Asian Development Fund loan, along with its terms and conditions with interest during the grace period accrued and capitalised, will be re-lent to GSE.

(iii)   Annual operation and maintenance costs of the system during the life of the project[34](analysis period), including interest, are to be paid by GSE.

(iv)   GSE will contribute taxes and duties for the project.

(v)   Physical and price contingencies will be in accordance with ADB's guidelines.

17.   Economic cost of project components: Costs and benefits are given in constant 2012 prices. Use is made of the domestic price numeraire; tradeable inputs are valued at their 'border price equivalent value' and are converted to their domestic equivalent using a standard exchange rate factor of 1.06×. Capital costs contain physical contingencies, but exclude (i) taxes, (ii) price contingencies and (iii) financial charges and duties during construction.

E.   **Weighted averagecost of capital and financial internal rate of return**

18.   To compute the WACC, the financing sources are assumed to comprise the GSE equity contribution financed through retained earnings. Based on Georgia's expected return on equity and 10-year borrowing rate, the cost of GSE's equity is calculated at 7 per cent in nominal costs. The other assumptions are a domestic inflation rate of 10.11 per cent and a tax rate of 15 per cent. The WACC for the project is 0.2 per cent (Table A4.1).

19.   The FIRR for the project is calculated at 3.8 per cent. The revenue stream mainly comes from the wheeling charges from additional exports that are possible with the project. Other revenues include the additional transmission of electricity to the expanding region of Poti. The FIRR compares favourably with the estimated WACC of 0.19 per cent, substantiating the financial viability of the project (Table A4.2).

20.   A separate analysis was carried out to examine the sensitivity of the FIRR and the financial net present value to adverse changes in key variables: (i) a 14 per cent increase in

Table A4.1 Weighted average cost of capital

| | Financing component | | |
| | ADB loan | Government | |
Description	(ADF) (%)	Equity (%)	Total (%)
A. Weighting	71.64	28.36	100
B. Nominal cost	1	10	–
C. Tax rate	20	0	–
D. Tax-adjusted nominal cost [B × (1 – C)]	0.8	10	–
E. Inflation rate	0.5	10.11	–
F. Real cost [(1 + D)/(1 + E) – 1]	0.3	0	–
G. Weighted component of WACC [F × A]	0.21	0	–
Weighted average cost of capital	–	–	0.19

Source:Asian Development Bank. Retrieved from: https://www.adb.org/sites/default/files/linked-documents/44183-013-geo-efa.pdf Accessed on 27 November 2017.

Note

a   ADB = Asian Development Bank; GSE = Georgian State Electro-system; WACC = weighted average cost of capital.

Table A4.2 Detailed financial internal rate of return (thousands of dollars)

Year	Total revenues	Capital investments	Operation costs	Net revenue	Year	Total revenues	Capital investments	Operation costs	Net revenues
2013	–	(21,210)	–	(21,210)	2031	7,052	–	(3,946)	3,105
2014	–	(20,992)	–	(20,992)	2032	7,052	–	(3,972)	3,080
2015	2,797	(10,277)	(559)	(8,040)	2033	7,052	–	(3,997)	3,054
2016	3,901	–	(2,965)	936	2034	7,052	–	(4,023)	3,029
2017	5,301	–	(3,266)	2,034	2035	7,052	–	(4,049)	3,002
2018	5,301	–	(3,288)	2,012	2036	7,052	–	(4,076)	2,976
2019	7,051	–	(3,661)	3,390	2037	7,052	–	(4,102)	2,950
2020	7,051	–	(3,683)	3,368	2038	7,052	–	(4,129)	2,923
2021	7,051	–	(3,706)	3,345	2039	7,052	–	(4,157)	2,895
2022	7,051	–	(3,729)	3,322	2040	7,052	–	(4,184)	2,868
2023	7,051	–	(3,752)	3,299	2041	7,052	–	(4,212)	2,840
2024	7,051	–	(3,776)	3,276	2042	7,052	–	(4,240)	2,812
2025	7,051	–	(3,799)	3,252	2043	7,052	–	(4,268)	2,784
2026	7,051	–	(3,823)	3,228	2044	7,052	–	(4,297)	2,756
2027	7,051	–	(3,847)	3,204	2045	7,052	–	(4,326)	2,727
2028	7,051	–	(3,872)	3,180	2046	7,052	–	(4,355)	2,698
2029	7,051	–	(3,896)	3,155	2047	7,053	–	(4,384)	2,668
2030	7,052	–	(3,921)	3,130					

Source: Asian Development Bank. Retrieved from: https://www.adb.org/sites/default/files/linked-documents/441 83-013-geo-efa.pdf Accessed on 27 November 2017.

Table A4.3 Financial results of sensitivity analysis

Item	FIRR (%)
Base case	3.8
Increase capital costs by 14%	2.9
Decrease revenue by 10%	2.2
Increase O&M costs by 30%	0.6

Source: Asian Development Bank. Retrieved from: https://www.adb.org/sites/default/files/linke d-documents/44183-013-geo-efa.pdf Accessed on 27 November 2017.

Note
a FIRR = financial internal rate of return; O&M = operation and maintenance.

capital costs, (ii) a 10 per cent decrease in revenues and (iii) a 30 per cent increase in operation and maintenance costs (Table A4.3).

In all cases, the rates compare favourably with the estimated WACC value of 0.2 per cent, substantiating the financial viability of the project.

F.   **Economic model and evaluation**
   i.   **Economic internal rate of return**
21.   To test economic viability, the EIRR was calculated based on the incremental cost and benefit streams associated with each project. The economic analysis evaluated the economic performance of the proposed components by comparing the with-project and without-project scenarios, i.e. the economic value of the supply of electricity that the components will provide compared with the existing patterns of energy use in non-electrified areas.
22.   Economic returns are acceptable for all components. This stems from the capacity of the substation enhancements and the development of the Khorga substation to bring

urban-quality power to new industrial areas in the southwest of Georgia and facilitate electricity exports to Turkey.

23. The EIRR is calculated at 13.4 per cent for the project. This rate compares favourably with the 12 per cent benchmark, substantiating the economic viability of the project (Table A4.4).

    iii. **Sensitivity analysis**

Table A4.4 Detailed economic internal rate of return (thousands of dollars)

Year	Revenue from induced benefits	Economic benefits		Total	Capital costs	Operation and maintenance	Net benefits
		Incremental transmission revenues	Non-incremental revenues				
2013	–	–	–	–	(23,432)	–	(23,432)
2014	–	–	–	–	(23,190)	–	(23,190)
2015	1,052	2,629	761	4,442	(11,353)	–	(6,911)
2016	1,052	2,629	765	4,446	–	(2,413)	2,033
2017	2,172	5,429	769	8,370	–	(2,437)	5,933
2018	2,172	5,429	773	8,374	–	(2,462)	5,912
2019	3,572	8,929	777	13,278	–	(2,486)	10,791
2020	3,572	8,929	781	13,281	–	(2,511)	10,770
2021	3,572	8,929	784	13,285	–	(2,536)	10,749
2022	3,572	8,929	788	13,289	–	(2,562)	10,728
2023	3,572	8,929	792	13,293	–	(2,587)	10,706
2024	3,572	8,929	796	13,297	–	(2,613)	10,684
2025	3,572	8,929	800	13,301	–	(2,639)	10,662
2026	3,572	8,929	804	13,305	–	(2,666)	10,639
2027	3,572	8,929	808	13,309	–	(2,692)	10,617
2028	3,572	8,929	812	13,313	–	(2,719)	10,594
2029	3,572	8,929	816	13,317	–	(2,746)	10,571
2030	3,572	8,929	820	13,321	–	(2,774)	10,547
2031	3,572	8,929	825	13,325	–	(2,802)	10,524
2032	3,572	8,929	829	13,330	–	(2,830)	10,500
2033	3,572	8,929	833	13,334	–	(2,858)	10,476
2034	3,572	8,929	837	13,338	–	(2,887)	10,451
2035	3,572	8,929	841	13,342	–	(2,915)	10,427
2036	3,572	8,929	845	13,346	–	(2,945)	10,402
2037	3,572	8,929	850	13,350	–	(2,974)	10,377
2038	3,572	8,929	854	13,355	–	(3,004)	10,351
2039	3,572	8,929	858	13,359	–	(3,034)	10,325
2040	3,572	8,929	862	13,363	–	(3,064)	10,299
2041	3,572	8,929	867	13,368	–	(3,095)	10,273
2042	3,572	8,929	871	13,372	–	(3,126)	10,246
2043	3,572	8,929	875	13,376	–	(3,157)	10,219
2044	3,572	8,929	880	13,381	–	(3,189)	10,192
2045	3,572	8,929	884	13,385	–	(3,220)	10,165
2046	3,572	8,929	889	13,389	–	(3,253)	10,137
2047	3,572	8,929	893	13,394	–	(3,285)	10,109
						EIRR=	13.4%
						ENPV@12%=	7,209

Source: Asian Development Bank. Retrieved from: https://www.adb.org/sites/default/files/linked-documents/441 83-013-geo-efa.pdf Accessed on 27 November 2017.

Note
a EIRR = economic internal rate of return; ENPV = expected net present value.

Table A4.5 Sensitivity of the economic
internal rate of return

Item	EIRR (%)
Base case	13.4
Increase capital costs by 14%	12
Decrease benefits by 10%	12
Increase O&M costs by 30%	12.4

Source: Asian Development Bank. Retrieved
from: https://www.adb.org/sites/default/files/
linked-documents/44183-013-geo-efa.pdf
Accessed on 27 November 2017.

Note
a EIRR = economic internal rate of return;
O&M = operation and maintenance.

24. The sensitivity of the project's economic performance to changes in key variables was
    tested. The results are summarised in Table A4.5. The combined EIRR values remain
    at around the 12 per cent cut-off for all scenarios.
25. The sensitivity analysis shows that the cases with an increase in capital costs of
    14 per cent and a benefit decrease of 10 per cent are most critical. However, the EIRR
    is still at the 12 per cent threshold for the entire project.

Notes

1 The basic reference material for this chapter was ADB. (2017)*Guideline for the Economic Analysis
  of Projects*. Asian Development Bank, Manila. Retrieved from: https://www.adb.org/sites/default/f
  iles/institutional-document/32256/economic-analysis-projects.pdf Accessed on 18 October 2019.
2 *See* http://bca.transportationeconomics.org/benefits/economic-effects Accessed on 18 October
  2019.
3 *See* http://bca.transportationeconomics.org/benefits/economic-effects Accessed on 18 October
  2019.
4 *See* https://hasrelopacu.bandcamp.com/album/cost-benefit-analysis-in-healthcare-pdf-download
  Accessed on 19 October 2019.
5 *See* https://www.adb.org/sites/default/files/institutional-document/32256/economic-analysis-pro
  jects.pdf Accessed on 18 October 2019.
6 *See* http://ec.europa.eu/regional_policy/sources/docgener/guides/cost/guide02_en.pdf Accessed on
  23 May2017.
7 However, according to ADB's *Guidelines for the Economic Analysis of Projects*, 'for social sector
  projects, selected poverty-targeting projects (such as rural roads and rural electrification), and
  projects that primarily generate environmental benefits (such as pollution control, protection of
  the ecosystem, flood control, and control of deforestation), the minimum required EIRR can be
  lowered to six per cent. Where a project's benefits cannot be adequately quantified in monetary
  terms, its cost-effectiveness must be demonstrated as part of alternative analysis'.
8 This figure is not necessarily a precise reflection of the opportunity cost of capital in borrower
  countries.
9 *See* http://siteresources.worldbank.org/INTTRANSPORT/Resources/336291-1227561426235/5
  611053-1231943010251/trn-6EENote2.pdf Accessed on 19 October 2019.
10 *See* http://siteresources.worldbank.org/INTTRANSPORT/Resources/336291-1227561426235/5
  611053-1231943010251/trn-6EENote2.pdf Accessed on 21 October 2019.
11 *See* http://www.eib.org/attachments/thematic/economic_appraisal_of_investment_projects_en.pdf
  Accessed on 21 October 2019.
12 *See* http://www.eib.org/attachments/thematic/economic_appraisal_of_investment_projects_en.pdf
  Accessed on 24 October 2019.

13 *See* https://www.giz.de/en/downloads/giz2014-en-hydropower-economic-development-mekong .pdf Accessed on 16 November 2017.

14 *See* https://www.google.co.in/url?sa=t&rct=j&q=&esrc=s&source=web&cd=5&cad=rja&uact=8 &ved=0ahUKEwi337O10MLXAhXCGpQKHUUDAP4QFgg_MAQ&url=https%3A%2F %2Fwww.tmr.qld.gov.au%2F-%2Fmedia%2Fbusind%2Ftechstdpubs%2FProject-delivery-and-maintenance%2FCost-benefit-analysis-manual%2FCBAManualPart2.pdf%3Fla%3Den&usg =AOvVaw2UaaHVq20EVZyRxMcoe3rc Accessed on 16 November 2017.

15 *See* https://www.vse.cz/polek/download.php?jnl=aop&pdf=558.pdf Accessedon 16 November 2017.

16 In low-income and lower-middle-income developing member countries.

17 *See* Unlimited Consulting and Auditing Partnership.(2018) Sensitivity Analysis.Retrieved from: https://www.readyratios.com/reference/analysis/sensitivity_analysis.html Accessed on 18 December 2019.

18 *See* ADB (2017)*Guidelines for the Economic Analysis of Projects*, op. cit., p. 54.

19 *See* https://www.adb.org/sites/default/files/institutional-document/32256/economic-analysis-pro jects.pdf Accessed on 18 October 2019.

20 *See* https://ac.els-cdn.com/S0263786398000350/1-s2.0-S0263786398000350-main.pdf?_tid=d11 cbf8c-cb4d-11e7-8f5d-00000aab0f26&acdnat=1510892254_7e8fbc90dd5ceea1c7398cdde88 0a028 Accessed on 16 November 2017.

21 *See* http://www.pmi.org/learning/library/2016/06/11/07/23/fundamentals-project-sustainability-9369 Accessed on 16 November 2017.

22 *See* http://siteresources.worldbank.org/INTRANETFINANCIALMGMT/Resources/FMB-Notes/ FM-Partners/Good-Practices-Guide-FM-Analysis-Revenue-Generating-Entities-February-2003.p df Accessed on 21 November 2017.

23 *See* https://www.adb.org/sites/default/files/institutional-document/32256/economic-analysis-pro jects.pdf Accessed on 23 November 2017.

24 *See* https://www.adb.org/sites/default/files/linked-documents/44183-013-geo-efa.pdf Accessedon 27 November 2017.

25 This appendix was retrieved from the Asian Development Bank's website (RRP GEO: Regional Power Transmission Enhancement Project) and presented here to provide more clarity to the readers. Retrieved from: https://www.adb.org/sites/default/files/linked-documents/44183-013-geo -efa.pdf Accessed on 23 November 2017.

26 The export-related benefits are considered in conjunction with the substation rehabilitation component.

27 Conservatively, this additional load should eventually equate to 1,051 gigawatt-hours/year of additional energy supplied, calculated as 300 MW × 8,760 hours/year × 0.4 load factor. From a financial perspective, serving the additional load will provide additional transmission and dispatch revenue of more than $4 million per annum based on existing tariffs.

28 Conservatively, an additional 100 kilometres or $25 million of transmission lines (based on the cost of $250,000/kilometre) would be required just to partially replace the distribution functionality of the Khorga substation and supply the load at Poti.

29 Exports from new generating facilities are estimated at 751 gigawatt-hours (GWh) in 2014, increasing to approximately 2,500 GWh in 2018. The capacity of the line is assumed to be capped at 2,500 GWh per annum for the duration of the analysis period.

30 An example of problems that can arise without contractual comfort of this nature was seen on the interconnection between Tajikistan and Uzbekistan, where in August 2009 Uzbekistan suspended imports from Tajikistan due to the Tajikistan network's poor reliability caused primarily by inadequate protection equipment on substations. This had cascading impacts on the Uzbekistan side of the interconnector.

31 Financial benefits will also arise for Energotrans and generating companies, including ones operating in the Russian Federation, Azerbaijan and Armenia; and exporting to Turkey via Georgia. However, these benefits are not considered in this assessment.

32 This is based on extrapolating data collated by GSE for January–June 2008 over a full year.

33 The financial benefits also allow for some small increase in revenue to GSE due to the additional volume of energy transmitted.

34 Assumed at 4 per cent in the beginning, increasing by 1 per centper annum due to the decreasing efficiencies of the project assets.

References

ADB. (1997) *Guidelines for the Economic Analysis of Projects*. Asian Development Bank, Manila.

ADB. (1999a) *Economic Analysis of Sub-Regional Projects*. Asian Development Bank, Manila.

ADB. (1999b) *Handbook for the Economic Analysis of Water Supply Projects: Guidelines, Handbooks, and Manuals*. Asian Development Bank, Manila. Retrieved from: https://library.pppknowledgelab.org/documents/4174/download Accessed on 24 May 2017.

ADB. (2004, June) *Key Areas of Economic Analysis of Projects: An Overview*. Asian Development Bank, Manila.Retrieved from: https://www.adb.org/sites/default/files/institutional-document/149709/key-areas-economic-analysis-projects-overview.pdf Accessed on 24 May 2017.

ADB. (2005) *Financial Management and Analysis of Projects*. Asian Development Bank, Manila.

ADB. (2017) *Guidelines for the Economic Analysis of Projects*, pp. 25–26. Retrieved from: https://www.adb.org/sites/default/files/institutional-document/32256/economic-analysis-projects.pdf Accessed on 18 October 2019.

Belli, Pedro, Anderson, Jock R., Barnum, Howard N., Dixon, John A. and Tan, Jee-Peng. (2001) *Economic Analysis of Investment Operations: Analytical Tools and Practical Applications*. World Bank Institute, The World Bank, Washington, DC. Retrieved from: http://siteresources.worldbank.org/INTCDD/Resources/HandbookEA.pdf Accessed on 18 April 2018.

Davis, H. Craig. (1990) *Regional Economic Impact Analysis and Project Evaluation*. UBC Press, Vancouver. Retrieved from: https://search.proquest.com/docview/2131403164/bookReader?accountid=34791 Accessed on 30 May 2017.

EIB. (2013) *The Economic Appraisal of Investment Projects at the EIB*. European Investment Bank, Luxembourg, p. 12. Retrieved from: http://www.eib.org/attachments/thematic/economic_appraisal_of_investment_projects_en.pdf Accessed on 23 May 2017.

Jovanovic, Petar. (1999) Application of sensitivity analysis in investment project evaluation under uncertainty and risk. *International Journal of Project Management*, 17(4), pp. 217–222. Retrieved from: https://ac.els-cdn.com/S0263786398000350/1-s2.0-S0263786398000350-main.pdf?_tid=d11cbf8c-cb4d-11e7-8f5d-00000aab0f26&acdnat=1510892254_7e8fbc90dd5ceea1c7398cdde880a028 Accessed on 17 November 2017.

Keating, Barry P. and Keating, Maryann O. (2017) *Basic Cost Benefit Analysis for Assessing Local Public Projects*, 2nd edition. Business Expert Press, LLC, New York.

Kee, J.E. (2005) Cost benefit analysis. In: Kempf-Leonard, K. (Ed) *Encyclopaedia of Social Measurement*, Vol. 1, pp. 537–544. Elsevier, London.

Konstantin, P. and Konstantin, M. (2018) Financial and economic analysis of projects. In: *Power and Energy Systems Engineering Economics*, pp. 65–76. Springer, Cham. Retrieved from: https://link.springer.com/chapter/10.1007/978-3-319-72383-9_5 Accessed on 28 May 2017.

Tangvitoontham, Nantarat and Chaiwat, Papusson. (2012) Economic feasibility evaluation of government investment project by using cost benefit analysis: A case study of domestic port (Port A), Laem-Chabang port, Chonburi Province. *Procedia Economics and Finance*, 2, pp. 307–314.

5   Tendering and bidding

5.1 Introduction

Once the need for a project has been established, the sequential processes involve careful consideration of possible project alternatives and their socioeconomic, financial and environmental evaluation, the establishment of project feasibility and recommendation of the most feasible proposal. Subsequently, project cost estimation, development of detailed design and contract documentation followed by tendering and bidding will take place. Competitive tendering for public works is often introduced when public services are put forward to private contractors with the intention to lower the costs of the publicly provided service compared to if the public entity produces the service on its own. Tendering is very similar to buying and selling goods and services. Project authorities or clients or, in the international terminology, the employer, is the buyer in a tender, and the contractor is the seller. The goals of both parties are opposite to each other; the employer wants the lowest possible price while the contractor wants to gain the greatest profits. The process of tendering attempts to ensure that good quality work is delivered at a reasonable price. The Indian Contract Act, 1872 and the Arbitration and Conciliation Act, 1996 (as amended in 2015) are major legislations governing contracts for procurement (both private and public) in general. The current chapter outlines the conventional tendering and bidding procedure.

5.2 Contract and the tender

When two or more persons communicate a common intention to create some obligation between them, this is said to be an agreement. An agreement[1] which is enforceable by law is a 'contract' (Pollock and Mulla, 2018).

The essential elements of a contract are:(a) an offer made by one person (the promisor); (b) the acceptance of an offer made by the other person (the promisee); (c) the undertaking of an act or abstaining from undertaking a particular act by the promisor for the promisee (called consideration); (d) the offer and acceptance should relate to something which is not prohibited by law; (e) the offer and acceptance constitute an agreement, which when enforceable by law, becomes a contract; and (f) to formulate a valid and binding agreement, the parties entering into such an agreement should be competent to make such an agreement.

For the purposes of an agreement, there must be a communication of intention between the parties thereto. Therefore, in the forms of a contract there are: (a) a proposal; (b) communication of the proposal; and (c) communication of the acceptance of the proposal.

The communication of acceptance of a proposal completes the agreement. An offer may lapse for want of acceptance or be revoked before acceptance (CPWD Works Manual 2014). Acceptance produces something which cannot be recalled or undone. A contract emerges as soon as the offer is accepted and imposes an obligation upon the person making the offer. It has been opined by the Ministry of Law that before communication of acceptance of an offer the tenderer would be within his right to withdraw, alter and modify his tender before its acceptance, unless there is a specific promise to keep the offer open for a specific period backed by a valid consideration. (GOI, 2014)[2]

A tender usually means an offer, also known as a bid, by a tenderer or a bidder to undertake a work in accordance with a signed documented contract against payment. A tender does not create a legal right until it is accepted in a recognised manner. Hence, one is not legally bound to accept a tender. Similarly, there is no legal obligation to accept the lowest tender and mere submission of a tender does not create a right to property. Even if there is a breach of the rules/regulations in accepting/rejecting a tender, no writ will lie for enforcement of contractual rights. Further, all the expenses in tendering will be borne by the tenderer and each bidder can submit only one bid per tender. All bidding costs, costs of site visits/survey and attending pre-bid meetings will be borne by the bidder.

5.3 Selecting a contractor

Tendering in the construction industry is the process of selecting a construction contractor to carry out specific packages of construction work. Contractors for construction contracts are selected on the basis of competitive bidding. It is a process whereby an employer/client invites bids for projects that must be submitted within a predetermined deadline. A tender, also known as a bid,[3] is a submission or an offer made by a prospective contractor or construction firm in response to an invitation to tender to carry out a specific package of construction work. The bid document outlines the offer that the bidder is making, its price and schedules as well as his competencies, experience and eligibility to bid for the project. A tender does not create a legal right until it is accepted and in a recognised manner. Therefore, one is not legally bound to accept a tender. Mere submission of a tender does not create a right to property. All expenses in tendering will be borne by the tenderer. The accepted bids are subsequently evaluated with regard to defined criteria. Generally, to ensure adequate competition, at least three tenders need to be received for a project to make the selection. Instructions to bidders (ITB) provide bidders with all information needed to prepare their proposals Tenders and contracts are governed by the Indian Contract Act.

Competitive bidding, whether international or domestic, helps an employer/client in his selection of a contractor after obtaining a good market response. The major elements of this comprehensive method are:

1. The employer/client invites the contracting firms to quote for the construction work.
2. The employer/client first decides what it wants, and informs the prospective contracting firms (bidders) about the specifications of the goods, inspection methods, competence of bidders, payment terms and delivery schedules.
3. Prospective bidders are usually informed through advertising in newspapers or specialist trade journals. Suppliers are then requested to submit their bids by a given date.
4. The buyer informs the suppliers about the terms and conditions of the contract for the purchase of goods.

5. The employer/client ensures the secrecy of the bids received, until they are publicly opened.
6. The employer/client evaluates the bids and makes a final selection.

In short, once the bids have been received, details of the bids received will be compiled and analysed with respect to specifications, financial soundness, technical ability, etc. Subsequently, the strengths and weaknesses and the previous performance records of the lowest three bidders will be examined. Then, the firm that submitted the lowest bid will be considered for contract award. In some cases, a bid evaluation committee/team may recommend negotiating with the bidder who submitted the lowest conforming tender for a better price with a view to awarding the tender to the bidder. Only prequalified bidders or firms are allowed to bid for the work.

Prospective bidders can contact the client/employer in writing for clarification on the bidding document. They can even examine the site and its surroundings to obtain necessary information for the bid preparation. The employer conducts a pre-bid meeting to clarify issues and to answer queries on any issue that arise at that particular stage. The process of project procurement is given in Figure 5.1.

5.3.1 Select types of tendering processes

The employer is required to confirm that all the necessary tender documents have been prepared and funding sources have been identified prior to opening the tendering process.

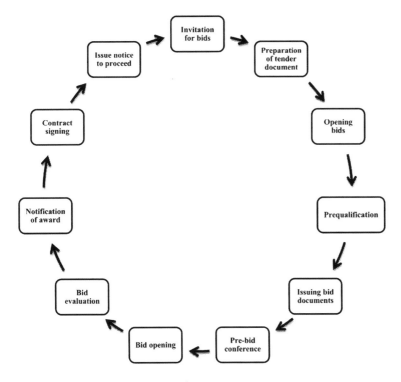

Figure 5.1 Project procurement cycle.

This is a primary requirement for the smooth management of the tendering procedures in the subsequent stages. The type of tendering to be adopted depends on the complexity of the project and its construction, level of expertise required, etc.

The majority of large projects rely on open tender to ensure that the procurement and works to be carried out are conducted in a fair and equitable manner without prejudice. However, it is only one among several tendering methods. The following section discusses different types of tendering methods including open tender.

5.3.1.1 Single-stage process

A single-stage process involves a request for proposal (RFP) only. The single-stage process is appropriate for smaller projects when there is an eminent and comparatively small group of private entities that are likely to bid, and when the project scope and service delivery options can be clearly specified in advance.[4] Invitation to tender documents are issued to various competing contractors who are given the chance to bid for the project based on the same tender documentation.

5.3.1.2 Two-stage tendering process

This process is generally followed for larger projects. Two-stage tendering obtains bids in two stages with receipt of financial bids after receipt and evaluation of technical bids. First, the technical bids are evaluated. The financial proposals remain sealed and secured at this stage. Firms attaining the minimum technical qualifying mark or greater, as stated in the invitation for proposals, are subsequently invited to the public opening of their financial proposals. The Ministry of Finance (Government of India) has mandated this bidding process for central sector public-private partnership (PPP) projects. There are variants to this procedure also.

As per 'Sample Request for Proposals for Large Scale Power Projects'[5] of the World Bank: proposals shall be prepared in the English language in two parts, firstly, a technical and commercial proposal, and secondly a financial proposal. Bidders shall submit one original and three copies of the technical and commercial proposal and one original of the financial proposals.

5.3.1.3 Open tender

Open tendering is the most common tendering method used by employers, both government and private sector. An open tendering procedure is an invitation to tender by public advertisement. The employer advertises the tender offer in the newspapers providing the detail and key information of the proposed work and inviting interested contracting firms to submit their bids or offers. There are no restrictions on who can submit a bid; however, contractors are required to tender all the necessary information and are evaluated against the stated selection criteria. Earnest money deposit (EMD) is mandatory for bidding, which will be returned once a legitimate bid is selected. Open tendering guarantees competition; however, it does not promise the acceptance of any offers.

5.3.1.4 Select tendering

A select tender is an alternative developed to address the limitations of the open tendering method, in which a short list of contractors is prepared to invite and submit bids. Often, the short list is sourced from an open tender or compiled by the client's consultant. This method can improve the quality of the bids received. It can also ensure that only bidders with the required experience and competence are provided the opportunity to submit the necessary bids. Usually, security reasons or the urgency of the work involved warrant this type of more manageable tendering procedure.

5.3.1.5 Multi-stage tendering process

The multi-stage tendering method is used when there are a large number of bidders. At each stage in the process, the most suited bidders to the specific contract requirements are selected. It has distinct stages: (i) expression of interest (EOI); (ii) request for qualification (RFQ); (iii) request for technical proposals (RTP); and (iv) request for proposal to short-list bidders and to seek their financial quotes (Figure 5.2).

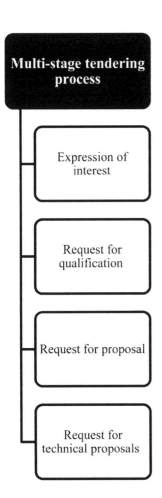

Figure 5.2 Multi-stage tendering process.

5.3.1.5.1 EXPRESSION OF INTEREST

An expression of interest is a request for interested bidders to register their interest in undertaking a specific contract. It is used to identify firms interested and available to bid for a project. Therefore, it is not an invitation to bid and itis not binding on either party. The EOI project-sponsoring authorities determine the level of interest and availability of potential partners and identify a preliminary list of firms who will be sent RFQs or RFPs.[6] The EOI results in a list of all interested firms.

5.3.1.5.2 REQUEST FOR QUALIFICATION

A request for proposal enables project-sponsoring authorities to fairly evaluate competing proposals while reviewing the broadest possible range of potential solutions. At this stage, potential bidders are known and identified, but the number of interested bidders can be large. The project-sponsoring authorities will narrow down the list of qualified firms that will be invited to bid. It is an invitation for contractors to submit a proposal. Unlike an EOI, 'RFQ submissions are evaluated and firms are eliminated on the basis of pre-determined qualifying criteria' (GOI, 2011). Each RFQ bidder should have the technical capability, experience and expertise to construct the facility. Appropriate experience in similar projects is essential. The RFQ bidder should give evidence showing strong credit backing as well as the ability to arrange financing and the required security to perform and complete the project. It must also prove that is has the requisite skill sets and the ability to execute a project of the same nature and scope. After an evaluation of RFQs, the received bidders will be shortlisted. These potential bidders are then invited to submit their proposals for the project at the RFP stage.

5.3.1.5.3 REQUEST FOR TECHNICAL PROPOSALS

In the case of exceptionally complex projects, the project-sponsoring authorities determine that bidders must submit their technical proposals at the RFQ stage, either together with the initial application or at an intermediate stage preceding the RFP stage.[7]

In the request for technical proposals documents, in prescribed format, the short-listed applicants (prospective bidder) submit their technical proposals for construction of the project. Further, modifications/suggestions, if considered necessary by the applicants, will be submitted along with technical proposals. The RTP primarily indicates a proper understanding of the scope of the work, transferring the onus of feasibility from the project authority to the prospective bidder. The consultant prepares the criteria for evaluating the technical proposal before the due date of submission for making an objective assessment.[8]

5.3.1.5.4 REQUEST FOR PROPOSALS

An RFP document is issued to interested bidders inviting them to participate in the bid process. The RFP is the formal bid document issued by the project sponsor and includes the project details and draft concession agreement. An RFP invites technical and financial proposals from interested entities (in case it follows an EOI) or qualified entities (in case it follows an RFQ) or from the market in general (in case of a single stage process).[9] The National Highways Authority of India (NHAI) and similar project sponsors require proposals in three parts: (Part 1) firm's credentials; (Part 2) technical proposal; and (Part 3) financial proposal.

5.4 Preparation of tender documents as per *CPWD Works Manual*

The *CPWD Works Manual* (GOI, 2003) stipulates that an exhaustive estimate showing: (i) the quantities; (ii) rates and amounts of the various items of work; and (iii) the specifications to be adopted, should be prepared prior to invitation to tender for work.[10] Prior to sanctioning, a draft of the detailed estimate, for works involving an architect, should be sent to the senior architect for examination concerning the specifications of various items provided by him.

5.4.1 Elements of tender documents

Tender documents should be prepared and approved by an authority empowered to approve the notice inviting tenders (NIT) before such notice is issued (GOI, 2014). The tender documents must comprise the following[11]:

i. Notice inviting tender in Form PWD 6.
ii. Form of tender to be used together with a set of conditions. Particular specifications and special conditions should not be repetitive and in contradiction to each other. Additional conditions should be decided by the NIT approving authority and he should be responsible for the same.
iii. Schedule of quantities of work.
iv. Set of drawings referred to in the schedule of quantities of work.
v. Specification of the work to be done.

Similarly, prior to approval of a NIT: i) availability of site, funds and approval of local bodies to plans; ii) confirmation that materials to be issued to the contractor would be available; and iii) arrangement for issue of drawing above slab at level 2 well in advance of actual requirement at site as per the programme of construction should be ensured.

As far as possible, the materials stipulated in the tender should be the materials that are either available at the time of inviting the tender or are likely to be obtained before the beginning of the work, if there are no other bona-fide reasons. Further, care must be taken to ensure that a description of the materials to be used is adequately specified in order to prevent any disputes.

'Tenders must be invited in the most open and public manner possible, by advertisement in the press and by notice in English/Hindi and the written language of the district, posted in public place' (GOI, 2003). A notice to tender generally contains: (i) name and address of client; (ii) name of work; (iii) estimated cost; (iv) construction period; (v) earnest money amount; (vi) cost of tender document; (vii) source of funding; and (viii) last date of submission of bid.

5.5 Bidding documents: World Bank–funded projects

World Bank–funded bidding documents contain three parts: (i) bidding procedures; (ii) works requirements; and (iii) conditions of contract and contract forms.

The bidding procedures of such projects consist of five sections: (i) instructions to bidders; (ii) bid data sheet; (iii) evaluation criteria and qualification criteria; (iv) bidding forms; and (v) eligible countries. The conditions of contract and contract forms contain: general conditions (GC); particular conditions (PC); and an annex to the particular conditions – contract

Content of bidding documents	- Instructions to bidders
	- Forms of bid and qualification information
	- Conditions of contract
	- Contract data
	- Specifications
	- Drawings
	- Bill of quantities
	- Forms of securities

Figure 5.3 Contents of bidding documents.

forms. According to World Bank standard bidding documents (SBD), the ITB issued by the employer is not part of the bidding documents.

In general, the set of bidding documents comprises: (1) instructions to bidders; (2) forms of bid and qualification information; (3) conditions of contract; (4) contract data; (5) specifications; (6) drawings; (7) bill of quantities (BQ); and (8) forms of securities. In addition, there are forms for bank guarantees and authorisation forms from manufacturers of specialised equipment (Figure 5.3). Three copies of bid and qualification information, contract data and bill of quantities are supplied to the prospective bidder. The number of copies to be completed and returned with the bid is specified in the bidding data (Figure 5.3).

5.6 Prequalification

Selecting the appropriate contractor can be a challenge for employers. Prequalification (PQ) may be necessary for large or complex projects/contracts. The employer invites notice for prequalification. Prequalification is a method in which the bidder is appropriately examined by analysing the bidder's experience in the same type of work, financial soundness, technical capabilities, managerial soundness, etc. Prequalified construction firms are allowed to bid for a tender.

The success of a project largely depends on the competence of the contractor. The selection of a contractor for construction works is usually through open advertisements. Prequalification proposals are invited through open advertisement. For that, the nature of the job to be executed is indicated in detail, and the contractor is requested to give the firm's registration; annual turnover for the previous 5 years; details of experience as the prime contractor on similar works in the previous 5 years, works in hand and contractual commitments; the construction equipment proposed to be deployed on the job which is either available with the contractor or through fresh procurement; key personnel; subcontracting works; balance sheet for 5 years and projections for the next 2 years; line of credit available; authority to seek information from bankers; details of any current litigation; and construction methodology, quality assurance and quality control methods.

These proposals are evaluated and a short list of contractors capable of executing the project is prepared for issue of tender documents. Short-listed contractors are universally known as prequalified contractors. Tender documents are only sold to such prequalified contractors so that the competition is only between competent agencies. International lending institutions such as the World Bank, the Asian Development Bank, etc., insist on the prequalification of both contractors and consultants so that proposals, both technical and financial, are only invited from the short-listed firms/individuals. Prequalification (short listing) is, thus, an international practice.

The procurement procedures laid down by the Ministry of Finance of the Government of India (GOI, 2006)[12] state the following:

1. Criteria for prequalification together with the evaluation system should be clearly spelt out in detail.
2. The client may engage in (PQ) proceedings with a view to identifying, prior to the submission of tenders, proposals or offers in procurement proceedings, suppliers and contractor that are qualified.
3. The PQ documents (GOI, 2006) shall include the following information[13,14]: (i) instructions for preparing and submitting PQ applications; (ii) a summary of the principal required terms and conditions of the procurement contract to be entered into as a result of the procurement proceedings; (iii) any documentary evidence or other information that must be submitted by contractors to demonstrate their qualifications; (iv) the manner and place, date and time for the submission of applications to prequalify and the deadline for the submission and allowing sufficient time for contractors to prepare and submit their applications, taking into account the reasonable needs of the procuring entity.
4. If the client engages in PQ proceedings, it shall provide a set of PQ documents to each contractor that requests them in accordance with the invitation to prequalify.
5. The client shall respond to any request by a supplier or contractor for clarification of the PQ documents that is received by the project authority within a reasonable time before the deadline for the submission of applications to prequalify. The response by the client shall be given within a reasonable time so as to enable the supplier or contractor to make a timely submission of its application to prequalify. The response to any request that might reasonably be expected to be of interest to other contractors shall, without identifying the source of the request, be communicated to all the contractors to which the project authority provided the PQ documents.
6. The client shall make a decision with respect to the qualifications of each supplier or contractor submitting an application to prequalify. In reaching that decision, the procuring entity shall apply only the criteria set forth in the PQ documents.
7. The procuring entity shall, upon request, communicate to suppliers or contractors that have not prequalified the grounds for its decision; however, the procuring entity is not required to specify the evidence or give the causes for its finding.

5.6.1 Evaluation

An application shall be evaluated in two stages:

(i) evaluation of technical documents submitted; and
(ii) evaluation of credentials (for the applications short listed after Stage I).

A sample evaluation format is given in Table 5.1.

The Government of India's *Manual for Procurement of Works* (GOI, 2019) on the construction experience of the firm and available bid capacity states as follows:

Table 5.1 Sample credential evaluation format

Sr. No.		Description	Points allotted	Max. points
1.		*Past experience of the firm in construction of similar types of work*[a]		
	(a)	Number of years of experience – minimum 3 years	5	7
		More than 3 years	1 per year	–
	(b)	Past experience of projects completed in last 7 years of similar nature	–	30
		One completed project of value 80% or more	12	–
		Every additional completed project of value 80% or more	3 each	–
		One completed project of value 60% or more	6	–
		Every additional completed project of value 60% or more	2 each	–
		One completed project of value 40% or more	4	–
		Every additional completed project of value 40% or more	1 each	–
2.		*Experience of key personnel*[a]		
	(a)	Graduate civil engineers with minimum years' experience	5	7
		Every additional graduate civil engineers with minimum 5 years' experience	1 each	–
	(b)	Graduate electrical engineers with minimum 3 years' experience	2	3
		Every additional quality engineer with minimum 3 years' experience	1 each	–
	(c)	Quality engineer with minimum 3 years' experience	2	3
		Every additional quality engineer with minimum 3 years' experience	1 each	–
3.		*Financial strength*[a]		
	(a)	Average turnover for last 3 years – minimum 100% of estimated value	7	15
		Every additional 10%	2	–
4.	(a)	Presently working in the project area	5	5
	(b)	Within the project state	2	–
5.		Bidders need to submit a proposed construction methodology to execute the job with details such asmanpower/timeline/ quality and similar type of executed works[a]		10
6.		*Performance on works (time over run)*[a](at least three completed work certificate)[a]	10 marks	

		Parameter	Score	Maximum marks
		Calculation for points		10
		If TOR =	1.00 2.00 3.00 >3.50	
	(i)	Without levy of compensation	10 7 5 5	
	(ii)	With levy of compensation	10 5 0 –5	
	(iii)	Levy of compensation not decided	10 7 0 0	

TOR=AT/ST, where AT = actual time; ST = stipulated time.
Note: Marks for value in between the stages indicated above are to be determined on a straight-line variation basis

7.	*Performance of works(quality)*[a](at least three completed work certificate)[a]	10 marks	
	(i)Outstanding	10	
	(ii)Very good	7	
	(iii)Good	5	
	(iv)Poor	0	
	Total		100

Source: Adapted from an application for prequalification for a project invited by National Projects Construction Corporation Ltd., Govt. of India. Retrieved from: http://www.npccindia.com/Writereaddata/data/Tender/JNV-Ch hatarpur.pdf Accessed on 14 March 2019.

Note
a Documentary evidence to be submitted

Particular Construction Experience and Key Production Rates
   The applicant should have:

1. Successfully completed or substantially completed similar works during last seven years ending last day of month previous to the one in which applications are invited should be either of the following:
    1.1 Three similar completed works costing not less than the amount equal to 40 (forty) per cent of the estimated cost; or
    1.2 Two similar completed works costing not less than the amount equal to 50 (fifty) per cent of the estimated cost; or
    1.3 One similar completed work costing not less than the amount equal to 80 (eighty) per cent of the estimated cost; and
2. The applicant should also have achieved the minimum annual production value of the key construction activities (e.g. dredging, piling, or earthworks etc.) stipulated.

The similarity of work shall be pre-defined based on the physical size, complexity, methods/technology and/or other characteristics described, and scope of works. Substantial completion shall be based on 80 (eighty) per cent (value wise) or more works completed under the contract. For contracts under which the applicant participated as a joint venture member or sub-contractor, only the applicant's share, by value, shall be considered to meet this requirement. For arriving at cost of similar work, the value of work executed shall be brought to current costing level by enhancing the actual value of work at simple rate of seven per cent per annum, calculated from the date of completion to the date of Bid opening.

Available Bid Capacity

The bidder should possess the bidding capacity as calculated by the specified formula. The formula generally used is:

   Available bid capacity $= A \times M \times N - B,$

where

A = Maximum value of engineering (Civil/Electrical/Mechanical as relevant to work being procured) works executed in any one year during the last five years (updated at the current price level), taking into account the completed as well as works in progress.
M = Multiplier Factor (usually 1.5)
N = Number of years prescribed for completion of the work in question.
B = Value (updated at the current price level) of the existing commitments and on-going works to be completed in the next 'N' years.[15]

5.6.2 *Sale of tender documents*

Tender documents should be prepared and kept ready for sale to the contractors prior to the notice beingsent to the press or pasted on a notice board. Every contractor desiring to tender shall be asked to make a written application. It is the duty of the executive engineer/assistant executive engineer/assistant engineer to ensure that the tender documents are made available to the contractors as soon as the application is made by the contractors in the prescribed form. Tender documents will essentially contain: (i) instructions to bidders, (ii) general conditions of contracts, (iii) conditions of particular application, (iv) technical specifications and method of measurements, (v) form of bid, (vi) appendix to bid, (vii) bid security, (viii) bill of quantities and rates, (ix) form of agreement, (x) forms of securities, (xi) performance bond, (xii) bank guarantee for mobilisation of advance payment, (xiii) drawings and (xiv) soil data.

5.6.3 Clarification of bidding documents

Multilateral lending agencies allow a bidder to seek clarifications. The bidder shall thoroughly examine the bidding documents in all respects and if any conflict, discrepancy, error or omission is observed, the bidder may request clarification. A clarification, in writing, may be asked for 28 days before the deadline for bids. The employer has to answer queries and questions, and inform each of the buyers of bidding documents about the nature of the clarifications sought and replies given. The employer's response shall also be put on his website.

5.6.4 Pre-bid meeting

After the tender documents have been issued to short-listed contractors and sufficiently ahead of the date of submission of tenders, a pre-bid conference is held by the employer, to which all intending bidders are invited. Before this conference, the intending bidders are also encouraged to visit the site and see in person the site conditions, the facilities available at the site and its vicinity and examine all data on investigations, etc., that are available with the project authority so that the contractors have a clear picture of what is expected of them and the conditions under which the works are to be executed. During the conference, the job requirements issues of bidders will be clarified and contract conditions explained wherever an explanation is sought.

If the estimated cost of work is Rs. 2 crore and above, as per the *CPWD Works Manual*, a pre-bid conference shall be held by the engineer-in-charge. The pre-bid conference shall be held about 10 days before the last date of submission of tenders to clarify any of the contractors' issues. Any additional suggestions proposed by the contractors will also be discussed during the pre-bid conference. The minutes of the meeting shall be circulated to all prospective tenderers attending the conference. For works with a contract value less than Rs. 2 crore, where necessary, a pre-bid conference may be conducted in the manner explained above at the discretion of the notice invitingtender approving authority.

This pre-bid conference allows the contractors to get a clear idea of the conditions of the contract and they are also free to ask for clarification on any of the job requirements, design specifications, etc., and the various contract clauses including their legal implications. A summary record of the conference will be prepared as an addendum and sent to all the intending bidders, to whom the contract documents would have been sent, along with modifications and amendments to the contract documents resulting from this conference. Representation from each tenderer should be limited two or three persons. Non-attendance of the meeting does not disqualify a bidder.

5.7 Preparation of bids

The level of detail of the project at the project preparation phase should be driven by the amount of information required by the prospective bidders for preparing bids. The client/ employer developing the project should provide adequate information for the preparation of bids. Bidders are expected to observe all instructions as well as study the drawings, specifications and schedule.

5.7.1 *Language of bid*

All documents relating to the bid shall be in the language specified in the contract data. English can be used as per the World Bank documents.

5.7.2 *Documents comprising the bid*

Documents comprising the bid should be as per the tender document referred to in the notice inviting tender (refer to Box 5.1).

BOX 5.1 DOCUMENTS COMPRISING THE BID AS PER THE STANDARD BIDDING DOCUMENT FOR HYDROPOWER PROJECTS, THE MINISTRY OF POWER, GOVERNMENT OF INDIA

a) The duly filled in bid form.
b) The demand draft for the cost of the bidding documents, placed in a separate cover, marked 'Cost of Bidding Document'.
c) The earnest money deposit in a separate cover marked 'Earnest Money Deposit'.
d) Bidder's address of communication:
   Telephone no(s):
   Office no.:
   Mobile no.:
   Facsimile (fax) no.:
   Electronic mail identification (E-mail ID):
e) Qualification information and documents.
f) Drawings prepared by the bidder (other than tender drawings provided by the employer).
g) Priced bill of quantity/price schedule.
h) Any other information/documents required to be completed and submitted by bidders in accordance with instructions as per ITB.

The documents marked 'ORIGINAL' listed under Sections 2 (Parts I & II), 4 and 7 of ITBClause 9 shall be duly filledin, signed, stamped and submitted without exception.

The following documents, which are not submitted with the bid, will be deemed to be part of the bid.

Section	Particulars
Invitation for bids (IFB)	
I.	Instruction to bidders (ITB)
II.	Information to bidders (INFB)
III.	Conditions of contract (GCC & SCC)
IV.	Specifications/drawings

The bidder shall treat the bid documents and contents thereof as confidential.

Source: Adapted from: http://www.cea.nic.in/reports/hydro/bidding_doc_hyd_sbd .pdf Accessed on 7 October 2019.

5.7.3 *Bid security*

Bidders are required to submitthe bid security along with the bids. This is to ensure that the contractor does not withdraw from his tender when it is under consideration or from executing the work after being awarded the work – the bid validity period.[16] The amount of bid security should ordinarily range between 2 per cent and 5 per cent of the estimated value of the goods to be procured.[17] Common forms of security include a letter of credit, a bank guarantee, a cashier's certificate and a crossed demand draft/pay order in favour of the employer from any nationalised or scheduled bank. The bid bond will remain valid for a period of 28 days beyond the validity period of the bid.

Prompt return of the bid security to unsuccessful bidders is assured. The bid security of the successful bidder will be released as soon as the bidder has signed the agreement and handed over the required performance security. The general financial rules of the Ministry of Finance, Government of India (GOI, 2005) states that the bid securities of the unsuccessful bidders are 'required to be returned to them at the earliest after expiry of the final bid validity and latest on or before the 30th day after the award of the contract'.[18]

5.7.4 *Format and signing of bid*

As per the procurement policies and procedures[19] prescribed by the World Bank, the bidder shall prepare an original of the documents comprising the bid as described in the clause 'Documents Comprising the Bid' of the ITB given above, bound with the volume containing the form of bid, and clearly marked 'ORIGINAL'. In addition, the bidder shall submit copies of the bid, in the number specified in the bidding data, clearly marked as 'COPIES'. In the event of a discrepancy between them, the original shall prevail.

The bid shall be typed or written in indelible ink and shall be signed by a person or persons duly authorised to sign on behalf of the bidder (Sub-Clauses 4.3(a) or 4.4(b)).

Sub-Clauses 4.3(a): Copies of original documents defining the constitution or legal status, place of registration, and principal place of business; written power of attorney of the signatory of the Bid to commit the Bidder.
Sub-Clauses 4.4(b): The Bid shall be signed so as to be legally binding on all partners.[20]

Each page of the bid shall be signed by the person or persons signing the bid.

5.8 Submission of bids

5.8.1 *Sealing and marking of bids*

This clause states information on how to seal and mark the bids as an original or a copy. Place the sealed bid, both original and copies of the bid, in two inner envelopes and one outer envelope, duly marking the inner envelopes as 'ORIGINAL' and 'COPIES' (refer to Box 5.2).

The inner and outer envelopes shall have: (a) the name and address of the employer; and (b) the identification details of the contract as defined in the bidding and contract data. It must carry the warning 'not to open before the specified time and date for bid opening' as defined in the bidding data. Each of the envelopes shall indicate the name and address of the bidder to enable the bid to be returned unopened in case it is declared late or is declared non-responsive.

BOX 5.2 SEALING AND MARKING OF BIDS AS PER STANDARD BIDDING DOCUMENT FOR HYDROPOWER PROJECTS, MINISTRY OF POWER, GOVERNMENT OF INDIA

The documents comprising the bid will be submitted in stages as stated in Clause 13 of the ITB as mentioned by the employer. The bid may be need to undergo a prequalification process or a post-qualification process as per the exigent needs of the employer with respect to the work package, and shall be submitted, as per the submission schedule laid down by the employer. The bid shall be submitted in separately sealed envelopes duly marked Envelope 1, Envelope 2a, Envelope 2b and Envelope 3, and each envelope shall contain the following documents:

Envelope 1: Envelope 1 will be named, marked and sealed as 'Bid Security'. This envelope shall contain the documents listed in ITB Clause 13.1(b), (c) and (d) and placed in separate covers as stated therein.

Envelope 2a: Envelope 2a will be named, marked and sealed as 'Bidders Qualification' and shall contain documents listed in ITB Clause 13.1(e).

Envelope 2b: Envelope 2b will be named, marked and sealed as 'Techno-Commercial Unpriced Bid'. This envelope shall contain documents listed in ITBClause 13.1(a) unpriced, (f), (h) and (i), bid addendum/corrigendum (if any) and blank formats of bill of quantities/price schedule placed in 'Bidders Copy' (without filling in any rate or amount).

Envelope 3: Envelope 3 will be named, marked and sealed as 'Price Bid' containing duly filled in rates and prices in the format listed in ITB Clause 13.1(a) priced, (g) regarding bill of quantities/price schedule marked 'ORIGINAL' giving the unit price and amount against each item with the grand total at the end in figures and in words. If the bid is submitted by a joint venture/consortium (JV/C), the percentage share of each partner of the JV/C shall be stated in the 'Price Bid'.

Source: Adapted from: http://www.cea.nic.in/reports/hydro/bidding_doc_hyd_sbd .pdf Retrieved on 7 October 2019.

5.8.2 Deadline for submission of bids

The deadline for submission of the bids clause is regarding the last date and time for receiving bids specified in the bidding data. The employer may extend the time for submission of bids by issuing an amendment in line with the clause on 'Amendment of Bidding Documents', in which case all rights and obligations of the employer and the bidders previously subject to the original deadline will then be subject to the new deadline.

5.8.3 Late bids

Bids received by the employer after the deadline will be returned unopened to the bidder.

5.8.4 Modification and withdrawal of bids

The World Bank permits withdrawal or modification of a bid before the deadline. Each bidder's modification or withdrawal notice shall be prepared, sealed, marked and delivered, with the outer and inner envelopes additionally marked 'MODIFICATION' or 'WITHDRAWAL', as appropriate. Although a bid may be withdrawn by notice dispatched before the deadline, it cannot be modified if the modification is received after the deadline for the submission of bids.

5.9 Processing of bids

Documents invariably specify the date, time, place and designation of the officer to whom the completed documents are to be submitted. As per the *CPWD Works Manual* (GOI, 2003), bidders are also expected to provide an earnest money deposit (EMD),[21] a fixed amount stipulated in the tender, along with the tender, in the form of cash or bond or bank guarantee. The EMD is to ensure that the contractor does not withdraw from his tender when it is under consideration or from executing the work after being awardedthe work. Tenders, unaccompanied by an EMD are normally rejected as being non-responsive. It is also a contractual requirement that the tender should be submitted exactly as per the provisions of the documents. 'Conditional tenders' or tenders with any deviation are summarily rejected as non-responsive.

5.9.1 Bid opening and evaluation

Bids are opened in the presence of the bidders or their authorised representatives who wish to be present at the notified time and place of opening. Once bids are opened, the officer opening the bids reads out the tendered amounts for the information of all those present. The tenders received are entered in a register and signed by all those present. Bids received after the due date and time of receipt are not considered. They should neither be opened nor entered in the tender opening register. They are returned unopened, giving the date and time of receipt.

Bids are opened at the time and place announced in the bidding documents. Bidders have a right to attend the bid opening. Bid evaluation is confidential and canvassing by bidders is prohibited. An employer has a right to seek clarification from a bidder; however, the response should not change the substance or price of the bid. A bidder can bring additional information if requested, only in writing. These provisions are clearly stated for World Bank–aided project, but are not spelt out in the CPWD or other such domestic-funded projects. Additionally, all bids should be substantially responsive and conform to all terms, conditions and specifications. The World Bank standard bidding document's clause prohibits any attempt by the bidder to turn a substantially non-responsive bid into a responsive one, through clarification, correction or withdrawal of an offending deviation. The CPWD also follows this practice but there are no published guidelines for such deviations.

a. Bids will be openedin the presence of the bidders and the bid prices and the total amount of each bid with discounts, if any, will be read out. Any bid price, discounts or an alternative bid price not declared and recorded will not be considered in an evaluation.
b. The employer prepares the minutes of the bid opening.
c. The processrelating to examination, clarification, evaluation and comparison shall not be disclosed to the bidders.
d. The employer may seek clarificationincluding a breakdown of prices, in writing; however, no change in bid prices will be permitted except confirmation on arithmetical errors.

e. Determination of responsiveness: (i) a bid meetsthe eligibility criteria of the International Bank for Reconstruction and Development IBRD; ii) a bid has been properly signed; (iii) a bid is accompanied by the required securities; (iv) a bid is substantially responsive to the requirements of the bidding documents; and (v) a bid has provided the clarification and/or substantiation required by the employer.

f. A substantially responsive bidis one that conforms to all the terms, conditions and specifications of the bid documents without material deviation or reservation. A material deviation/reservation is one (i) which affects the scope, quality or performance of works in a substantial way; (ii) which limits employer's rights or bidder's obligations under the contract; and (iii) whose rectification would unfairly affect the competitive position of other bidders presenting substantially responsive bids.

g. A bid not substantially responsive shall be rejected and no subsequent corrections or withdrawals will be allowed.

h. Corrections and errors: Some arithmetical errors in the bidding are permitted to be corrected, involving unit rate and item value. Discrepancies between amounts and figures in words are resolved by accepting the value in words. The words will govern discrepancies between words and figures. The unit rates will govern the discrepancy between unit rates and quantity, unless there is an obvious gross decimal mistake in the unit rates, in which case the line item total will govern and the unit rate corrected.

i. A bid amount will be adjustedwith the concurrence of the bidder; however, if the bidder does not accept this, then the bid will be rejected and the bid security may be forfeited.

j. Conversion to single currency: (i) The bid price will be broken down into respective amounts payable in various currencies at exchange rates payable in the contract. (ii) The amounts in various currencies in which the bid price is payable (excluding provisional sums but including day work where priced competitively) will be converted to the employer's currency at the selling rates on the date specified in the bidding document.

5.9.2 Bid evaluation

The initial step in the evaluation of tenders is the rejection of all conditional tenders and those without EMD. Acceptable tenders are subsequently evaluated. Apart from arithmetical scrutiny, the tendered rates for individual items are compared with the estimated rates to spot those with exceptionally high or low rates. The employer reserves the right to accept or reject any offer without conveying any reason. Therefore, the employer is free to reject any tender with exceptionally high or low rates for individual items. Similarly, it is also not obligatory to reject all tenders. The impact of such items on the overall costs of the work can be assessed and if the variation in the total cost is marginal, such tenders can be considered for comparison. Generally, the lowest tender meeting the requirements stipulated in the tender documents is accepted. Earnest money given by all contractors except the lowest bidders should be refunded within a week from the date of receipt of tenders.

Bid security will be in the form as specified by the employer issued by a bank acceptable to the employer. The bid bond format cannot be changed without the prior approval of the employer. The format provides for encashment without justification and to be honoured on demand. A bid bond is valid for 28 days beyond the validity period of the bid. Bids received without a valid bid bond will be rejected as non-responsive. The bid security of unsuccessful bidders will be returned promptly but no later than 28 days after the expiry date of its validity. The bid security of the successful bidder will be returned after signing the agreement and the submission of the performance bond. The bid security will be forfeited if the bidder withdraws his bid during the period of validity or the bidder does not accept a correction of his bid price or a successful bidder fails, within a specified time limit, to sign the agreement or submit the required performance security or to furnish the domestic preference security.

A performance security should be furnished within the time, and in the amount, currency and form prescribed in the contract document. In the event of failure to produce a performance security according to the stated requirement, the contract is declared null and void and the contractor forfeits his tender security. The employer then awards the contract to the next lowest bidder. The EMD is returned once the performance security is returned. If the timing is such that it will be impossible for the contractor, in compliance with the terms of the contract, to provide the performance security before the validity period has expired, the employer should protect himself by asking the other bidders to extend the validity period of their tenders. Refusal by any bidders to extend the validity of his tender would not entitle the employer to call his tender security.

The employer should prepare the minutes of the bid opening. The processes relating to examination, clarification, evaluation and comparison shall not be disclosed to bidders. The employer may seek clarification including a breakdown of prices, in writing, but no change in bid prices will be permitted except toconfirm an arithmetical error.

Only substantially responsive bids of the bidders will be assessed. A substantially responsive bid can be defined as a bid that conforms to all the terms and conditions and specifications of the bid documents without material deviation or reservation. A material deviation/ reservation is one: (i) that substantially affects the scope, quality and performance of works; (ii) that limits the employer's rights or the bidder's obligations under the contract; and (iii) whose rectification would unfairly affect the competitive position of other bidders presenting substantially responsive bids. A bid not substantially responsive shall be rejected and no subsequent corrections or withdrawals will be allowed.

The World Bank and ADB-funded projects have a domestic price preference of an amount equal to 7.5 per cent of the bid amount that will be added to bids received from foreign contracting firms for the purpose of evaluating and comparingbids.

As per the *CPWD Works Manual*, the government has the right to forfeit 50 per cent of the earnest money if the bidder withdraws his tender within the validity period or makes any alterations to the terms and conditions of the tender which are not acceptable to the central public works department. In case the contractor fails to commence the work specified in the tender documents on the 15th day or such time period as mentioned in the letter of award after the date on which the engineer-in-charge issues written orders to commence the work, or from the date of handing over the site, whichever is later, the government shall, without discrimination to any other right or remedy, be at liberty to forfeit the whole of the EMD.

Different lending agencies suggest different methods of bid evaluation. The bid evaluation procedure of the Asian Development Bank is shown in Figure 5.4.

5.10 Award of work

Acceptance of tender is defined in Section 2 [interpretation-clause– (b)] of the Indian Contract Act, 1872 as: 'When the person to whom the proposal is made signifies his assent thereto, the proposal is said to be accepted. A proposal when accepted becomes a promise'.

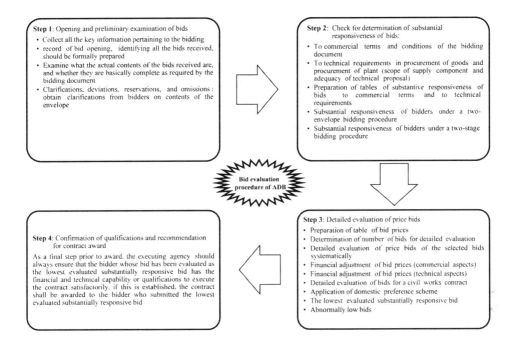

Figure 5.4 Guide on bid evaluation.

According to Section 4 of the Indian Contract Act, 1872 (communication when complete), the moment an acceptance of tender is posted (by the government), it is complete as against the tenderer and a valid and binding contract comes into existence.

The tender becomes 'the contract', a legal document, subsequent to acceptance and signing.[22] The contract will be awarded to the bidder whose bid is evaluated as substantially responsive to the bid documents[23] and has offered the lowest evaluated bid price and has been determined to be eligible and qualified in terms of the documents. Acceptance of a tender is communicated to the contractor through the letter of award, indicating the award of the work to the contractor for a specified sum and advisinghim to attend the office of the employer to sign the agreement. The date of commencement and the date of completion of work are also specified. Although the signing of the contract agreement does not affect the validity of the contract established by the issue of the letter of acceptance (LOA), it is very common for the contract to be formalised in this way. Such prescriptions may have a legal effect under certain legal systems. The initiative lies with the client, and the contract agreement, if required, should generally be signed together with, or shortly after, the issue of the letter of acceptance. Generally, the contract contains: (i) articles of agreement; (ii) form of tender; (iii) letter of acceptance; (iv) general and specific conditions of contract; (v) employer's requirements including the drawings referred to therein; (vi) contractor's proposals including a statement of the contract sum, the completed breakdown of the contractor's rates and prices and the drawings referred to therein; and (vii) correspondences and other agreed documents relating to the award as may be referred to in the letter of acceptance.

At this stage, the contractor is expected to furnish a performance guarantee which may be in the form of a bond or a bank guarantee for the notified amount. The guarantee of theearnest money deposit will be released at this stage. The notification by the letter of acceptance will constitute the formation of a contract. If, however, the contractor does not turn up to sign the contract or fails to provide the performance guarantee by the stipulated time and date, the work may be awarded to another contractor. Under this condition, the EMD of the contractor to whom the work was originally awarded, will stand forfeited to the employer. After the successful bidder has furnished a performance security, the employer will promptly notify other bidders that their bids have been unsuccessful. The employer must arrange for the return of tender securities to the unsuccessful bidders.

Where it is not immediately possible to issue a formal letter of acceptance to the successful bidder, the issue of a letter of intent (LOI) to enter into a contract may, in some circumstances, serve a useful purpose. Care should be exercised in formulating the LOI so that it is not construed as being the letter of acceptance. The LOI is not normally a legally binding document. It is only a request to the bidder to acknowledge receipt of the LOI and to confirm his acceptance.

5.11 Conclusion

In sum, once the project feasibility is established, the client will select the contractor to deliver the project at a reasonable price through tendering. A tender, also known as a bid,means an offer by a tenderer or a bidder to undertake a work in accordance with a signed documented contract against payment. A tender does not create a legal right until it is accepted in a recognised manner. Hence, one is not legally bound to accept a tender and mere submission of a tender does not create a right to property. The entire expense in tendering will be borne by the tenderer. The procedure of selectinga contractor is, in the simplest sense, tendering. Tendering and bidding are generally carried out in four stages: (i) prequalification; (ii) issue of tender documents; (iii) receipt and review of tenders; and (iv) award of contract.The above discussion presented the contract and the tender, project procurement cycle, contractor selection, types of tendering processes, preparation of tender documents, bidding documents, prequalification process, preparation of bids, submission of bids, processing of bids and award of work.

Notes

1 Section 10 of the Indian Contract Act, 1872 provides what agreements are contracts. It states that 'All agreements are contracts if they are made by the free consent of parties competent to contract, for a lawful consideration and with a lawful object, and are not hereby expressly declared to be void. Nothing herein contained shall affect any law in force in India, and not hereby expressly repealed, by which any contract is required to be made in writing or in the presence of witnesses, or any law relating to the registration of documents'.
2 This paragraph is extracted, edited and adapted from the Government of India: *CPWD Works Mannual*2014, Chapter III Contracts Section 13 Contracts and Forms, Sub-section 13.1.5, Central Public Works Department, New Delhi, p. 91.
3 Throughout this chapter, the words tender and bid are used interchangeably.
4 *See*   https://www.pppinindia.gov.in/toolkit/ports/module2-leapsfp-dotpp.php?links=ctbspm1bA ccessedon 18 March 2019.
5 *See*   http://siteresources.worldbank.org/INTINFANDLAW/Resources/RFPFORPPAVOLI.pdf Accessed on 12 March 2019.
6 *See*https://www.pppinindia.gov.in/toolkit/ports/module2-leapsfp-dotpp.php?links=ctbspm1b   Op. cit.
7 Ibid.

8 Abstracted from http://www.indianrailways.gov.in/railwayboard/uploads/directorate/land_amen/downloads/Approved_Model_RFP_Tech_Consultant_Rail_Stations.pdf Accessed on 16 March 2019.

9 *See* https://www.pppinindia.gov.in/toolkit/ports/module2-leapsfp-dotpp.php?links=ctbspm1b Accessed on 18 March 2019.

10 *See CPWD Works Manual* 2003, Section 15, pp. 81–86.

11 This paragraph is extracted, edited and adapted from the Government of India, *CPWD Works Manual* 2014, Chapter III Contracts Section 15: Preparation of Tender Documents, Sub-section 15.1.3, Central Public Works Department, New Delhi, p. 98.

12 *See*GOI.(2006) pp. 14–15).

13 *See*GOI. (2006)*Manual on Policies and Procedure for Procurement of Works*. Section 4.2.3.4, p. 14.

14 *See*https://uncitral.un.org/sites/uncitral.un.org/files/media-documents/uncitral/en/proc93.pdf.

15 *See*GOI. (2019)*Manual for Procurement of Works 2019*,pp. 28–29.

16 The World Bank document states the various acceptable forms of bid security and their amount.

17 *See* http://finmin.nic.in/the_ministry/dept_expenditure/GFRS/GFR2005.pdf Accessed on 7 October 2019.

18 *See*GOI. (2005) General Financial Rules, 2005, p. 31.

19 *See* http://web.worldbank.org/WBSITE/EXTERNAL/PROJECTS/PROCUREMENT/0,,contentMDK:20063456~pagePK:84269~piPK:84286~theSitePK:84266,00.html.

20 Ibid.

21 *See CPWD Works Manual* 2003, Section 10.

22 *See* http://cvc.nic.in/3%20Tender%20Stage.pdf.

23 An ambiguous agreement leads to poor contract performance and litigation and also gives a contractor the opportunity to make a profit out of ambiguous conditions.

References

GOI. (2003) *CPWD Works Manual 2003*. Central Public Works Department, Government of India, New Delhi, India.

GOI. (2005) General Financial Rules, 2005. Department of Expenditure, Ministry of Finance, Government of India. Retrieved from: https://doe.gov.in/sites/default/files/GFR2005.pdf Accessed on 10 October 2019.

GOI. (2006) Manual on Policies and Procedure for Procurement of Works.Ministry of Finance, Government of India. Retrieved from: https://doe.gov.in/sites/default/files/Structure%20CP%20OWG.pdf Accessed on 20 March 2019.

GOI. (2011) *PPP toolkit for improving PPP decision-making processes*. Department of Economic Affairs, Ministry of Finance, Government of India, New Delhi.

GOI. (2014) *CPWD Works Manual 2014*. Central Public Works Department, Government of India, New Delhi, India. Retrieved from: https://cpwd.gov.in/Publication/worksmanual2014.PDF Accessed on 18 March 2019.

GOI. (2019) *Manual for Procurement of Works 2019*. Government of India, Ministry of Finance, Department of Expenditure, New Delhi. Retrieved from: https://doe.gov.in/sites/default/files/Manual%20for%20Procurement%20of%20works%202019.pdf Accessed on 27 December 2019.

Pollock and Mulla. (2018) *The Indian Contract Act, 1872*, 15th edition.LexisNexis, Haryana, India.

6 Contracts and contract management

6.1 The contract

The construction industry is relatively unique among commercial endeavours where the 'output or final product is sold before it is made'. Preparing a contract document for a large project involves sufficient investigation of the project to include all aspects that influence its design, minimising uncertainties that may occur at the time of execution, studying possible technical alternatives and working out detailed designs, construction drawings and estimates. The contract document typically includes the notice inviting tender, information and instructions to tenderer, general conditions of contract, special conditions of contract, schedule of quantities and cost, tender drawings, specifications for works, forms for bank guarantee in respect of earnest money, performance guarantee, mobilization advance, etc. (Figure 6.1). The contract document defines the contract. Although conditions within different clauses of diverse parts of the document have their individual independent objective, significance and interpretation, no uniform contract document is available that is applicable to all situations. The contract for every project has to be drawn up on the basis of experience gained on similar projects so that uncertainties and pitfalls that occurred in earlier projects are not repeated. The appropriate management of a contract entails the interpretation of each clause in agreement with others for efficient and timely implementation of the work. It is extremely relevant to both parties, i.e. the contractor and the client.

Different sections of the Indian Contract Act, 1872 provide the essential elements of a contract. They are: (i) proposal and acceptance; (ii) consideration – lawful consideration with a lawful object; (iii) capacity of parties to contract – competent parties; (iv) free consent; (v) an agreement must not be expressly declared to be void; (vi) writing and registration if so required by law; (vii) legal relationship; (viii) certainty; (ix) possibility of performance; and (x) enforceability by law.

6.1.1 Contract document

The contract for a construction project is a complex legal document with a number of interconnected documents each of which plays a vital role in defining the obligations and responsibilities of the parties concerned and details of the works to be constructed. Contract management begins with the development of the contract document. The preparation of the contract document will be preceded by detailed preparatory activities. It contains a description of the civil engineering works that the client wants to carry out, such as the specifications; the contract value or the price to be paid to the contractor for executing the work; the roles, liabilities and responsibilities of concerned parties; the time frame, payment

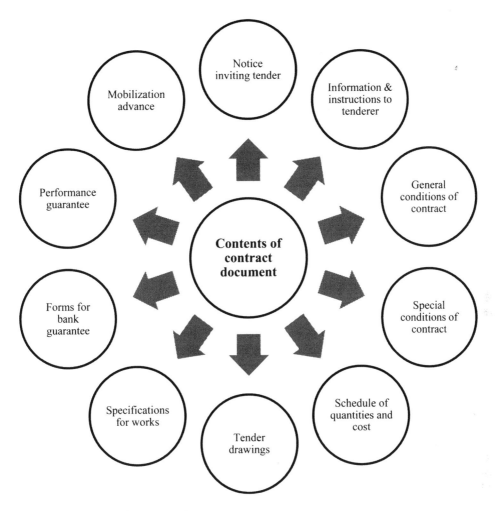

Figure 6.1 Contents of a contract document.

terms and change procedures. It should also give detailed information in respect of soil and other technical data and on the levy of additional taxes/duties during the execution of the works. Contract documents are prepared with great care and are examined for comprehensive coverage, precision and consistency with one another prior to invitation to tender. Large contracts often form part of a more complex contractual system which also includes the supply and erection of machinery, the transfer of technology and so on, where a large number of contractors operate simultaneously and are interdependent. If issues affecting the performance of a contract are given due recognition and the rights and responsibilities of the various parties to the contract are properly defined and made equitable, the work environment in the construction industry will improve and will lead to better performance. Today, model contract documents prepared by governmental agencies can be used, but these seldom suit the requirements of different types of works since no standard document is applicable to all situations. A reasonable delay in project execution for reasons beyond the control of a contractor may be deemed to be covered by a contract; for unreasonable delay, the contractor needs to be suitably compensated including losses suffered by him, in addition to an extension of time.

6.1.2 Time is essence of contract

If, in a contract, time is the essence of performance, the contract becomes voidable at the option of the party who is affected by such non-performance, if the contract is not executed within the predetermined time. However, if time is not the essence of the contract, as implied under Section 55 of the Act, and the promisor does not perform the contract within the stipulated time, the contract does not become voidable. However, the promisee can claim compensation for the delayed performance. The most important item in any contract is that if time is critically important to the contract, it should be clearly stated in the contract. In general, it depends on whether time was considered by the parties to be essential to the contract or not. In this respect, Section 55 of the Indian Contract Act, 1872 lays down the following:

When a party to a contract promises to do a certain thing at or before a specified time, or certain things at or before specified times, and fails to do any such thing at or before the specified time, the contract, or so much of it as has not been performed, becomes voidable at the option of the promisee, if the intention of the parties was that time should be of the essence of the contract.

6.1.3 Valid contract

A contract is an agreement and an agreement is a promise. The promise is an accepted proposal. As a result, every agreement is the result of a proposal from one side and its acceptance by the other (Figure 6.2). To be enforceable by law, an agreement must have certain essential elements (Figure 6.3). A contract is an agreement that is legally enforceable (Poole, 2014). It is an agreement between two or more parties which can be implemented by law. According to the Indian Contract Act, 1872[1] 'all agreements are contracts if they are made by the free consent of parties competent to contract, for a lawful consideration and with a lawful object, and are not hereby expressly declared to be void'. Agreements which are not legally enforceable are not contracts and remain void agreements. Such agreements are not enforceable at all or as voidable agreements which are enforceable by only one of the parties to the agreement. The Act states that all agreements are contracts if they are made with the free consent of parties competent to contract for a lawful consideration and with a lawful object and are not hereby expressly declared to be void. It also specifically states who are competent to contract. According to Section 11 of Indian Contract Act, 1872, every person is competent to contract who is of the age of majority according to the law to which he is subject and who is of sound mind and is not disqualified from contracting by any law to which he is subject. Sound mind means that at the time of making a contract, the person is capable of understanding it and of forming a rational judgement.

6.1.4 Offer, acceptance, agreement and consent

'An offer is an act on the part of one person whereby he gives to another the legal power of creating the obligation called contract' (Corbin, 1917). The accepted offer should create legal obligations between parties, i.e. an agreement which arises only if there is an 'offer'

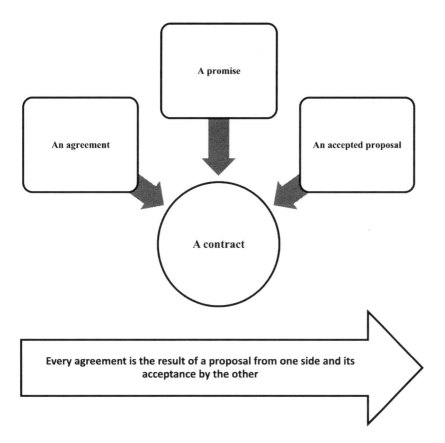

Figure 6.2 Valid contract.

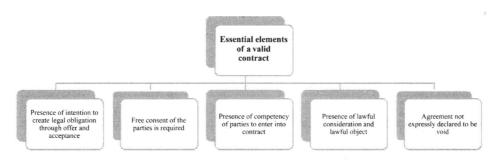

Figure 6.3 Essential elements of a valid contract.

by one party and the 'acceptance' of such offer by another party. Therefore, an agreement, between parties is one of the most important elements of a valid contract. A lawful offer and acceptance creates a binding legal contract. Offer and acceptance ascertain that an agreement exists between parties. When a person makes a proposal to another and the proposal is agreed upon, it is called acceptance. The acceptance must be given before a lapse of time or within a reasonable time. Offer and acceptance constitute the initiation of a legal contract. In Section 2(b) of the Indian Contract Act, 1872, acceptance is defined as 'when the person to whom the proposal is made signifies his assent thereto, the proposal is said to be accepted. A proposal when accepted becomes a promise'. When an offer is accepted, it results in an agreement. Two or more persons are in consent while the said persons agree upon the same thing in the same sense. Section 14 of the Act defines free consent, i.e. 'consent is free when it is not caused by coercion, undue influence, fraud, misrepresentation or mistake'. The steps involved in a contract are: (i) proposal and its communication; (ii) acceptance of proposal and its communication; (iii) agreement by mutual promises; (iv) contract; and (v) performance of contract.

6.2 Contracts on the basis of validity, formation and performance

The Indian Contract Act, 1872 lists different types of contracts on the basis of validity, formation and performance (Figure 6.4). Contracts on the basis of validity are: valid contracts, void contracts, voidable contracts and illegal agreements. An agreement which has all the essential elements of a contract is called a valid contract whereas an illegal agreement is one which is forbidden by law.

6.2.1 Void contract and voidable contract

Section 2(j) of the Act states that 'A contract which ceases to be enforceable by law becomes void when it ceases to be enforceable'. 'Sections 15, 16, 17, 18, 19 and 19A also deals with "voidable contracts", i.e., if they are vitiated by "coercion", "undue influence", "fraud", "misrepresentation", they all say that the "agreement" which is a "contract" is voidable. These deal with "procedural unfairness" of contracts'.[2] A void contract cannot be enforced by a court of law.

Section 2(i) of the Act states that 'An agreement which is enforceable by law at the option of one or more of the parties thereto, but not at the option of the other or others, is a voidable contract'. Where one of the parties to the agreement is in a position or is

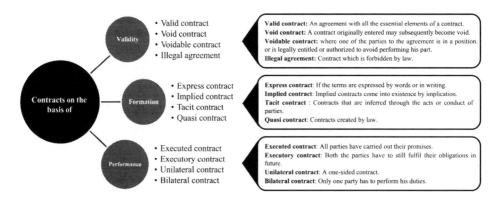

Figure 6.4 Types of contracts.

legally entitled or authorised to avoid performing his part, then the agreement is treated and becomes voidable.

6.2.2 Contracts on the basis of formation

Contracts on the basis of formation are: express contracts, implied contracts, tacit contracts and quasi contracts. Section 9 of the Act states, 'in so far as the proposal or acceptance of any promise is made in words, the promise is said to be express' (express contract). For example, A calls B and proposes to purchase his four-wheeler for a specified sum and B gives acceptance over the telephone. Section 9 of the Act further states, 'in so far as such proposal or acceptance is made otherwise than in words, the promise is said to be implied' (implied contract). Such contracts come into existence by implication. Tacit contracts are those that are inferred through the acts or conduct of parties. Under this contract, the terms and conditions are known to the concerned parties. An example is withdrawing cash from an ATM. Quasi contracts are created by law. By mistake, A leaves goods in B's house, who treats the goods as his own and consumes them.

6.2.3 Contracts on the basis of performance

Contracts on the basis of performance are: executed contracts, executory contracts, unilateral contracts and bilateral contracts. A contract becomes an executed contract after all parties have carried out their promises. An executory contract is a contract that has yet to be fully executed. A unilateral contract is a one-sided contract; only one party has to perform his duty or obligation in a unilateral contract. A bilateral contract is one in which the obligation or promise is outstanding on the part of both parties. Both parties have to fulfil their obligations in a unilateral contract.

6.3 Contract delivery formats/methods

Selecting contractors and ascertaining a price for each part of the work are necessary forerunners to the beginning of construction work. Clients and contractors' duties and obligations are defined in a contract. Many types of contracts are used in construction. The main types of contracts are categorised as: build only/construction contracts; design-build contracts; concession agreement contracts; and operations and maintenance contracts. Each category has different types of contracts. The build only/construction contract has different types of contracts such as item rate, cost plus and lump sum. The variants in design-build contracts are design and plant build, engineering, procurement and construction (EPC) and lump sum turnkey (LSTK). Concession agreement contracts have build operate transfer (BOT), build own operate (BOO), build own operate and transfer (BOOT), build operate lease and transfer (BOLT), design build operate and transfer (DBOT) and many other types of contracts (Figure 6.5). Operations and maintenance contracts have operation and maintenance (O&M) and operation, maintenance and transfer (OMT) contracts. Each type of contract has its advantages and disadvantages with respect to the owner and the contractor.

6.3.1 Item rate contracts

For item rate contracts, the client provides a detailed design as well as estimates of the quantities for different items of work (bill of quantities [BOQs]). Payments to the contractor are

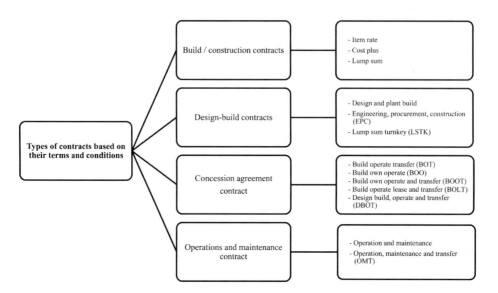

Figure 6.5 Types of contracts based on their terms and conditions.

made on the basis of measurements of the work done in respect of each item. The item rate contract is also known as the unit price contract or schedule contract.

Under these types of contracts, a project is divided into sections and each section is split into items and the quantity of each item. Rates are required to be quoted per unit of measurement so specified. Such contracts provide for provisional items which may or may not be executed and are, therefore, not included in the total price. For minor works required to be executed outside the scope of the items specified, day work rates for labour, materials and equipment are quoted with an increase in percentage for overheads. The BOQ provides for a lump sum amount to limit the operations of such items and this is included in the total bid price.

Item rate contracts are prone to excessive time and cost overruns, besides recurrent disputes involving large claims because of inadequate project preparation and estimation together with the allocating of several construction risks to the client.[3]

6.3.2 Cost plus contracts

In a cost plus contract, the purchaser agrees to pay the cost of all labour and materials plus an amount for the contractor's overheads and profit. Indian Accounting Standard (Ind. AS) 11 for construction contracts defines the cost plus contract as 'a construction contract in which the contractor is reimbursed for allowable or otherwise defined costs, in addition a percentage of these costs or a fixed fee'.[4] Cost plus contracts are used when the scope of the work cannot be estimated at the time of award. Cost needs to be defined on which the percentage will be charged. Furthermore, sometimes an incentive clause is included if the contractor completes the project before a certain specified time limit. These types of contracts are preferred for works where the scope of the task is indeterminate or highly uncertain and the kinds of labour, material and equipment needed are also uncertain.

6.3.3 Lump sum contracts

A lump sum contract is a contract in which the final quantities are computed and the final price to be paid is determined by adding to/deducting from the contractor's accepted tender price, the value of variations and other specified items (such as provisional quantities and contingency items). It is also customary for contracts for plant and equipment and small construction works to be awarded a lump sum based on specifications and drawings (rather than quantities). Traditionally, every civil engineering works is contracted on an estimated quantities and remeasurement basis. Lump sum contracts are used for standard plans such as houses with standard-type plans and multiple numbers where the contractor agrees to execute a complete work with all its contingencies in accordance with the drawings and specification for a fixed sum. In such contracts, all detailed drawings and specifications need to be specified to avoid future claims. Extra rates for foundation works below a certain depth are quoted separately.

6.3.4 Plant and design-build

The plant and design-build conditions of a contract (Yellow Book) are recommended for the provision of electrical and/or mechanical plants, and for the design and execution of building or engineering works. It is intended for use where the contractor designs and provides, in accordance with the employer's requirements, a fully functional plant or other works which may include any combination of civil, mechanical, electrical and/or construction works. Under these types of contracts, the contractor is responsible for the design of the works, as well as having significant responsibilities for the commissioning of the works (including training the employer's staff). The employer defines the scope of the works.

6.3.5 Engineering, procurement and construction

Item rate contracts were prone to enormous time and cost overruns, and recurrent disputes involving large claims. For these reasons, a new form of contract, i.e. engineering, procurement and construction, has emerged. The engineering, procurement and construction contract is suitable for complex infrastructure projects with the provision on a turnkey basis of a process or power plant, a factory or similar facility, or an infrastructure development project where (i) a high degree of certainty regarding the final price and time is required and (ii) the contractor assumes the entire responsibility for the design and execution of the project, with little involvement from the employer. Under the normal arrangements for turnkey projects, the contractor carries out all the engineering, procurement and construction, providing a fully equipped facility, ready for operation (at the 'turn of the key'). Besides delivering a complete facility, the contractor has to deliver the facility for a guaranteed price by a certain date and it has to perform as per the specified level of quality.

6.3.6 Lump sum turnkey

Generally, a lump sum turnkey contract gives single-point responsibility to the contractor, including but not limited to design, engineering, procurement, manufacture, packing, supply, transportation, customs clearance, insurance and all other systems and facilities required for the installation and commission, operation and maintenance of the project within a specified period of time. The contractor is accountable for the delivery of the entire project implementation (cost, time, quality) along with the contract specification. The specifications must scrupulously identify the specific requirements (scope, quality, time) of the owners, which may possibly include commissioning if specified in the contract. It is particularly useful when the contractor is also the holder of the process know-how and/or the supplier of the main equipment. The advantages include: (i) most of project risk is transferred to the contractor; (ii) requires less supervision from the owner; and (iii) can be extremely cost and schedule effective. Nevertheless, the time for project preparation, tendering and evaluation of the contract is likely to be long. Any amendments to the specification will be extremely expensive and also likely to bring about claims for extra time. In these types of contracts, compliance with the specification needs to be strictly monitored and there are high risks of contract disputes.

The following part discusses different public-private partnership (PPP) contractual schemes. The PPP approach entails a conventional mechanism of project financing through a private sector borrower who desires to have financing on a limited recourse or a non-recourse basis.

6.3.7 Build operate transfer

In the roads sector, BOT is a common PPP mode, with revenues for the private operator often earned from tolls (BOT tolls contract) or from a fixed annual/semi-annual payment (BOT annuity contract). Common examples of different modes of BOT family projects are management contracts, lease contracts, build own operate contracts and build operate transfer contracts.

Build operate transfer is a contractual arrangement whereby the project developer/private operator undertakes the construction (including financing) of the infrastructure facility, and the operation and maintenance thereof. The operator (company) generally earns its revenues through a fee charged to the utility under a BOT project.[5] The developer operates the project over a predetermined term during which he is permitted to charge facility users appropriate tolls, fees and charges not beyond those proposed in the bid, or as negotiated and incorporated in the contract, to allow the project developer to recover his investment and operating and maintenance expenses from the project. The project developer or contractor transfers the facility to the client at the end of a fixed term. Major participants in a BOT project are shown in Figure 6.6.

Before transferring the constructed facility, the project developer finances, builds, operates and maintains the facility as per the concession agreement. The investor collects the user fee throughout the concession period to recover the construction cost, debt servicing and operation cost. At the end of the concession, the facility reverts back to the government that has given the concession. Generally, a concession period is the time period granted by the government to the private company within which the private company is responsible for the financing, construction and operation of a BOT project (Shinya and Palapus, 2012). Figure 6.7 shows a general BOT framework.

The consortium of companies undertaking the financing and development of a project is known as the concessionaire. The organisation created for a BOT project to design, build, finance and maintain the asset is known as a special purpose vehicle (SPV).

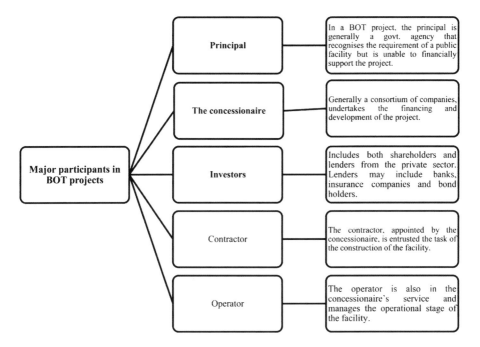

Figure 6.6 Major participants in a BOT project.

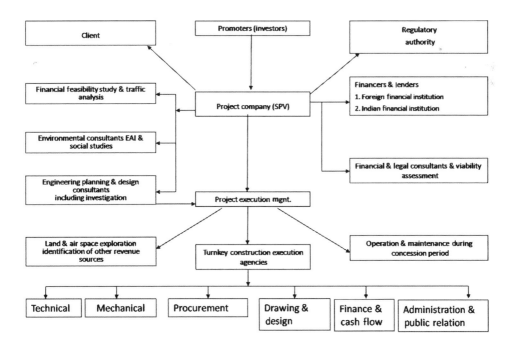

Figure 6.7 General BOT framework.

According to the Public-Private Partnership Legal Resource Center of the World Bank:

> A Concession gives a concessionaire the long term right to use all utility assets conferred on the concessionaire, including responsibility for operations and some investment. Asset ownership remains with the authority and the authority is typically responsible for replacement of larger assets. Assets revert to the authority at the end of the concession period, including assets purchased by the concessionaire. In a concession the concessionaire typically obtains most of its revenues directly from the consumer and so it has a direct relationship with the consumer. A concession covers an entire infrastructure system (so may include the concessionaire taking over existing assets as well as building and operating new assets). The concessionaire will pay a concession fee to the authority which will usually be ring-fenced and put towards asset replacement and expansion.[6]

6.3.8 Build own operate (BOO)

Under a build own operate structure the asset ownership is with the private sector as is the responsibility for the service/facility provision. The BOO is a contractual arrangement whereby a project developer is authorised to finance, construct, own, operate and maintain an infrastructure development facility, from which the developer or contractor is allowed to recover the entire investment, O&M costs, as well as a reasonable return thereon, by collecting tolls, fees and charges from facility users. Under these types of projects, the developer or contractor who owns the assets of the facility/project may assign its operation and maintenance to a facility operator.

6.3.9 Build own operate transfer

Under this type of contract, a developer/private firm designs and builds a complete project at little or no cost to the government or as a joint venture partner. The project company owns and operates the constructed facility/project as a business for a specified period (usually 10–30 years) and subsequently transfers it to the client at a previously agreed-upon price (Figure 6.8). The project company assumes equity and other commercial and construction risks.

As given in Merna and Smith (1994), a BOOT project is a project in which a government (client) grants a concession to a promoter (concessionaire) to construct and transfer a facility (project) to the client at no cost to the client. During the concession period, the concessionaire owns and operates the facility and collects revenues with the purpose of repaying the financing and investment costs, maintaining and operating the facility and making a reasonable profit. Design build finance operate (DBFO) is another variant of BOOT.

6.3.10 Build operate lease and transfer

The government grants a developer the right to finance and construct a project. The developer is entitled to design, build, own, operate and lease the facility and subsequently transfer

Figure 6.8 Role of project company.

ownership to the government at the end of the lease period. The government obtains a specific percentage of the revenue rather than ownership of the project at the end of the concession period.

6.3.11 Annuity contracts

In annuity projects, the special purpose vehicle constructs the project and obtains the right to receive fixed payments from the client or state government authorities throughout the concession period. It has an obligation to maintain the quality of the project. However, unlike toll road projects, the demand risk is mitigated due to the availability of a steady stream of assured payments from the concessioning authority. These types of projects are common in the road construction sector.

6.3.12 Hybrid annuity model

In 2016, the hybrid annuity model (HAM) was introduced by the government (National Highways Authority of India [NHAI]) to revive public-private partnership in highway construction in India. Hybrid annuity means that the government makes a fixed amount of payments for a considerable period and then a variable amount of payments for the remaining period. In the case of annuity projects, the concessionaire will be entitled to a fixed 'annuity' during the entire concession period. After the concession period, the ownership of the facility is transferred to the employer.

The hybrid annuity model is a mix of pure EPC and BOT models. Under the EPC model, the contractor executes the order book and gives it back to the government, whereas under the BOT model, the developer/contractor has to build, operate and maintain the project for a specific period before it is given back to the government. In hybrid annuity projects, the concessionaire receives payments during the project implementation phase to fund a certain percentage of the project cost, and also receives annuity payments during the operational phase, which are inflation adjusted. The National Highways Authority of India provides 40 per cent of the construction cost during the project stage. This reduces the funding risk due to the lower equity contribution by the developer during the initial stage. The remaining 60 per cent of the project cost is given by the NHAI using a fixed stream of annuity payments, over the concession period.[7] The toll rights are vested with the NHAI and not with the developer. The pace of growth of HAM projects has reduced due to the banks' reluctance to fund HAM projects as a sizeable number of construction companies have been unable to provide bank guarantees.

6.4 Special purpose vehicle

In the initial stages of public-private participation projects, a consortium of private sector sponsors (sometimes a governmental agency) is formed to examine the proposal, to prepare an initial feasibility and viability study and to submit a bid. Subsequently, the selected tenderer will form the special purpose vehicle for equity contributions and borrowing funds to finance the project.

An SPV, a limited liability company, is formed for a single, well-defined and narrow purpose such as to undertake a project created by a firm (known as the originator), whose ownership and management are independent of the originator. The project company is known as a 'concessionaire' since its rights and obligations are defined in the 'concession agreement'.

The SPV raises funds by collateralising future receivables and generally utilise as a means of securitisation of assets without disturbing the managerial relationships. An SPV enters into the primary contract with the government/public sector entity which typically involves providing both assets and services over the contract duration. Figure 6.9 illustrates the contractual structure of a typical BOT project. An SPV enters into contract with the government, the construction contractor, the operations and maintenance agency, the equipment/material supplier, etc., and executes, operates and maintains the project. Therefore, it is basically a partnership involving public sector organisations and private sector investors mainly in the form of a stand-alone business venture. The concession agreement defines the rights and obligations of an SPV and the government as well as risk allocation.

The contractor has to decide the debt–equity ratio and finalise the legal structure of the entire arrangement by way of agreements, sureties, etc. An escrow account (independent account) mechanism is required for the timely and proper release of funds and will enable easy checks, monitoring and auditing. For an escrow account, the bank and the contractor prepare an agreement and a trustee, with well-defined instructions on releasing the funds, is appointed to the account. After the formation of an escrow account, the next steps prior to commencement of execution are the preparation of execution/quality/maintenance/ expenditure manuals and set procedures thereto. On reaching this stage, the contractor should complete all procedures with regard to project financing and confirm making the

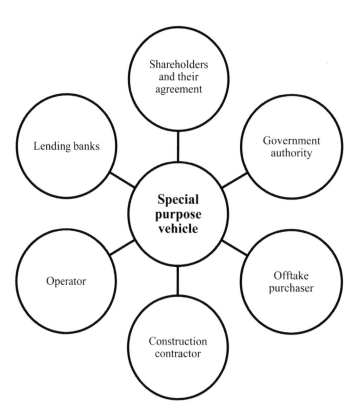

Figure 6.9 Contractual structure of a typical BOT project.

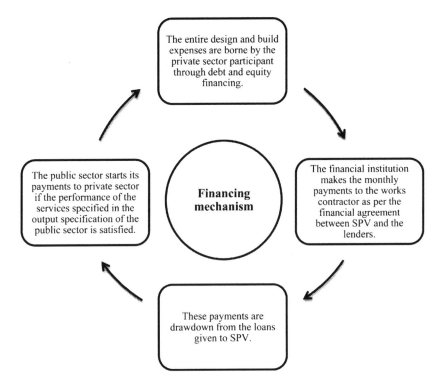

The entire design and build expenses are borne by the private sector participant through debt and equity financing.

Financing mechanism

The financial institution makes the monthly payments to the works contractor as per the financial agreement between SPV and the lenders.

The public sector starts its payments to private sector if the performance of the services specified in the output specification of the public sector is satisfied.

These payments are drawdown from the loans given to SPV.

Figure 6.10 General financing processes of a BOT project.

same available as and when required as per the terms of the agreement and achieve financial closure. The general financing processes are outlined in Figure 6.10.

In general, the contractor divides the project into different parts/activities and awards each one to competent, experienced and resourceful subcontractors/agencies. The project execution is monitored and controlled efficiently. Before substantial completion/completion certificate and permission to proceed with toll collection from the commercial operation date, the contractor needs to fix the toll collection mechanism, system and control over the activities. The trustees supervise the financial activities including depositing cash. At the end of the concession period, the project asset is to be handed over to the owner and the SPV will then be dissolved. The SPV will legally dissolve subsequent to completion of the necessary formalities, by issuing the required legal notices, declaration and termination of all agreements made by the SPV for the purposes of the project and termination of the escrow arrangement.

6.5 Employer, contractor, architect and engineer in contract management

The contract spells out the rights and obligations of the parties that guide and manage their roles. The employer ensures that the engineer properly performs the assigned duties as an independent and impartial body. The engineer has a big role to play in large-scale infrastructure development projects while architectural firms may have a limited role to play in larger projects. The responsibilities of the major parties in a contract, as described in the NHAI works manual (NHAI, 2006, July), is shown in Figure 6.11.

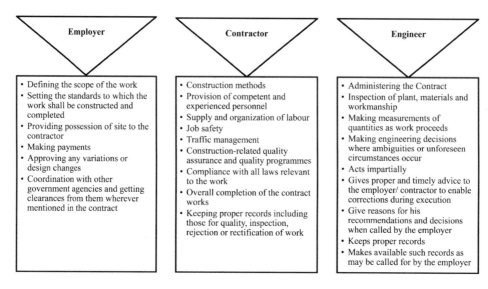

Figure 6.11 Responsibilities of the major parties in a contract.

The employer must consider that a construction project is a total team effort involving himself, the contractor and management teams from both sides. His approach should be to ensure that the coordination of diverse skills and interests results in a concrete and coordinated effort that meets his expectations in respect of quality, time and cost. The client must demonstrate that he can make positive and irrevocable decisions in a timely manner and at the same time not impede the contractor's rights, such as possession of a site on commencement; facilities as promised under the contract; timely supply of drawings, details and information required during execution of the contract; freezing the changes at the early stage of a project; prompt payment of dues; prompt and fair settlement of extra items/claims; and releasing payments.

6.5.1 The employer

The employer/client is an individual, a limited company, a corporation, a local authority or a government department requiring construction work. The major task for clients is to ensure that their project is duly managed, ensuring the health and safety of all who might be affected by the work, including members of the public.[8] Contracts are awarded on behalf of the clients and they are expected to perform the following duties in discharging their obligations under the contract: (i) give uninterrupted possession of the site immediately after the order to commence work has been given to the contractor; (ii) appoint an engineer and an architect before the contractor starts the work so that the work can be completed under the direction and satisfaction of a skilled person and the process of payment and valuation of extra items is carried out as per the terms and conditions of the contract; (iii) supply the necessary details and instructions regarding requirements for the project to the contractor to carry out the work as per the contract provision; (iv) approve the nomination of subcontractors and suppliers in a timely fashion; (v) make interim and final payments to the contractor on receipt of architect/engineer certificates within a reasonable time; (vi) pay the architect/engineer's fee as per their agreement of appointment; (vii) supply instructions as to the carrying out of the work; and (viii) permit the contractor to carry out the whole of the work. One of the major duties of the employer is to make the site easier to work in, to conduct surveys, take bore holes and make other investigations. In the absence of any express term, the obligation to supply drawings and information must be done within a reasonable time. Non-performance of the employer's responsibilities will result in time and cost overruns to the contractor which, in turn, may have to be borne by the employers.

6.5.2 The contractor

The contractor is the party who undertakes a project for a predetermined price for the owner/employer/client. The contractor procures and assembles all of the resources required to construct the building in accordance with the documented design. The contractor is the party who makes use of all the required skills; makes all the materials, plant and equipment available; and undertakes to construct what the client has intended. The contractor controls the execution of the work awarded to him under a written legal contract. The contractor has to ensure that any materials supplied by him under the contract are up to the specified quality standards prescribed in the contract and, if not, at least reasonably fit for their intended purpose. The contractor acts in accordance with other terms and conditions given in the contract; organises, commences, plans and executes the work within the stipulated time; and commences the work on time. The contractor should provide the necessary information to the engineer and employer if (i) works are expected to be delayed for want of drawings; or (ii) further instruction is not issued by the engineer within a reasonable time. Further, it is required that the contractor should notify the employer and the engineer of any error, omission or defect in the design and specification provided to him by the client. The contractor should also prepare quality assurance manuals, employ quality control personnel and perform quality assurance tests to ensure quality works as per the specifications and standards.

6.5.3 The engineer

An engineer is a person who possesses expert knowledge on engineering design and the execution of the work and should be adequately qualified to perform the duties and functions of an engineer. An engineer could be an employee/consultant of the client or a separate consultant employed for the purpose. Engineers in government departments are entrusted with the overall responsibility for planning, designing and executing schemes. The International Federation of Consulting Engineers (FIDIC) makes it clear that the engineer [also project management consultant (PMC)] who represents the employer shall carry out his duties as specified in the conditions of a particular application. The engineer must clarify issues on which he must seek prior approval from the employer, such as cost increases or time extensions. The engineer has no authority to relieve the contractor of any of his obligations under the contract, but has full authority to act in an emergency. He can delegate any of his duties and authorities to his representative. The engineer should operate impartially on all matters entrusted to him.

6.5.4 The architect

An architect (architectural firm) may be hired early in the pre-design phase. An architect identifies and surveys the client's needs regarding the new facility for construction. An architect is expected to possess the qualifications specified in the schedule of the Architect's Act of 1972. The architect should perform in accordance with the professional practices, standards, codes and local laws prevailing in the country. The architect's duties in general include to prepare plans, drawings, specifications and designs according to the owner's requirement and within the cost limitations with proper cost planning and cost control; to prepare tenders and finalise contractual arrangements for the execution of the projects; to help the client in selecting the main and nominated subcontractors; to supervise the construction work and settle all matters which arise during construction; and to issue interim and final certificates for finalising interim and final payments.

6.6 Management of contracts

Contract management is the process of systematically and efficiently making and executing contracts with minimum financial and operational risks. The subject matter of contract management is negotiating the terms and conditions of contracts to ensure compliance with the technical specifications, terms and conditions as well as documenting and agreeing on any changes or amendments that may arise during their implementation or execution. The management of a contract begins with identifying the contract and its pertinent documents to support the contract's purpose. The objective of contract management is to ensure that all the parties meet their obligations and that the project is delivered on time and to the required specifications, standards, quality and price. Adherence to national and international standards to the greatest extent minimises the chances of misinterpreting the requirements of quality construction materials and works, thereby avoiding disputes.

According to the Word Bank (2018), the principles of good contract governance include: 1. clearly defined roles and responsibilities at all levels; 2. each role has sufficient representation and authority to fulfil its responsibilities; 3. disciplined governance arrangements supported by appropriate systems and controls (especially tracking progress, monitoring against service levels/quality standards, financial controls (budget, invoicing, forecasts etc.) and reporting); 4. a multi-tiered decision-making framework which provides for escalation from operational to management to governance; 5. decisions made at appropriate authorization points are recorded and communicated; 6. independent scrutiny of contract progress, outputs and outcomes undertaken on a regular basis; 7. a comprehensive contract management plan which is agreed and communicated to all parties; 8. strict control of change management decisions; 9.comprehensive communications strategy; 10. stakeholders are engaged at a level that is commensurate with their relevance to the contract and in a manner that fosters trust; and 11. culture of candid feedback and improvement.[9]

6.6.1 Contract administration

Contract administration generally involves: (i) records management, including setting up and operating a records management system to record correspondence, claims, meeting minutes, performance reviews and other records; (ii) managing and tracking of payments to the contractor against invoices and/or claims and the timeliness thereof; (iii) managing performance; (iv) conducting review meetings; (v) managing changes, i.e., variations and projected costs against budget; and (vi) managing disputes (Figure 6.12).[10]

6.6.2 Management of key contract information

The management of key contract information records must be kept up to date. The performance and progress of awarded contracts should be managed throughout the contract

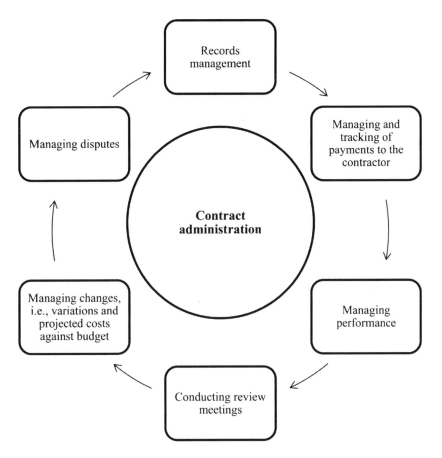

Figure 6.12 Contract administration.

duration through regular site inspections and tests by an engineer or any other personnel of the client (refer to FIDIC Sub-Clause 4.21: progress reports).

6.6.3 *Managing payments*

Managing payments made, monitoring actual and planned payments and calculating future payments are the main aspects of managing payments. Delayed payments coupled with the recovery of interest on mobilisation (where so provided in the contract) even for such periods of delay further adds to the financial problems of the contractor with the result that progress is adversely affected. The contractor is not expected to bear the cost for long delays in the payment of progress bills. In such situations, the employer must realise that this constitutes a breach of contract on his part which entitles the contractor to claim for costs and compensation for other losses. To effectively control funds, the employer should also establish a system that will enable him to get an updated financial position at regular intervals covering authorised expenditure, details of savings and extras, actual expenditure incurred and estimates of final cost.

6.6.4 Managing changes

The efficient and effective administration of the processing of contract variations by the client is one of the most crucial elements of contract management. Certain changes in quantities arising from changes in the scope of the work, changes in the construction drawings and changes necessary to meet functional and safety requirements, unforeseen events, the settlement of a claim arising from the contract and to correct errors and omissions are inescapable in construction contracts. Changes at a late stage not only increase the cost to the contractor but also disrupt the smooth execution of the work and cause loss of labour and construction plant productivity. The contractor's cash flow position is adversely affected if there are drastic changes in items of work included in the contract due to changes in the drawings given to the contractor at the contract stage, and if those changes are not decided in a timely and equitable manner by the employer. Consequently, the project is delayed and disputes arise between parties. Another area which does not receive proper or fair attention from the employer in certain cases relates to extra items.

In handling a variation claim (ADB, 2018, June), the client should: (i) assess the reasons for the variation, and whether this may show an emerging or actual performance problem; (ii) assess the impact of the variation on the contract deliverables; (iii) determine the effect the variation will have on the overall contract price and schedule; and (iv) follow the terms and conditions stated in the contract for review and approval of variations (Figure 6.13).[11]

6.6.5 Claim and dispute settlement

Due to shortcomings in the preparation of contracts, a situation arises that was not anticipated, the employer's control team generally feels impelled to defend their documents and deny the contractor and his rights. According to ADB's 'Contract Management Guidance Note on Procurement' (ADB, 2018, June), there are a variety of reasons for contract variations

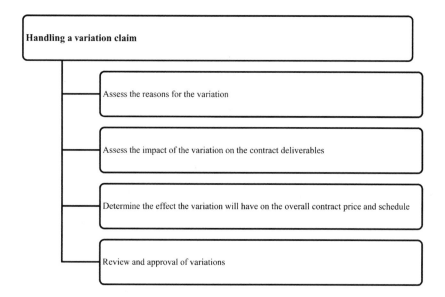

Figure 6.13 Handling a variation claim.

including: '(i) changes in the scope of the work; (ii) unforeseen events, e.g., ground conditions, climatic conditions; and (iii) settlement of a claim arising from the contract'.[12] These can lead to delays and disputes. When the pricing of drastic changes to drawings is not decided in a timely and equitable manner on the employer's side or when extra items or delayed pricing of quantities exceed a variation limit, these can affect a contractor's cash flow and lead to disputes between parties. If the fault lies with the employer, he must face the consequences instead of arbitrarily passing it on to the contractor. Similarly, he should be judicious in assessing the circumstances in which the contractor was placed, and whether it was reasonably possible for him to foresee the situation he has actually encountered. If he is convinced that the contractor could not have foreseen the situation and where the claim is due to the employer's default, the contractor's claim should be given fair treatment and settled quickly. There may be certain complex claims where the employer should evolve some method in agreement with the contractor so that they are capable of settling the claim. Entering into negotiation under the guidance of a neutral advisor with vast experience in the relevant field and also experienced in conciliation/mediation can sort these types of situations. Arbitration should only be resorted to in rare cases when an amicable resolution is not possible. Therefore, specifying a dispute resolution mechanism when drafting the contract is very important.

6.6.6 *Preparing a contract management plan*

The essential elements of a contract management plan (CMP), as given in the World Bank's Contract Management Practice (2018), include contract management roles and responsibilities for contract closure procedures (see Figure 6.14). The CMP is a tool that monitors

Figure 6.14 Essential elements of a contract management plan.

the contractor's performance. The CMP sets out how the obligations of all the parties should be effectively and efficiently implemented. It should be shared with the contractor and all parties involved in contract implementation, management, administration and governance. The World Bank also prescribes the initiation of a risk register and including it within the CMP as a practical tool to support effective contract management.

6.6.7 *Facilitating contract start-up*

To facilitate a contract start-up process, the client should discuss the CMP with the contractor and ensure that a fully equipped contract management team capable of undertaking its tasks is in place and familiar with: (i) the CMP; (ii) the contract management systems and processes; (iii) supporting the contractor in obtaining the necessary documentation including work permits; and (iv) all of the actions necessary for contract start-up.

The client should accurately comprehend the contract's functional performance and technical requirements and generate key performance indicators (KPIs). At this stage, it is necessary to confirm that the contractor has submitted valid and acceptable environmental, social, health and safety (ESHS) performance security and to check the adequacy of any insurance policies taken out by the contractor and advance payment security prior to making the advance payment to the contractor.

Building permits or similar permissions for the works should be obtained before entering into a contract. It must be ensured that the contractor is given access to, and possession of, the site within the time stated in the contract. Similarly, if a letter of credit is required, the client must arrange it in a timely manner.

The employer/client must consider that a construction project is a total team effort involving himself, the architect and engineering team, the owners and investors, and the contractors. The client is responsible for providing the project's scope and requirements, as well as funding for the project. The design team includes architects, engineers, and consultants, who generate the construction documents for the owner. The contractor builds the project in a viable setting and relies heavily on subcontracted and sub-subcontracted labour. Composition construction teams vary from stage-to-stage of a project's lifecycle. Even though, all team members share the common goal of completing the project, they may also have conflicting priorities. Owners are always after completing the project timely with quality and within budget, whereas the architect and engineers are concerned with aesthetics and safety. Aligning these interests and completing a project on time and on budget require teamwork from all participants. Effectiveness of the team depends on timely and prompt decision making by the client, project manager and the leaders of each construction team.

The success of any project depends on the selection of a project manager and his team. The project manager and other members of his team must be persons of proven competence in the relevant field with adequate experience of tackling and solving problems in a timely manner, and who can work in a spirit of goodwill and honesty of purpose.

6.7 Factors to be controlled in contract administration

Contract administration, the day-to-day routine administrative requirements of a contract, begins at contract award and continues all through the contract execution period. Contract administration is generally the responsibility of an individual or consulting firm designated by the client.

To complete a project as expected, it should be controlled in four ways: (i) the final product/project must perform as originally intended (performance); (ii) it must cost no more than originally planned (cost); (iii) it must be completed as originally anticipated (time); and (iv) above all, the final product/project must meet specified quality standards (Figure 6.15). The performance, cost and time depend on three conditions. First, the parties to the contract are required to have a clear understanding of and a commitment to their obligations. Second, clients must select a project team with adequate experience, good exposure to project problems, the courage to provide prompt and fair decisions and fully dedicated to work in a spirit of goodwill and honesty of purpose, to make prompt payments and to settle extra items/ claims judiciously and promptly. Thirdly, the contractor should select a competent project manager and other members of his team, make correct assessments and deploy the required resources, by proper forward planning, keeping in view time, cost and quality control.

'Time is the essence of contract', i.e. completing the work included in a contract within a fixed time is an essential condition of a contract and, in the event of failure to complete the works within the fixed time, will enable the employer (client) to terminate the contract and dismiss the contractor from the site.

Effective contract management ensures that: (i) agreed-on strategic priorities are delivered in a timely and cost-effective manner; (ii) issues of non-compliance are dealt with or appropriately escalated for resolution; (iii) costs and risk are managed well; (iv) reviews are carried out; and (v) experience gained is fed back into the commissioning and procurement

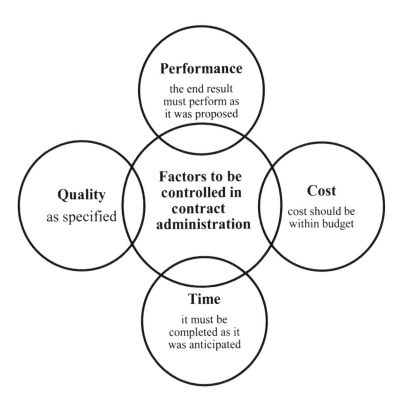

Figure 6.15 Factors to control in contract administration.

process to ensure continuous improvement (Basingstoke and Deane, 2013). The contract decision strategies during the development of a contract affect cost, time and the quality of the project by influencing: (a) the responsibilities of the parties; (b) the control of design, construction and commissioning and, consequently, the coordination of the parties; (c) risk allocation and risk management policies; and (d) the extent of control transferred to contractors. Therefore, they affect cost, time and quality.

6.8 Contractors' reporting to the employer[13]

The contractor regularly submits reports on the progress and performance of the contract to the client. They include: (i) initial contract area condition report; (ii) performance measures conformance report; (iii) monthly report; (iv) pre-construction report; (v) post-construction report; (v) asset damage report; (vi) structure inspections report; (vii) end of contract handover report; and (viii) final report (Figure 6.16).

6.8.1 *Report on initial contract area condition*

The contractor submits, within 1 month of the commencement date of the contract, his first report to the employer, i.e. the initial contract area condition report. The content of the initial contract area condition report is given in Figure 6.17.

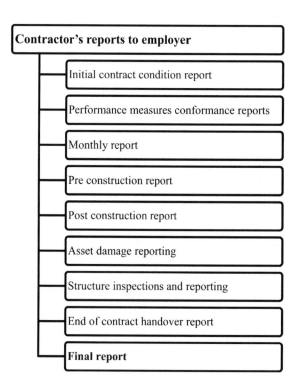

Figure 6.16 Contractor's reports to the employer.

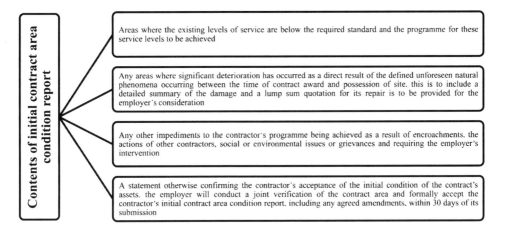

Figure 6.17 Report on initial contract area condition.

6.8.2 Performance measures conformance reports

The performance measures conformance report of each month should be submitted by the tenth calendar day of the succeeding month. It includes: (i) contractor's self-documented conformance with the management performance measures; (ii) road user safety & comfort performance measures; and (iii) road durability performance measures.

6.8.3 Daily construction progress record

A generally accepted form of construction progress record (Figure 6.18) is in the form of a daily construction report completed by the resident project representative. The daily construction progress report in combination with an inspector's daily diary or log discloses the true progress of the work. Throughout the progress of the work, it may be advisable to submit the daily diary at regular intervals to the project manager of the design organisation or owner to allow inspection of its contents. In this way, the project manager can be advised of all the dealings that have been taking place in the field.

6.8.4 Monthly report

Monthly reports include: (i) monthly progress of construction works; (ii) the list of minor uncompleted works and a programme for their completion; (iii) all the required quality control reports on the monthly completed works; (iv) cash flow estimates; (v) pavement repair work summary; (vi) emergency works reports; (vii) network performance inspection form; (viii) bridge and other structures inspection form; (ix) road signs report; (x) stakeholders relationship/ meetings; and (xi) environmental and social management framework compliance reporting.

6.8.5 Pre-construction report

Generally, for each section, such as improvement works, rehabilitation, resurfacing works or any other agreed on additional works, a separate report is compiled. The pre-construction

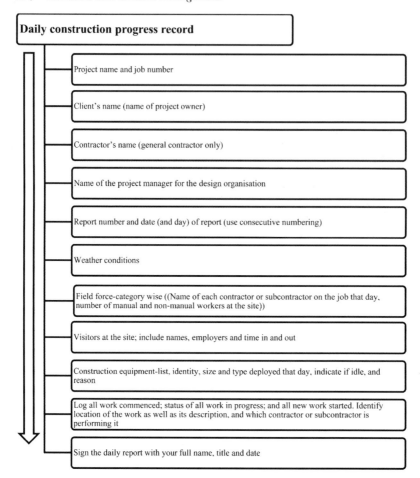

Figure 6.18 Daily construction progress record.

report includes all the required: (i) investigation, (ii) survey plans, (iii) design information, (iv) construction drawings, (v) environmental and social assessment outcomes, (vi) environmental management plan and (vii) supporting documentation for the required construction works, both conforming and non-conforming designs. Similarly, all the necessary traffic diversions and detours as well as all associated ancillary works, for instance, temporary embankments, foundations, pavements, surfacing, drainage, signage and delineations. According to the World Bank, the pre-construction report must be submitted within 2 months of the 'start date of the contract'.

6.8.6 Post-construction report

A post-construction report must be given separately for each completed part of improvement, rehabilitation or resurfacing works. This report should include the details given in Figure 6.19. Along with a summary of future inspection and maintenance requirements, for instance, required frequency of inspections, ongoing environmental monitoring

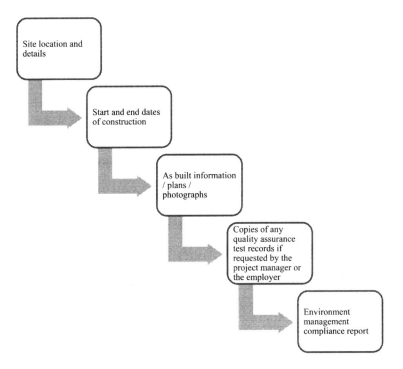

Figure 6.19 Post-construction report.

requirements, specific ongoing maintenance requirements, dated photographs, etc., will become part of a post-construction report.

6.8.7 Road asset damage reporting

The contractor shall report damage from vehicle accidents, vandalism, theft, deterioration of minor road structures, such as parapets, guardrails, handrails, etc., deterioration of signs and other delineation, in his monthly report. The report should also include significant or persistent damage to assets within the right of way.

6.8.8 Structure inspections and reporting

Bridges, culverts, drains and other structures require routine visual inspections and an annual detailed structural inspection along with associated reporting to the project manager. The contractor shall refer to the relevant Indian Roads Congress (IRC) codes for structure inspections and reporting.

6.8.9 End of contract handover report

The contractor is also required to provide an end of contract handover report to allow a smooth transition to the next contract and ensure that the next contractor and employer are aware of any outstanding issues, within 1 month prior to the end of the contract. In addition, a 'concise summary' of the condition of the road assets, including the residual life of the pavements, being handed back should be provided. This handover report should include: summaries of any unresolved issues; a schedule of outstanding defects and liabilities and a programme for their correction; unresolved issues; sensitive issues with the potential to impact upon the employer; a year-by-year trend analysis of the roughness, rutting and deflection results over the duration of the contract; updated databases and all 'as built' drawings; and any ongoing special monitoring/maintenance needs to be included in subsequent contracts.

6.8.10 *Final report*

The final report must contain a concise summary of the entire contract, outcomes achieved, lessons learned, good practices and recommendations to the employer for the improvement of future projects. The final report has to be presented to the employer within 3 months of the contract completion date (after the start of the defect liability period).

6.9 Contract closure

The term 'contract closure' refers to the process of carrying out all tasks and terms that are mentioned as deliverable and outstanding as per the contract.

6.9.1 *Contract closing*

This is the last phase of the contract life cycle which combines handing over the project deliverables and documenting the lessons learned. Subsequent to the parties agreeing that the contract has been completed, the closure process (particularly for more complex works) typically requires (ADB, 2018) the parties to: (i) complete all administrative matters; (ii) confirm that all the goods, services, and technical inputs have been received, including as-built drawings for works, operating manuals, user training, and guarantees or warranties; (iii) test, install, inspect, commission, and hand over the works, goods, and, where appropriate, services; (iv) prepare a final defects list; (v) determine the extent of any liquidated damages to be deducted from the contract price; (vi) prepare the completion certificate; (vii) finalize payments; (viii) during the liability period, follow up remedial works on the defects list; (ix) record the date and details of the final inspection and issuance of the performance certificate; (x) record the end of the retention and guarantee periods, and the date of release of retention monies and/or guarantees; (xi) handle warranties, indemnities, and insurance; (xii) summarize any claims made against or received from the contractor by the executing agency; (xiii) record final contract payments, and reconcile all payments; and (xiv) record any transfer of assets, asset verification, and disposals.[14] Documenting (i) the lessons learned from the project, (ii) important issues faced in the project, and (iii) their resolution helps in planning for such type of issues in the early stages of future projects.

6.10 Conclusion

Construction contract management is a complex task involving the communication and coordination of the skills and efforts of a number of agencies along with managerial staff, labour, material resources and equipment. Time, cost, quality and contractual risks are the constraints within which a contract is required to be delivered. Contract managers engaged for a particular project should understand the requirements of the project in its entirety. With proper understanding of the basic requirements for the formation of a contract enforceable by law, the contractor will be better able to avoid drawbacks, legal complications and liabilities. Contract management is the process of systematically and efficiently making and executing a contract with minimum financial and operational risks.

Notes

1 Act No. 9 of Year 1872, dated 25 April 1872.
2 *See* http://lawcommissionofindia.nic.in/reports/rep199.pdf p.195 Accessed on 26 April 2019.
3 *See* http://planningcommission.gov.in/sectors/ppp_report/1.Model%20Concession%20Agree ment%20Overview/22-Model-Agreement-for-EPC-Civil-works.pdf Accessed on 29 April 2019.
4 *See* http://www.mca.gov.in/Ministry/pdf/Ind_AS11.pdf Accessed on 26 April 2019.
5 *See* https://ppp.worldbank.org/public-private-partnership/agreements/concessions-bots-dbos Accessed on 26 April 2019.
6 *See* https://ppp.worldbank.org/public-private-partnership/agreements/concessions-bots-dbos Accessed on 12 May 2019.
7 *See* https://www.crisil.com/mnt/winshare/Ratings/SectorMethodology/MethodologyDocs/criteria/ CRISILs%20criteria%20for%20rating%20annuity%20roads.pdf Accessed on 12 May 2019.
8 *See* http://www.hse.gov.uk/construction/cdm/2015/responsibilities.htm Accessed on 12 May 2019.
9 Adapted from the Word Bank (2018), pp. 8–9.
10 *See* Adapted from ADB (2018, June), p. 10.
11 Adapted from ADB (2018, June), p. 14.
12 Ibid, p. 13.
13 This section is based on the Government of Punjab (Public Works Department Buildings and Roads) (2011). Bidding documents for procurement of contract under output and performance based road contract (OPRC) (asset management contract) for improvement, rehabilitation, resurfacing and routine maintenance works of roads of Sangrur-Mansa-Bathinda Contract Area, Punjab Roads and Bridges Development Board. Retrieved from: http://prbdb.gov.in/files/oprc/OPRC%20f inal.pdf Accessed on 23 May 2019.
    The reporting processes and procedures given here are of a road project extracted from the aforementioned document except the daily construction progress record. Also see Government of Gujarat, Roads and Buildings Department, Second Gujarat State Highway Project (GSHP-II) (Under Assistance From World Bank) Bidding documents (Part-2), International Competitive Bidding (ICB), pp. 47–50. Retrieved from: http://gshp2.gov.in/sites/default/files/notices/Part%202 %20Section%20VI%20A.%20Specifications.pdf Accessed on 23 May 2019.
14 Adapted from ADB (2018) Contract Management Guidance Note on Procurement, p. 15.

References

ADB. (2018, June) Contract management guidance note on procurement, Asian Development Bank, Philippines, Manila, p. 13. Retrieved from: https://www.adb.org/sites/default/files/contract-manage ment.pdf Accessed on 30 October, 2019.

Basingstoke and Deane. (2013) *Procurement and Contract Management Strategy 2013–2017*. Retrieved from www.basingstoke.gov.uk/content/doclib/350.pdf Accessed on 25 April, 2019.

Corbin, Arthur L. (1917) Offer and acceptance, and some of the resulting legal relations. *The Yale Law Journal*, 26(3), p. 171. Retrieved from: https://digitalcommons.law.yale.edu/cgi/viewcontent .cgi?referer=https://www.google.com/ &httpsredir=1&article=2527&context=ylj Accessed on 26 April 2019.

Hanaoka, Hazel Shinya and Palapus, Perez. (2012) Reasonable concession period for build-operate-transfer road projects in the Philippines. *International Journal of Project Management*, 30(8), pp. 938–949. Retrieved from: https://www.sciencedirect.com/science/article/pii/S0263786312000294 ?via%3Dihub Accessed on 30 April 2019.

Merna, A. and. Smith, N.J. (1994) Concession contracts for power generation. *Engineering, Construction and Architectural Management, 1*(1), p. 17. doi:10.1108/eb020990.

NHAI. (2006, July) *NHAI Works Manual.* National Highways Authority of India, Government of India, p. 86. Retrieved from: http://www.nhai.org/writereaddata/Portal/Document/NHAI%20Works%20Manual%202006_new.pdf Accessed on 26 May 2019.

Poole, Jill. (2014) *Casebook on Contract Law*, 12th edition. Oxford University Press, UK. Retrieved from: http://www.pbookshop.com/media/filetype/s/p/1405700635.pdf Accessed on 25 April 2019.

Word Bank. (2018) *Contract Management Practice*, 1st edition. The Word Bank, Washington, DC. Retrieved from: http://pubdocs.worldbank.org/en/277011537214902995/CONTRACT-MANAGEMENT-GUIDANCE-September-19-2018-Final.pdf Accessed on 28 October 2019.

7　FIDIC conditions of contract

7.1 Introduction

The growth and development of the international construction industry gave rise to the development of standard conditions of contract by the International Federation of Consulting Engineers (FIDIC). This chapter introduces the FIDIC conditions of contract for construction for building and engineering works, in general major amendments on the 1992 document, some new features of the 1999 document and new provisions of the new Red Book favouring the contractor and the employer.

A contract is a document that defines the 'scope of services' and is composed of two interdependent parts: the legal and commercial section, titled the form of contract, and the technical section. The technical section is where the 'scope of services' or the 'product' is fully described in terms of layout, specifications, requirements, planning and risks. The form of contract states the contract management rules, the role and interfaces of each party and where the inherent contract risk is allocated.

To complete a project on time and within budget, it is essential that each phase of its preparation and execution, starting with the feasibility assessment and terminating with the handing over of the completed project by the contractor to the owner, is formulated with precision. Careful contract formulation can avoid delays, disputes and unforeseen additional costs. It can also ensure timely completion of the project, provide comfort to the investors, mitigate risks, fix liabilities and responsibilities in the construction phase and ultimately ensure revenue streams to the project in the operational phase. The contract conditions, therefore, are central to successful project development.

The World Bank imposes the use of 'standard bidding documents' (SBDs) in the procurement of civil engineering works for all financed projects. SBDs include not only a model for the contract, both form of contract and technical section, but also all other bidding and/or non-contractual documents such as invitation for bids, instructions to bidders, etc. In the SBDs, the World Bank adopts the FIDIC's model form of contract. In addition, other organisations, institutions and even governments use this model in their construction contracts, particularly in the case of invitations for international tenders of large projects. FIDIC's form of contract is widely used and has been tested and tuned by years of experience.

7.1.1 Fédération Internationale des Ingénieurs-Conseils

The majority of governments around the world use or have a model form of contract for construction works. The models are usually elaborated by associations of consulting engineers. One of these associations is the Fédération Internationale des Ingénieurs-Conseils (FIDIC).

FIDIC is the international federation of consulting engineers focused in the definition and regulation of the role of many parties involved with the international construction industry. 'Founded in 1913, FIDIC is charged with promoting and implementing the consulting engineering industry's strategic goals on behalf of its Member Associations and to disseminate information and resources of interest to its members'.[1] It publishes standard forms of contract documents related to the procurement of engineering and construction works. FIDIC represents most private practice consulting engineers around the world. Only associations of consulting engineers are eligible for FIDIC membership.

FIDIC is best known for publishing standard forms of contract documents related to the procurement of civil engineering works, electrical/mechanical works, engineering, procurement, construction (EPC), etc. These standard forms include those for use by consulting engineers and other forms that are suitable for the procurement of building and engineering works. FIDIC's previous standard forms of contract documents for building and engineering works have been known by the colours of their respective covers.[2]

In 1999, FIDIC created a core 'rainbow suite' of four contracts: (i) the Red Book (Conditions of Contract for Construction for Building and Engineering Works); (ii) the Yellow Book (Conditions of Contract for Plant and Design-Build); (iii) the Silver Book (Conditions of Contract for EPC/Turnkey Projects); and (iv) the Green Book (Short Form of Contract). The books in the 1999 suite are known as first editions while books published in 1998 are known as test editions. In December 2017, FIDIC came up with a new suite of rainbow contracts, by publishing amended Red, Yellow and Silver books.

First editions deal with: (i) Conditions of Contract for Construction, which are recommendations for building or engineering works where the employer provides most of the design. However, the works may include some contractor-designed civil, mechanical, electrical and/or construction works; (ii) Conditions of Contract for Plant and Design-Build, which are recommendations for the provision of electrical and/or mechanical plants, and for the design and execution of building or engineering works. The scope of these conditions thus embraces both the old Yellow and Orange Books, for all types of contractor-designed works; (iii) Conditions of Contract for EPC/Turnkey Projects, which are suitable for projects on a turnkey basis of a process or power plant, of a factory or similar facility or of an infrastructure project or other type of development, where (a) a high degree of certainty concerning the final price and completion time is required, and (b) the contractor takes total responsibility for the design and execution of the project; and (iv) Short Form of Contract, which is recommended for building or engineering works of relatively small value; depending on the type of work and the circumstances, this form may also be suitable for contracts of greater value, particularly for relatively simple work.

This chapter deals with the FIDIC harmonised edition (FIDIC, 2006) which is commonly in use, i.e. Conditions of Contract for Construction for Building and Engineering Works Designed by the Employer.

There are 20 clauses in FIDIC's Conditions of Contract for Construction for Building and Engineering Works Designed by the Employer. The following presents a concise summary of all the clauses and important sub-clauses.

7.2 Clause 1: General provisions

The general provisions define the important constituents of a contract such as the contract, the parties involved in the contract, payments and contract agreement. There are 15 sub-clauses to the general provisions clause. The important sub-clauses are given below.

7.2.1 Contract

This sub-clause (*Sub-Clause 1.1.1.1*) includes the contract agreement, letter of acceptance (LOA), letter of tender, specifications, drawings, schedules and other documents listed in the contract agreement or letter of acceptance.

7.2.2 Contract agreement

An agreement signed in a standard form annexed to a tender document within 28 days of the letter of acceptance (*Sub-Clause 1.6*). Stamp duty and other expenses will be borne by the contractor.

7.2.3 Letter of acceptance

The letter of formal acceptance of the 'letter of tender' submitted by the tenderer, including any annexed memoranda comprising the agreement between and signed by both parties (*Sub-Clause 1.1.1.3*). If there is no such agreement, then 'letter of acceptance' means date of signing the contract agreement.

7.2.4 Letter of tender

By 'letter of tender' the FIDIC conditions of contract refers to an offer by the tenderer to the employer (*Sub-Clause 1.1.1.4*).

7.2.5 Engineer

A person so appointed by the employer to act as an engineer and named in the Appendix to Tender/Contract data, having the authority to perform certain obligations under the contract, in terms of the contract agreement (*Sub-Clause 1.1.2.4*). For replacement of the engineer, see Sub-Clause 3.4.

7.2.6 Contractor's representative

Sub-Clause 1.1.2.5 is referring to Sub-Clause 4.3 for details of the contractor's representative. It defines the contractor's representative as a person so named by the contractor to act as his authorised representative for the performance of the contract and to whom all notices will be issued by the engineer.

7.2.7 DB (generally known as the dispute adjudication board [DAB])

This refers to the dispute adjudication board of one or three persons so named in the contract under Sub-Clause 20.2 (appointment of the dispute board) (*Sub-Clause 1.1.2.9*).

7.2.8 Dates, tests, periods and completion (Clause 1.1.3)

Base date connotes 28 days prior to submission date of tender (*Sub-Clause 1.1.3.1*). Commencement date indicates date of commencement of work as specified in the contract agreement under Sub-Clause 8.1 (*Sub-Clause 1.1.3.2*). Time for completion is the period mentioned in the Appendix to Tender/Contract data for completion of works (*Sub-Clause 8.2*). Tests on completion refer to the tests to be carried out on completion of works or sections of works before taking over (*Clause 9*). Tests after completion refer to tests (if any) specified in the contract, after the works are taken over by the employer (*Sub-Clause 1.1.3.6*).

A taking-over certificate is a certificate issued on taking over the works as per Clause 10 (employer's taking over) (*Sub-Clause 1.1.3.5*). A defects notification period (under Sub-Clause 11.1 – completion of outstanding work and remedying defects) refers to the period specified in the Appendix to Tender (generally 12 months) calculated from the date of taking over the works under Sub-Clause 10.1 (extension of defects notification period) and Sub-Clause 11.3 (extension of defects notification period) (*Sub-Clause 1.1.3.7*). A performance certificate means the certificate of performance of the contract including the defects notification period issued by the engineer under Clause 11.9 (to be issued within 28 days of completion of the defect liability period). Bank guarantees will subsequently be released (*Sub-Clause 1.1.3.8*).

7.2.9 *Money and payments (Clause 1.1.4)*

An accepted contract amount refers to the amount accepted in the letter of acceptance for the execution and completion of the works and remedying defects (*Sub-Clause 1.1.4.1*). The contract price (*Sub-Clause 1.1.4.1*) (see also Clause 14.1 – the contract price) is defined as the price determined and evaluated (vide Clause 12.3) by multiplying the quantities of each item of work at the appropriate rate and includes adjustments in terms of the contract (price determined based on the actual measurement of works executed).

Costs are defined as all reasonable expenditure incurred by the contractor including overheads but excluding profits (*Sub-Clause 1.1.4.3*).

The final payment certificate refers to the payment certificate issued under Sub-Clause 14.13 (issue of final payment certificate) for final payments due to the contractor, as certified by the engineer (*Sub-Clause 1.1.4.4*). Such payments are to be paid within 56 days of receiving the performance certificate and after receipt of application from the contractor for the final certificate (Clause 14.11). An interim payment certificate is a payment certificate issued under Clause 14 (contract price and payment) other than the final payment certificate (*Sub-Clause 1.1.4.7*). The provisional sum is the amount specified in the bill of quantities (BOQ) for the execution of any part of the works or for supply of materials, plant or services under Sub-Clause 13.5 (provisional sum) (*Sub-Clause 1.1.4.10*).

The retention money (RM) is an amount retained from interim payments under Sub-Clause 14.3 and released under Sub-Clause 14.9 (payment of retention money) (*Sub-Clause 1.1.4.11*).

7.2.10 *Law and language*

The contract shall be governed by the law of the country (or other jurisdiction specified in the contract data [*Clause 1.4*]). The ruling language is as specified in the contract data and the language for communication shall be as specified in the contract data. If no language for communication is stated there, the language for communication shall be the ruling language of the contract.

Priority of documents: The sequence of the priority of the documents is as follows:

(i) Contract agreement (if any)
(ii) Letter of acceptance

   (iii) Letter of tender
   (iv) Particular conditions – Part A
   (v) Particular conditions – Part B
   (vi) General conditions
  (vii) Specification
 (viii) Drawings
   (ix) Schedules (BOQ) and any other documents forming part of the contract

The engineer shall issue clarifications or instructions for any ambiguity found in the documents.

7.2.11 Contract agreement

The parties shall enter into a contract agreement within 28 days after the contractor receives the letter of acceptance based on the contract agreement annexed to the particular conditions. The employer will bear all the duties and statutory payments to be made in connection with the contract agreement (*Clause 1.6*).

7.2.12 Delayed drawings or instructions

The contractor should give notice, with details of the necessary drawing or instruction and implications of delay, to the engineer whenever the works are likely to be delayed (*Clause 1.9*). If the engineer fails to issue the notified drawing or instruction within a reasonable time, the contractor can claim and receive the cost subject to Sub-Clause 20.1 (contractor's claims) and receive an extension of time under Sub-Clause 8.4 (extension of time for completion). In such a situation, the engineer can proceed in accordance with Sub-Clause 3.5 (determinations). If the delay is due to failures on the part of the contractor, the contractor is not entitled to such an extension of time, cost or profit.

7.2.13 Joint and several liability

In a joint venture (JV) or consortium contract, constituted under applicable laws, there is joint responsibility for each party and they are severally responsible to the employer for the performance of the contract. The name of the principal of the JV should be notified to the employer because the JV principal will represent the JV and receive all notices and shall have the authority to bind the contractor to each member of the JV. The prior consent of the employer is required for altering the composition or legal status of the JV (*Clause 1.14*).

7.3 Clause 2: The employer

The employer shall give the contractor right of non-exclusive access and possession of a site within the time (or times) specified in the Appendix to Tender only after giving the performance security. The contractor's right of access to a site is not exclusive and is subject to the performance security being provided (*Sub-Clause 2.1: Right of Access to the Site*). The contractor's claims are applicable in case of delay by the employer.

    Delay to handover the site as per the programme will entitle the contractor to claim for extension of time and compensation (cost plus profit) (Sub-Clause 8.4 – extension of time for completion) after notice by the contractor under Sub-Clause 20.1 (contractor's claims) and processing the claim under Sub-Clause 3.5 (determinations). Failure of the contractor to submit the claims and documents in a timely manner shall not entitle him to such extension of time, cost or profit.

The employer shall provide reasonable assistance to the contractor for obtaining: (i) copies of the laws of the country; (ii) obtaining permits and licenses for complying with the laws of the country; (iii) delivery of goods through customs; and (iv) export of contractor's equipment (*Sub-Clause 2.2: Permits, Licences or Approvals*).

The employer will ensure the cooperation of the contractor with the personnel of other contractors and their own staff and ensure safe procedures (Sub-Clause 4.8 – safety procedures) and the protection of the environment (Sub-Clause 4.18 – protection of the environment) (*Sub-Clause 2.3: Employer's Personnel*).

The employer has to provide reasonable assistance to the contractor to obtain local permits/approvals and to confirm financial arrangements for payment to the contractor within 28 days of request for information (*Sub-Clause 2.4: Employer's Financial Arrangements*), and also to convey any changes in the employer's financial arrangement. The employer's claims are ratified by the engineer and can be deducted from the payment certificate issued to the contractor (*Sub-Clause 2.5: Employer's Claims*).

7.4 Clause 3: The engineer

This clause provides the engineer's duties and authority, delegation by the engineer, instructions of the engineer, replacement of the engineer and determinations.

'Engineer' refers to the person appointed by the employer to act as the engineer for the purposes of the contract and is named in the contract data, or other person appointed from time to time by the employer and notified to the contractor. The employer shall appoint the engineer who shall carry out his duties as assigned in the contract.

The engineer has no authority to amend the contract (*Sub-Clause 3.1: Engineer's Duties and Authority*). The engineer's authority and limitations can be defined in the contract. Prior approval of the employer in exercising his authority shall only be as stated in particular conditions. The employer shall not impose further constraints on the engineer's authority without the consent of the contractor. The engineer shall be deemed to act for the employer in the exercise of his authority under the contract. The engineer's approval does not relieve the contractor from any responsibility he has under contract. The engineer shall consult both parties to reach agreement on claims or otherwise proceed to make a fair determination in accordance with the contract.

The engineer can delegate his authority (or revoke the same) to his staff which shall be done in writing and a copy of same shall be given to the contractor (*Sub-Clause 3.2: Delegation by the Engineer*). Assistants to whom the authority has been delegated by the engineer shall only be authorised to issue instructions to the contractor to the extent of such delegation and shall have the same effect as if issued by the engineer. Any failure to disapprove the work, plant or material shall not constitute approval and shall not prejudice the right of the engineer to reject the same.

The contractor can question and refer the decision of the assistant to the engineer for further determination. The contractor shall receive instructions only from the engineer or his assistant to whom such authority has been delegated (*Sub-Clause 3.3: Instructions of the Engineer*). If instructions constitute a variation, Clause 13 (variations and adjustments) shall apply. The contractor shall comply with the instructions given by the engineer or his authorised assistant. Instructions shall, as far as possible, be in writing. Oral instructions shall be confirmed by the contractor in writing within 2 days of giving instructions and if the same are not rejected by the engineer or his assistant within 2 days of receipt of such confirmation, then the confirmation shall constitute written instructions.

The employer can replace the 'engineer' with any other person against whom the contractor does not have any reasonable objections (*Sub-Clause 3.4: Replacement of the Engineer*). The employer can replace the engineer by giving 21 days' notice to the contractor in writing, giving details of the new engineer. The contractor can object in writing to such a replacement by a new appointee against whom the contractor has reasonable objection with supporting particulars. The employer has to give full and fair consideration to this objection.

The engineer shall be fair in his determination of all matters under the contract and shall do so in consultation with the contractor and take cognisance of all relevant matters (*Sub-Clause 3.4: Determinations*). The engineer shall give notice to both parties of such determination or agreement with all supporting particulars unless and until revised under Clause 20 (claims, disputes and arbitration).

7.5 Clause 4: The contractor

This clause outlines the contractor's general obligations, performance security, contractor's representative, subcontractors, assignment of benefit of subcontract, cooperation, setting out, safety procedures, quality assurance, site data, sufficiency of the accepted contract amount, unforeseeable physical conditions, rights of way and facilities, avoidance of interference, access route, transport of goods, contractor's equipment, protection of the environment, electricity, water and gas, employer's equipment and free-issue materials, progress reports, security of the site, contractor's operations on site and fossils.

7.5.1 General obligations

The contractor shall design if required to the extent specified in the contract (*Sub-Clause 4.1*). The contractor shall design (to the extent specified), execute and complete the works in accordance with the contract and with the engineer's instructions, and remedy all defects in the works. He shall be responsible for the supply of all materials, plant and equipment, personnel and all services of a temporary or permanent nature, for completion and remedying all defects.

The contractor is responsible for the safety of operations and methods of construction, all documentation, as specified in the contract, but shall not be responsible for the design and specifications of permanent works. If the contract specifies (in particular conditions) his responsibility for the design of any part of the permanent works, the contractor shall submit all documentation in accordance with the specifications and drawings and other criteria and shall be responsible for the same. He shall furnish as built drawings before tests and all operation and maintenance manuals and hand over all works to the full satisfaction of the engineer.

7.5.2 Performance security

The contractor shall deliver the performance security to the employer within 28 days of receipt of the letter of acceptance (*Sub-Clause 4.2*). Within 28 days of receiving the letter of acceptance, the contractor shall furnish to the employer a performance security from an approved bank in the country, in the format specified and for the amount and currencies specified in the particular conditions. The performance security shall be valid and enforceable until the specified expiry date and after the end of the defects liability period. If the contractor is not entitled to receive the performance certificate by the date so specified in the contract, then 28 days prior to this expiry date, the contractor shall extend the validity until the works have been completed and all defects rectified. The employer shall return the performance security to the contractor within 21 days of the performance certificate being issued on completion of the contract.

7.5.3 *Contractor's representative*

The contractor shall appoint the contractor's representative and shall give him all authority necessary to act on the contractor's behalf under the contract (*Sub-Clause 4.3*). If consent is withheld or subsequently revoked or the contractor's representative fails to act as his representative, the contractor shall appoint another person with the consent of the employer. The contractor shall not, without the prior consent of the engineer, revoke the appointment of the contractor's representative or appoint a replacement.

7.5.4 *Subcontractors*

As per Sub-Clause 4.4, the whole works cannot be subcontracted. The subcontractor can be named in the tender for other subcontractors proposed during execution; the engineer's approval is required. The contractor shall be responsible for the acts or defaults of any subcontractor, his agents or employees, as if they were the acts or defaults of the contractor (*Sub-Clause 4.4*). The contractor shall give the engineer not less than 28 days' notice of the intended date of commencement of each subcontractor's work, and of the commencement of such work on site.

7.5.5 *Assignment of benefit of subcontract*

If a subcontractor's obligations extend beyond the expiry date of the relevant defects notification period and the engineer, prior to this date, instructs the contractor to assign the benefit of such obligations to the employer, then the contractor shall do so (*Sub-Clause 4.5*). Unless otherwise stated in the assignment, the contractor shall have no liability to the employer for the work carried out by the subcontractor after the assignment takes effect.

7.5.6 *Cooperation*

The contractor shall cooperate with the employer's personnel or other contractors working in and around the contractor's site (*Sub-Clause 4.6*). However, if the contractor incurs expenditure, he will be entitled to compensation under the variation clause.

7.5.6 *Setting out*

The contractor shall be responsible for the correct setting out of works at his cost and shall be responsible for all levels and alignment (*Sub-Clause 4.7*). The employer shall be responsible for any errors in the specified items of reference but the contractor shall make all reasonable efforts to verify their accuracy. Any delay on this account shall entitle the contractor for an extension and compensation under Clauses 8.1 and 20.1, and the engineer shall, if satisfied, take necessary steps under Clause 3.5 (determinations).

7.5.7 *Safety procedures*

The contractor shall be responsible for observing all safety procedures, the safety of all personnel and the work site, and provide any temporary works (including roadways, footways, guards and fences) which may be necessary (*Sub-Clause 4.8*).

7.5.8 *Quality assurance*

The contractor shall submit a quality assurance manual which will be audited by the engineer (*Sub-Clause 4.9*). Compliance with the quality manual shall not relieve the contractor of his responsibilities under the contract.

7.5.9 *Site data*

The employer shall make available to the contractor all site data and hydrological data in his possession including environmental aspects before the base date and also make it available to the contractor when it comes into his possession after the base date (*Sub-Clause 4.10*). The contractor will be responsible for interpreting such data. However, before submitting a tender, and following a site inspection, the contractor should have satisfied himself of all relevant aspects which would affect his tender costs, including (without limitation): (i) form and nature of the site (including subsurface condition); (ii) hydrological and climatic conditions; (iii) extent and nature of works necessary for execution of works and remedying defects; (iv) laws and labour practices in the country; and (v) contractor's requirements of all services.

7.5.10 *Sufficiency of accepted contract amount*

The contractor shall satisfy himself of the correctness of the accepted contract amount based on all site conditions referred to in Sub-Clause 4.10 (site data) and includes all the contractor's obligations (including those under provisional sums) necessary for the proper execution of works and remedying defects (*Sub-Clause 4.11*).

7.5.11 *Unforeseeable physical conditions*

Physical conditions mean natural physical conditions including subsurface conditions which the contractor encounters at a site other than adverse unforeseen conditions and excluding climatic conditions (*Sub-Clause 4.12*). The contractor shall give notice to the engineer in writing of an inspection of the unforeseen conditions giving reasons for it, and the engineer, after inspection and on satisfaction, gives instructions to the contractor to carry out works which, as necessary, will entitle the contractor to an extension of time and compensation vide Clauses 8.4 (extension of time for completion), 20.1 (contractor's claims) and 3.5 (determinations). However, if favourable conditions are met to the engineer's satisfaction, he shall have the right to assess the reduction in cost and effect such reductions provided the overall effect of such additions and reductions do not result in a reduction in the contract price.

7.5.12 Rights of way and facilities

The contractor shall bear all costs of works for access to a site including that for outside the site of works (*Sub-Clause 4.13*).

7.5.13 Avoidance of interference

The contractor shall not cause inconvenience to public access to roads/footpaths and shall indemnify the employer against all damages, losses, etc. (*Sub-Clause 4.14*). The contractor shall indemnify and hold the employer harmless against and from all damages, losses and expenses (including legal fees and expenses) resulting from any such unnecessary or improper interference.

7.5.14 Access route

The contractor shall be responsible for the maintenance of and damage to all access routes and shall indemnify the employer for all damages and bear all costs (*Sub-Clause 4.15*).

7.5.15 Transport of goods

The contractor, unless otherwise specified, shall bear all costs for the transportation of all goods and shall give 21 days' notice to the engineer of the arrival of such goods to the site (*Sub-Clause 4.16*). The contractor shall indemnify and hold the employer harmless against and from all damages, losses and expenses (including legal fees and expenses) resulting from the transport of goods.

7.5.16 Contractor's equipment

The contractor is responsible for the transportation of all equipment required for works and shall not remove major items of equipment without the engineer's consent (*Sub-Clause 4.17*). But no such consent is required for the movement of transport vehicles and personnel off site.

7.5.17 Protection of the environment

The contractor is responsible for the protection of the environment on site and off site (*Sub-Clause 4.18*) and to ensure that his activities do *not* result in exceeding pollution values beyond those specified in the specifications and in the laws of the country.

7.5.18 Electricity, water and gas

The contractor shall bear all costs of electricity, water and gas and arrange for measurements of their use on works and the employer shall recover these from payments due to the contractor (*Sub-Clause 4.19*). The quantities consumed and the amounts due for supplies of electricity, water, gas and other services shall be agreed or determined by the engineer in accordance with Sub-Clause 2.5 (employer's claims) and Sub-Clause 3.5 (determinations).

7.5.19 Employer's equipment and free-issue materials

If so specified in the contract, the employer shall make available to the contractor equipment for his use on works at prices so fixed and shall be responsible for its maintenance if the same is operated by the contractor (*Sub-Clause 4.20*). The employer shall supply to the contractor 'free issue of materials' as specified in the contract and the contractor shall take over (quantities duly checked) and keep proper custody of the same. Shortages and defects should be identified immediately on taking over.

7.5.20 *Progress reports*

The contractor should submit six copies of a monthly progress report to the engineer before the seventh of each month (*Sub-Clause 4.21*). This will be continued until the taking-over certificate is issued. The report should contain charts and detailed progress on all activities, procurements, construction, testing and work of each subcontractor, together with photographs of works, details of each main item of equipment, percentage progress, contractor's personnel test certificates, manual submissions, details of claims, if any, safety aspects, environment aspects, any other item of importance to be brought on record together with a comparison of planned and actual progress with reasons for deficiencies and action proposed for making good the shortfall.

7.5.21 *Security of the site*

The contractor shall keep all unauthorised persons off the site and authorised persons shall be limited to the contractor's and employer's personnel and others as notified by the engineer/employer to the contractor (*Sub-Clause 4.22*).

7.5.22 *Contractor's operations on site*

The contractor shall confine his activities to the site and other areas specified. The site should be kept clean and clear of all obstructions (*Sub-Clause 4.23*). After issue of the taking-over certificate, the contractor shall clear the site of all surplus materials and equipment except those required for defect rectification.

7.5.23 *Fossils*

All items of geological and archaeological interest found on site shall be placed under the care and authority of the employer without damaging any of these findings (*Sub-Clause 4.24*). All costs incurred by the contractor in this connection shall be paid to the contractor duly following the procedures laid down. Further, the contractor will also be entitled to an extension of time, if any.

7.6 Clause 5: Nominated subcontractors

7.6.1 *Definition*

A nominated subcontractor is one who has been so stated in the contract and whom the engineer instructs the contractor under Clause 13 (variations and adjustment) to employ as a subcontractor (*Sub-Clause 5.1*).

7.6.2 *Objection to nomination*

The contractor is not obliged to accept the subcontractor so nominated by the engineer in which case he will submit his reasons thereof (*Sub-Clause 5.2*). The objections can be considered reasonable if they fall under categories so specified in Clause 5.2 unless the employer agrees to indemnify the contractor from any damages due to the acts of the subcontractor.

7.6.3 *Payments to nominated subcontractors*

Before issuing the payment certificate, the engineer may request the contractor to furnish evidence of payments received by the subcontractor as per their subcontract (*Sub-Clause 5.3*). In the absence of such evidence, the employer may make payments directly to the subcontractor.

7.6.4 *Evidence of payments*

The engineer may request the contractor to supply reasonable evidence that the nominated subcontractor has received the entire amount due in line with previous payment certificates less applicable deductions for retention or otherwise, before issuing a payment certificate which includes an amount payable to a nominated subcontractor (*Sub-Clause 5.4*).

7.7 Clause 6: Staff and labour

Engagement of staff and labour

The contractor shall make all arrangements for staff and labour and for their payments, housing, food and transport (*Sub-Clause 6.1*).

7.7.1 *Rate of wages and conditions of labour*

Rates of wages and conditions of labour shall not be less than are prevalent in the area (*Sub-Clause 6.2*).

7.7.2 *Persons in the service of employer*

The contractor shall not recruit persons in the service of the employer (*Sub-Clause 6.3*).

7.7.3 *Labour laws*

The contractor shall obey all labour laws applicable to his personnel and shall allow them all their legal rights (*Sub-Clause 6.4*). The contractor's employees shall obey all laws including safety at works.

7.7.4 *Working hours*

No work shall be carried out outside working hours or on holidays unless stated otherwise in the contract or approved by the engineer (*Sub-Clause 6.5*).

7.7.5 *Facilities for staff and labour*

The contractor shall provide all accommodation and welfare facilities to his personnel and also to the employer's personnel as specified in the contract. Quarters forming part of the permanent works shall not be utilised as staff quarters (*Sub-Clause 6.6*).

7.7.6 Health and safety

The contractor shall take care of the health and safety of the contractor's personnel in collaboration with the local health authorities (*Sub-Clause* 6.7). The contractor will ensure the availability of medical facilities for first aid and ambulance service availability and maintain good hygienic conditions at the site. The contractor shall appoint a qualified safety officer ('an accident prevention officer') to the site with adequate authority to ensure safety at the site. The contractor shall send regular safety reports to the engineer.

7.7.7 Contractor's superintendence

The contractor shall provide all necessary superintendence to plan and execute works using adequate numbers of personnel with knowledge of the local language for the satisfactory execution of works (*Sub-Clause* 6.8). The superintendence shall be given by an adequate number of persons with sufficient knowledge of the language for communications (defined in Sub-Clause 1.4 [law and language]) and of the operations to be carried out, for the satisfactory and safe execution of the works.

7.7.8 Contractor's personnel

The contractor's personnel shall be (i) appropriately qualified, (ii) skilled and (iii) experienced in their respective trades or occupations. The consultant can order the removal of any person from site for disobedience, incompetence and not observing safety (*Sub-Clause* 6.9).

7.7.9 Record of contractor's personnel and equipment

The contractor shall furnish monthly details of each type of equipment and category of staff on site until issue of the taking-over certificate for the works (*Sub-Clause* 6.10).

7.7.10 Disorderly conduct

The contractor is to take reasonable precautions to prevent unlawful and disorderly behaviour of his personnel, preserving the peace and protecting the property on site (*Sub-Clause* 6.11).

7.7.11 Foreign personnel

The contractor may bring into the country any foreign personnel who are necessary for the execution of the works to the extent allowed by the applicable laws by providing them with the required residence visas and work permits (*Sub-Clause* 6.12). The employer should assist the contractor in bringing in the contractor's personnel. However, the contractor shall be responsible for the return of these personnel to the place from where they were recruited.

7.7.12 Do's and don'ts

The contractor shall arrange for the provision of: (i) a sufficient supply of suitable food; (ii) an adequate supply of drinking and other water for the use of the contractor's personnel; (iii) necessary precautions to protect from insect and pest nuisance; (iv) respecting the country's recognised festivals, days of rest and religious or other customs; (v) keeping complete and accurate records of the employment of labour at the site; and (vi) recognising workers' rights to form and to join workers' organisations in countries where the relevant labour laws allow. Similarly, the contractor shall not: (i) make provision for the import or use of items not in accordance with the laws of the country; (ii) employ forced labour or employ children in a manner that is economically exploitative; and (iii) make employment decisions on the basis of personal characteristics unrelated to inherent job requirements (refer *Sub-Clauses 6.13–6.24*).

7.8 Clause 7: Plant materials and workmanship

7.8.1 *Manner of execution*

The contractor shall execute all works as specified in the contract in a workman-like manner and in accordance with recognised practices with properly equipped facilities (*Sub-Clause 7.1*).

7.8.2 *Samples*

All samples of materials shall be approved by the engineer before use in works at his costs, including additional samples as instructed by the engineer (*Sub-Clause 7.2*).

7.8.3 *Inspection*

The employer shall have right to access to places of source of materials during production, manufacture and construction at the site or elsewhere and check the workmanship and progress (*Sub-Clause 7.3*). No such activity shall relieve the contractor of his obligations and responsibility. The employer's personnel should be allowed to examine, inspect, measure and test the materials and workmanship, and check the progress of manufacture of plant and production and manufacture of materials. The contractor shall get his works inspected before the same are covered, failing which the engineer can get these reopened for inspection at the contractor's costs.

7.8.4 *Testing (prior to taking over)*

The contractor shall provide all plant and equipment and staff for all tests of plant and materials as required under the contract (*Sub-Clause 7.4*). The engineer, under Clause 13 (variation), can instruct the contractor to carry out additional tests at the cost of the contractor. The engineer will give more than 24 hours' notice of his intention to be present for tests. Any delays on this account shall entitle the contractor to an extension and compensation if any. The contractor shall forward all details of tests to the engineer. Tests carried out in the absence of the engineer, after giving him due notice, shall be deemed to have been carried out in his presence.

7.8.5 *Rejections*

The engineer shall reject all works/materials found defective and get the defects rectified at the contractor's cost (*Sub-Clause 7.5*). The engineer can get works re-tested and should such re-testing or rejections cause additional costs to the employer, the same will be recovered from the contractor under Clause 2.5 (employer's claims).

7.8.6 *Remedial works*

The engineer may instruct the contractor to remove/re-execute defective works and to execute works required for the safety of the works which shall be done by the contractor within a reasonable time, failing which the employer can get them executed and recover the costs under Clause 2.5 (employer's claims) (*Sub-Clause 7.6*).

7.8.7 *Ownership of plant and materials*

All plant and materials delivered on site or are entitled for payment under Clause 8.10 (payments in the event of suspension) shall be the property of the employer (*Sub-Clause 7.7*).

7.8.8 *Royalties*

The contractor shall pay all royalties for all materials, rents and other payments for natural materials obtained from outside the site and the disposal of demolished or surplus materials except to the extent that the disposal areas within the site have been specified (*Sub-Clause 7.8*).

7.9 Clause 8: Commencement, delays and suspension

7.9.1 *Commencement of works*

The engineer shall give the contractor not less than 7 days' notice of the commencement date which shall be within 42 days of receipt of the letter of acceptance, unless otherwise stated in the particular conditions (*Sub-Clause 8.1*).

7.9.2 *Time for completion*

The contractor shall complete all works including tests for completion, for the purpose of taking over vide Clause 10.1 (taking-over of works or sections) within the time for completion of the works stated in the contract (*Sub-Clause 8.2*).

7.9.3 *Programme*

The contractor shall submit a detailed programme of work to the engineer within 28 days of the receipt of notice under Clause 8.1 and shall revise the same, whenever the same is inconsistent with the actual programme (*Sub-Clause 8.3*).

Each programme shall include: (i) a time chart of work sequence, documentation, plant delivery, erection, testing, etc.; (ii) each stage of work performed by the nominated subcontractor (Clause 5); (iii) sequence and timings of tests specified in the contract; (iv) a report on the methodology of works and details of equipment and personnel at various stages of works; the engineer shall give notice, within 21 days of receipt of the programme, indicating modifications required to the programme, failing which the contractor will proceed with the programme; the contractor shall promptly issue notice to the engineer of any probable delays and extra costs likely to take place for work execution; and if the work does not proceed as per the programme, the engineer shall give notice to the contractor to revise the programme.

7.9.4 *Extension of time for completion*

The contractor shall be entitled to an extension of time for completion (subject to Sub-Clause 20.1: contractor's claims) if: (a) a variation or a substantial change in quantities occurs; (b) a cause for delay entitles extension under this clause; (c) exceptionally adverse climatic conditions; (d) unforeseeable shortages of goods and personnel caused by epidemic or by government action; and (e) any delay caused by the employer's actions or by other contractors on site (*Sub-Clause 8.4*). The engineer, on receipt of notice from the contractor, under Sub-Clause 20.1 (contractor's claims) shall proceed with his determination, and review previous determinations and may increase but shall not decrease the total extension of time.

7.9.5 *Delay caused by authorities*

Delay or disruption will be considered if caused by the procedures of and by public authorities and due to unforeseeable disruptions, and will entitle the contractor to an extension of time as per Sub-Clause 8.4 (*Sub-Clause 8.5*).

7.9.6 *Rate of progress*

If progress falls short of the planned programme due to causes other than those listed in Sub-Clause 8.4, the contractor shall revise the programme with revised methods to complete the works on time. The contractor shall increase his resources to complete the works on time, failing which the engineer will do so at the risk and costs to the contractor and recover costs vide Sub-Clause 2.5 (employer's claims) in addition to recovering damages vide Sub-Clause 8.7 below (*Sub-Clause 8.6*).

7.9.7 *Delay damages*

Delay on the part of the contractor to complete the works within time or extended time shall result in the recovery of damages, at rates stated in the Appendix to Tender, subject to the limitations of the total amount, if any, stated in the Appendix to Tender (*Sub-Clause 8.7*). Delay damages shall not relieve the contractor of his obligations under the contract and shall be the only damages to be recovered for such default other than those in the event of termination vide Clause 15.2 (termination).

7.9.8 *Suspension of works*

The engineer may, by notifying reasons, suspend part or the whole work (*Sub-Clause 8.8*). Clauses 8.9, 8.10 and 8.11 shall not apply if the cause of suspension is due to the contractor.

7.9.9 *Consequences of suspension*

The contractor shall be entitled to extensions of time and compensation for additional costs (Sub-Clauses 8.4: EOT and 20.1: claims), provided he complies with Clause 8.8 (suspension of works – protection of site and works) (*Sub-Clause 8.9*).

7.9.10 *Payment of plant and materials in event of suspension*

If the suspension is for more than 28 days and the plant and/or materials have been marked as the employer's materials as per the engineer's instructions, the contractor shall be entitled to payment of the value on the date of suspension of such plant and materials which have not been delivered on site (*Sub-Clause 8.10*).

7.9.11 Prolonged suspension

If the suspension (Sub-Clause 8.8) is for more than 84 days, the contractor can request the engineer to proceed with works (*Sub-Clause 8.11*). If no permission is given within 28 days of such notice, the contractor can claim suspension as an omission under Clause 13 (variation and adjustments) for the affected part. Similarly, if the whole work is affected, the contractor can give notice of termination to the engineer under Sub-Clause 16.2 (termination of contract).

7.9.12 Resumption of work

On resumption of works, the engineer will jointly assess the loss incurred by the contractor (*Sub-Clause 8.12*).

7.10 Clause 9: Tests on completion

7.10.1 Contractor's obligations

Tests on completion shall be carried out as per Clause 7.4 (testing), after providing documents as per the contractor's general obligations vide Sub-Clause 4.1(d). The contractor shall give 21 days' notice to the engineer to carry out tests (*Sub-Clause 9.1*). The contractor shall carry out tests within 14 days of the date so decided by the engineer. On completion of the tests, the contractor shall submit to the engineer the results of the tests.

7.10.2 Delayed tests clause

Sub-Clause 7.4 (testing – fifth paragraph) and Sub-Clause 10.3 (interference with tests) shall apply if the tests are delayed by the employer (*Sub-Clause 9.2*). The entitlement will be subject to Sub-Clause 20.1 (contractor's claims). If there is a delay on the part of the contractor to carry out tests, the engineer will give 21 days' notice to conduct tests, failing which the engineer shall carry out the tests at his risk and costs and these tests results shall be deemed as accepted.

7.10.3 Re-testing

If the works fail to pass tests (Sub-Clause 7.5 – rejection), the same work will be repeated under the same terms and conditions (refer *Sub-Clause 9.3*).

7.10.4 Failure to pass tests on completion

If the works fail to pass following repeated testing vide Sub-Clause 9.3 above, then: (a) order further repetition of tests vide Clause 9.3; (b) reject the works with remedies as provided in Sub-Clause (c) of 11.4 (failure to remedy defects); and (c) issue a takeover certificate, if the employer so requests. In this case, the contract price may be reduced as appropriate to be agreed on by both parties and recovered before the taking-over certificate is issued or determined and paid under Sub-Clause 2.5 (employer's claims) and Clause 3.5 (determinations) (refer *Sub-Clause 9.4*).

7.11 Clause 10: Employer's taking over

7.11.1 *Taking over of works and sections*

Works shall normally be taken over when they have been completed in agreement with the contract (*Sub-Clause 10.1*). The employer is required to take over the work when completed according to the contract within 14 days of notice from the contractor. The engineer is required to issue the 'taking-over certificate' to the contractor within 28 days of the above notice or reject the application giving reasons. If the engineer fails either to take over or to reject within 28 days, but the work is substantially complete, the certificate shall be deemed to have been issued on the last day of that period (Sub-Clause 10.1). The contractor shall give 14 days' notice to the engineer for such take over and the engineer shall issue a takeover certificate within 28 days of such notice when works have been substantially completed and the balance of the works does not affect the use of the works to be taken over, or reject the application if the works have not been substantially completed, giving reasons thereof. If the engineer fails to take action on the contractor's request for taking over within 28 days of such request, the takeover certificate will be deemed to have been issued on the last day of that period.

7.11.2 *Taking over of parts of the works*

The employer can take over parts of the works for use, only after the issuing of a takeover certificate for such parts (*Sub-Clause 10.2*). However, if any part is used before taking over, the same is deemed to have been taken over from the date of its use.

The contractor is not liable for any part used by the employer unless specified in the contract. Any costs incurred by the contractor after issue of taking over shall be paid for by the engineer. Recovery of delay damages shall be proportionately reduced corresponding to the value of the works taken over. The defects liability period for the parts taken over shall start from the date such parts are taken over.

7.11.3 *Interference with tests on completion*

If the contractor is prevented from carrying out tests for more than 14 days, then the takeover certificate will be issued by the engineer on the due date and these tests will be carried out during the defect liability period. Any delay and costs on this account shall be compensated by the engineer (*Sub-Clause 10.3*).

7.11.4 *Surfaces requiring reinstatement*

Except as otherwise stated in the taking-over certificate, a certificate for a section or part of works shall not be deemed to certify completion of any ground requiring reinstatement (*Sub-Clause 10.4*).

7.12 Clause 11: Defects liability

7.12.1 *Completion of outstanding works and remedying defects*

Complete all works outstanding on the date of the taking-over certificate and rectify all defects before the expiry date of the defects notification period (*Sub-Clause 11.1*). Any work by the contractor within the defects notification period, as pointed out by the employer, well within contract requirement is to be carried out at the risk and cost of the contractor.

7.12.2 Cost of remedying defects

All the works under Sub-Clause 11.1 shall be carried out at the cost of the contractor, so long as it is on the contractor's account (*Sub-Clause 11.2*). Otherwise, the contractor will be entitled to variation under Sub-Clause 13.3.

7.12.3 Extension of defects notification period

The employer can seek an extension of the defects notification period if a major item of the plant cannot be used for the purposes for which it was intended, but not exceeding 2 years in extension and is not rectified within the notified period (*Sub-Clause 11.3*). However, the contractor's obligations under this clause shall not apply to defects or damages occurring more than 2 years after the defects notification period, if the works were suspended under Sub-Clause 8.8 (suspension of works) or Clause 16.1 (contractor's entitlement to suspend works).

7.12.4 Failure to remedy defects

The contractor shall remedy defects within a reasonable period of extended time, failing which the employer can get the same executed at his costs and effect recovery and reduce the contract price (*Sub-Clause 11.4*). If as a result of non-rectification, any part of the works cannot be put in to use, the contract can be terminated and all costs recovered including financing and other costs.

7.12.5 Removal of defective work

If any part or equipment cannot be repaired on site, the employer may permit the contractor to replace it by removing it from site for which the contractor will have to provide additional bank security before removing it from site (*Sub-Clause 11.5*).

7.12.6 Further tests

The contractor shall be liable to carry out further tests on items of works so remedied as per standards laid down (*Sub-Clause 11.6*).

7.12.7 Right of access

The contractor will have access to the site until the performance certificate is issued consistent with security restrictions (*Sub-Clause 11.7*).

7.12.8 Contractor to search

If required by the engineer, the contractor shall search for the cause of the defect and shall be paid for doing so unless the defect is to be remedied at the cost of the contractor (*Sub-Clause 11.8*).

7.12.9 Performance certificate

The contractor's obligations under the contract shall be deemed to have been completed from the date of issue of the performance certificate by the engineer which shall be issued within 28 days from the expiry date of the defects notification period (*Sub-Clause 11.9*). The engineer shall issue a copy of this certificate to the employer. Only the performance certificate shall be deemed to constitute acceptance of works.

7.12.10 Unfulfilled obligations

After the issuing of the performance certificate, both parties shall determine the unperformed obligations until which the contract shall be deemed to remain in force (*Sub-Clause 11.10*).

7.12.11 Clearance of site

Within 28 days of receipt of the performance certificate, the contractor shall remove all materials and equipment from the site, failing which the employer may sell the same at his cost (*Sub-Clause 11.11*). Any balance money shall be paid to the contractor. Only the performance certificate constitutes acceptance of the work. Both parties are liable for any unperformed obligations and the contract remains in force to that extent.

7.13 Clause 12: Measurement and evaluation

7.13.1 Works to be measured

The engineer shall measure the works in the presence of the contractor after giving him notice to be present and supply the necessary documents, failing which the engineer will proceed to record measurements which shall be considered as accurate (*Sub-Clause 12.1*). The contractor shall be asked to examine and sign the measurements prepared by the contractor from records, failing which they would be considered as accurate. In case of disagreements, the contractor shall write to the engineer with details. The engineer shall examine the details after which they will be confirmed.

7.13.2 Method of measurement

All measurements shall be of net actual quantities and as per the bill of quantities, schedules or specifications (*Sub-Clause 12.2*). If no rate is quoted for a BOQ item, the same shall be deemed to have been covered under other item/items.

7.13.3 Evaluation

Price for payment shall be evaluated by applying the corresponding unit rate to the measurements so recorded for each item of work (*Sub-Clause 12.3*). A new rate shall be decided if: (a) the quantity for an item is changed by more than 25 per cent; (b) the variation in the amount of each item is more than 0.25 per cent of the contract price; (c) the change in quantity directly changes the cost per unit quantity by more than 1 per cent of the contract price; and (d) the item is not specified in the contract, or the work is instructed under Clause 13 (variations), and no rate or price is specified in the contract, and no rate specified in the contract is appropriate for application due to the nature of the work.

The new rate derived shall be based on a similar type of work in the contract and if no such work is in the contract then it will be based on actual cost plus reasonable profit (cost to include overheads). Until an appropriate rate is agreed and determined, a provisional rate is applied for interim payments.

7.13.4 Omissions

The contractor shall refer to the engineer regarding the omission of any work which forms part of a variation and the engineer shall decide the same under Clause 3.5 unless it is covered under any other item of BOQ (refilling of earthwork) (*Sub-Clause 12.4*).

7.14 Clause 13: Variation and adjustment

7.14.1 Right to vary

The engineer can vary any work prior to the taking-over certificate (*Sub-Clause 13.1*). The contractor shall execute such works unless he convinces the engineer of his inability to do so. Variation will include change in quantity, quality, levels, dimensions, omission of any work, any additional work or change in the sequence of work. The contractor shall not make any modification in work without the approval of the engineer.

7.14.2 Value engineering

The contractor can suggest improvements in works by value engineering subject to the engineer's acceptance (*Sub-Clause 13.2*). If accepted, the contractor shall be entitled to a fee of 50 per cent of the difference between: (i) a reduction in the contract price; and (ii) a reduction in the value to the employer due to a reduction in quality, anticipated life or operating efficiency. However, no fee is payable if item (i) is less than item (ii).

7.14.3 Variation procedure

Prior to instructing a variation, if the engineer requests the contractor for a variation, the contractor shall respond in writing giving reasons why he cannot comply or submit the details of the proposed work and a programme for its execution with modifications if any and his proposal for evaluation if any and the engineer shall respond (*Sub-Clause 13.3*). The contractor shall not delay the work while awaiting a response. The engineer shall issue instructions to execute the variation and record the costs. Each variation shall be evaluated under Clause 12 (measurement and evaluation) unless otherwise instructed under this clause.

7.14.4 Payment in applicable currencies

If the contract provides for payment in different currencies, the variation amount shall be paid in the same proportions of currencies as that provided in the contract (*Sub-Clause 13.4*).

7.14.5 Provisional sums

Provisional sums will only be used as instructed by the engineer and the contract price adjusted accordingly (*Sub-Clause 13.5*). It shall be utilised only for works for which it relates.

The engineer may instruct the execution of works and their value under Clause 13.3 (variation procedure) and/or works to be done through a nominated subcontractor which shall be paid on costs and profits as specified in the Appendix to Tender (all quotations etc. shall be produced by the contractor).

7.14.6 Day works

The engineer can direct variations to be executed at rates specified in the day work schedule (*Sub-Clause 13.6*). The contractor is to produce invoices, vouchers, etc., and all records of this work shall be jointly signed on the basis of which the price statements are issued to the engineer under interim payments.

7.14.7 Adjustments for change in legislation

The contract price shall account for a 'decrease or increase' in cost due to a change in law (*Sub-Clause 13.7*). In addition to an extension of time, any additional costs due to a change in law after the base date shall be paid for and determined by the engineer vide Sub-Clause 3.5 (determinations).

7.14.8 Adjustment of change in cost

An adjustment of a change in cost is to be calculated as per the formula given in the Appendix to Tender (*Sub-Clause 13.8*). It shall be paid in the currencies applicable (at selling rate). Escalation is not applicable if the contract is extended due to a failure on the contractor's part. The formula given in Sub-Clause 13.8 of FIDIC is as follows:

$$\mathrm{Pn} = a + b\frac{\mathrm{Ln}}{\mathrm{Lo}} + c\frac{\mathrm{En}}{\mathrm{Eo}} + d\frac{\mathrm{Mn}}{\mathrm{Mo}} + \cdots$$

where:

'Pn' is the adjustment multiplier to be applied to the estimated contract value in the relevant currency of the work carried out in period 'n', this period being a month unless otherwise stated in the contract data;

'a' is a fixed coefficient, stated in the relevant table of adjustment data, representing the non-adjustable portion in contractual payments;

'b', 'c', 'd' are coefficients representing the estimated proportion of each cost element related to the execution of the works, as stated in the relevant table of adjustment data; such tabulated cost elements may be indicative of resources such as labour, equipment and materials;

'Ln', 'En', 'Mn' are the current cost indices or reference prices for period 'n', expressed in the relevant currency of payment, each of which is applicable to the relevant tabulated cost element on the date 49 days prior to the last day of the period (to which the particular payment certificate relates);

'Lo', 'Eo', 'Mo' are the base cost indices or reference prices, expressed in the relevant currency of payment, each of which is applicable to the relevant tabulated cost element on the base date.[3]

7.15 Clause 14: Contract price and payment

7.15.1 *The contract price*

Unless otherwise specified in the particular conditions, the contract price is the price agreed by the parties or determined by the engineer under Clause 12.3 (evaluation) and is subject to adjustment according to contract Clause 13.7 (adjustments) (*Sub-Clause 14.1*). It includes all taxes, duties and fees payable as per the contract (*Sub-Clause 14.1*). The quantities in the BOQ are estimated and are subject to actual cost. Within 28 days of the commencement date, the contractor shall give a breakdown of the lump sum prices included in the BOQ and may make use of it for interim payments.

7.15.2 *Advance payments*

Advance payments denotes an interest-free loan for mobilisation against a bank guarantee and after payment of the performance security as laid down in the contract (*Sub-Clause 14.2*). The total amount of advance payments, stages if any and the currencies and proportions shall be as mentioned in the Appendix to Tender. The recovery of advance payments shall be at a percentage of the interim bills as stated in the Appendix to Tender and the bank guarantee can be released in stages to the extent of such recoveries. All recoveries shall be completed within the validity period of the performance security.

7.15.3 *Application for interim payment certificates*

The contractor shall submit a monthly statement in six copies in the form specified by the engineer, together with supporting documents and a report on progress under Clause 4.21 (progress report) (*Sub-Clause 14.3*). The interim certificate shall state: (i) the gross amount payable in the currencies of the contract for works executed and so payable up to the end of the month including all accepted variations; (ii) amounts due under Clauses 13.7 and 13.8 (adjustments for changes in law and changes in cost); (iii) deductions for retention money; (iv) advance payment recovery (Clause 14.2); (v) recoveries for plant and materials (Clause 14.5); (vi) any other deductions including under Clause 20 (claims); and (vii) deduction of amounts certified up to the end of the previous bill.

7.15.4 *Schedule of payments*

If the contract includes a schedule of payments specifying instalments, these instalments shall be paid consistent with actual progress (*Sub-Clause 14.4*). If the contract does not includes a schedule of payments, the contractor must submit a non-binding estimate of payment during each quarterly period, starting from 42 days after the commencement date.

7.15.5 *Plant and materials intended for works*

Part payment is applicable if the plant and equipment and materials are to be incorporated in the permanent works on their arrival on site (*Sub-Clause 14.5*). This clause shall not apply if such payment is not included in the Appendix to Tender. Part payment is also admissible on shipment subject to the submission of all documentation and a bank guarantee valid till the arrival of the goods on site and such payments shall be at 80 per cent of the actual costs. The payment shall be in the same currencies as mentioned in the contract.

7.15.6 *Issue of interim payment certificates*

No amount shall be paid under interim certificates until the employer approves the performance security (*Sub-Clause 14.6*). The engineer shall issue an interim payments certificate to the employer within 28 days of receipt of such statements and documents from the contractor, indicating the amounts so payable. The net amount payable shall not be less than that stated in the Appendix to Tender. Interim payments can be withheld if the payments are not as per the contract or if the contractor fails to perform his obligations under the contract. The engineer can also modify and correct the certificate for the amount payable which shall be notified to the contractor. The payment certificate does not deem to indicate approval of the contractor or his consent or satisfaction.

7.15.7 *Payment*

The first instalment of the advance payment shall be made within 42 days of the issuing of the letter of acceptance, or within 21 days of receipt of the performance security, whichever is later (*Sub-Clause 14.7*). Interim payments shall be made within 56 days of receipt of it by the engineer. Final bill payment shall be done within 56 days of the date of receipt of the performance security by the engineer. All payments in each currency shall be made in the account nominated by the contractor in the currency of the country specified in the contract.

7.15.8 *Delayed payment*

The contractor is entitled to financing charges compounded monthly for all delayed payments due vide Clause 14.7 above (*Sub-Clause 14.8*). Financing charges for delayed payments will be calculated at 3 per cent above the discount rate of the central bank in the country of the currency of payment.

7.15.9 *Payment of retention money*

Fifty per cent of the retention money shall be released on issue of the taking-over certificate for the whole or for parts (in proportion) and the balance after expiry of the defects notification period (*Sub-Clause 14.9*). If the taking-over certificate is issued for part of the section then the refund of the retention money shall be 40 per cent of the pro rata amount (and not 50 per cent). On issue of the defect liability certificate for part of the section, another 40 per cent on a pro rata basis can be released.

7.15.10 *Statement of completion*

Within 84 days of receiving the taking-over certificate, the contractor shall submit in six copies a statement of completion with supporting documents (as per Sub-Clause 14.3: application for interim payment certificates) and the value of all works up to completion as per the contract (*Sub-Clause 14.10*). For further sums considered by the contractor as due, and any other amounts as will become due, the engineer will certify the issuing of the interim payment certificate (Sub-Clause 14.6: issue of interim payment certificates).

7.15.11 *Application for final payments*

Within 56 days of receiving the performance certificate, the contractor shall issue six copies of the draft final statement with supporting documents indicating the final payments
due and any further payments due. Following discussions with the engineer, the contractor
will make a final statement for the amounts agreed by the engineer and he will issue to the
employer an interim payment certificate and the disputed items will be settled under a dispute adjudication board under Clause 20 or 20.5 (amicable settlement) and will then submit
a final statement to the employer (*Sub-Clause 14.11*).

7.15.12 *Discharge*

On the above final statement, the contractor shall certify final settlement of all dues under
the contract and this discharge will be effective on receipt of the performance security by
the contractor and the outstanding balance, in which event the discharge will be effective
from that date (*Sub-Clause 14.12*).

7.15.13 *Issue of final payment certificate*

Within 28 days of receiving the final statement and written discharge under Clauses 14.11
and 14.12, the final payment certificate shall be issued by the engineer after giving all credit
due to the employer, indicating the amount due (*Sub-Clause 14.13*). If the contractor does
not apply for the final payment certificate vide Clauses 14.11 and 14.12, the engineer shall
request the contractor to do so within 28 days, failing which the engineer will determine the
amounts due and issue the final payment certificate.

7.15.14 *Cessation of employer's liability*

The employer shall not be liable to the contractor for any amounts after the issuing of the
final statement certificate, except for matters arising after the issuing of the taking-over
certificate and also in the statement of completion described in Clause 14.10 (*Sub-Clause
14.14*). However, this clause does not limit the employer's liability under indemnification
liability, in case of fraud or misconduct by the employer.

7.15.15 *Currencies of payment*

The contract price shall be paid in the currencies mentioned in the Appendix to Tender
(*Sub-Clause 14.15*). Unless otherwise stated in particular conditions, a payment in multiple
currencies shall be made as:

(a) If the accepted contract amount is expressed in the local currency only: (i) payments in
     foreign currencies shall be made at a fixed rate of exchange as stated in the Appendix to
     Tender unless agreed otherwise; (ii) payments and deductions under Clauses 13.5 and
     13.7 (provisional sums and adjustments for changes in law) shall be made in the appro-
     priate currencies; and (iii) all interim payments under Sub-Clauses 14.3(a–d) shall be
     made as per this clause (14.7)(a)(i).
(b) Payments of damages stated in the Appendix to Tender shall be made in currencies as
     specified therein.

(c) Other payments to the employer by the contractor shall be made in currencies expended by the employer or as agreed by both parties.

(d) If the amount payable to the employer in a currency exceeds the amount payable to the contractor in that currency then it can be recovered in the other currency.

(e) If no rate of exchange is stated in the Appendix to Tender, then it shall be that prevailing on the base date and determined by the central bank of the country.

7.16 Clause 15: Termination by employer

7.16.1 Notice to correct

The engineer may issue notice to the contractor to make good his failures to perform his obligations under the contract (*Sub-Clause 15.1*).

7.16.2 Termination by employer

The employer is entitled to terminate the contract if the contractor: (a) fails to comply under Clause 4.2 (performance security) or a notice under Clause 15.1 (notice to correct); (b) abandons the works or demonstrates intentions not to continue the performance of his obligations; (c) without reasonable excuse, fails to proceed with works in accordance with Clause 8 (commencement, delays and suspension) or to comply with notice within 28 days issued under Clause 7.5 or 7.6 (rejections, remedial works); (d) subcontracts the whole works or assigns the contract without a required agreement; (e) becomes bankrupt or insolvent or goes into liquidation etc., as specified in this clause; (f) gives or offers to give (directly or indirectly) a bribe, gift, commission or other things of value as an inducement or reward. Termination under clause (a) to (d) shall be after giving 14 days' notice while no notice is required under clause (e) or (f); and under (g) the contractor shall leave the site, hand over all documents and comply with other instructions as notified. After termination, the employer will proceed to do the work through another agency (*Sub-Clause 15.2*).

7.16.3 Valuation at date of termination

On termination of a contract under Clause 15.2, measurements shall be taken by the engineer to determine under Clause 3.5 dues payable to the contractor under the contract (*Sub-Clause 15.3*).

7.16.4 Payment after termination

After termination of the contract, the employer will proceed to access the claims of the employer under Clause 2.5 and no payments shall be made to the contractor until the works rate is completed and costs to be recovered are accessed (*Sub-Clause 15.4*). After a notice of termination under Sub-Clause 15.2 has taken effect, the Employer may recover from the contractor, from his dues, as accessed, all extra costs damages and losses to the employer and return balance, if any.

7.16.5 Employer's entitlement to termination for convenience

The employer can terminate the contract at his convenience on receipt of 28 days' notice from the contractor (*Sub-Clause 15.5*). This clause cannot be used by the employer to do the work himself or through any other agency. After termination, the contractor will proceed under Clause 16.3 (cessation of works and removal of contractor's equipment) and Clause 19.6 (optional termination, payment and release).

7.16.6 *Corrupt or fraudulent practices*

If the employer determines that the contractor has engaged in a fraudulent act, under Sub-Clause 15.6, he may terminate the contract and expel the contractor from the site by giving 14 days' notice.

7.17 Clause 16: Suspension and termination by contractor

7.17.1 *Contractor's entitlement to suspend work*

The contractor can suspend the work or reduce the progress of the work after giving 21 days' notice to the employer if the engineer fails to issue interim payment certificates (Sub-Clause 14.6: issue of interim payment certificates), comply with Clause 2.4 (employer's financial arrangements) or Clause 14.7 (payments) (*Sub-Clause 16.1*). This action shall not prejudice his entitlement under Sub-Clause 14.8 (delayed payments) and later termination under this clause. The contractor shall be entitled to an extension of time and compensation subject to Sub-Clauses 20.1 (contractor's claims) and 8.4 (extension of time for completion). After receiving this notice from the contractor, the engineer can proceed in accordance with Sub-Clause 3.5 (determinations) to agree or determine these matters.

7.17.2 *Termination by contractor*

The contractor can terminate the contract if the contractor is not provided with the following (*Sub-Clause 16.2*):

(a) failure of the employer to comply within 42 days to Clause 2.4 (employer's financial arrangements);
(b) non-issue of the payment certificate within 56 days of receipt of documents from the contractor;
(c) non-receipt of the interim payment certificate within 42 days after expiry of time under Clause 14.7 except employer's claims;
(d) employer's substantial failure to perform his obligations under the contract;
(e) employer fails to comply with Clause 1.6 (contract agreement) and Clause 1.7 (assignment);
(f) prolonged suspension by employer under Clause 8.11; and
(g) employer becomes bankrupt.

In all such cases, the contractor can terminate the contract after giving 14 days' notice except under item (f) or (g) in which case it can be terminated immediately.

7.17.3 *Cessation of work and removal of contractor's equipment*

If a notice of termination by the employer under Clauses 15.5 and 16.2 has taken effect, then the contractor shall cease all further work except that required for protection and safety, and hand over all documents, plant and materials for which the contractor has received payments and remove all goods from the site except those required for safety (*Sub-Clause 16.3*).

7.17.4 *Payment on termination*

After termination under Clause 16.2 (termination by contractor), the employer shall return the performance security, pay the contractor as per Clause 19.6 and pay the contractor any loss of profit or other loss or damages suffered by the contractor as a result of such termination (*Sub-Clause 16.4*).

7.18 Clause 17: Risk and responsibility

7.18.1 *Indemnities*

The contractor indemnifies and holds harmless the employer as follows: the contractor shall indemnify the employer for all losses including legal fees; for injuries to any person due to his negligence; for damage to any property: (i) arising during the execution of works and (ii) due to the negligence of the contractor (*Sub-Clause 17.1*).

It is mandatory to the Contractor to indemnify the Employer, the Employer's Personnel and their respective agents against personal injury to the extent that the former arises otherwise than as a result of negligence, wilful acts or breach of the Contract by the Employer and loss or damage to any property (other than the Works) to the extent that the latter arises as a result of the Contractor's design or the Contractor's negligence, wilful act or breach of the Contract.

7.18.2 *Contractor's care of the works*

The contractor shall take full responsibility for the care of the works from the commencement date to the taking over of the works except those covered under Clause 17.3 (employer's risks), after which it is the responsibility of the employer except outstanding works so specified (*Sub-Clause 17.2*).

7.18.3 *Employer's risks*

The employer shall cover all consequences due to war, rebellion, riots, war explosives, pressure waves caused by aircraft, occupancy by employer on a part of the permanent works, employer's designs of works and unforeseeable forces of nature (*Sub-Clause 17.3*). The employer is required to meet the risks arising from:

(a) war, hostilities and invasion;
(b) rebellion, terrorism and civil war;
(c) riot, commotion or disorder (provided it is not caused by the contractor's personnel);
(d) explosives, ionising radiation or contamination by radioactivity except for those caused due to use of such materials by the contractor for construction purposes;
(e) pressure waves from aircraft caused by sonic or supersonic speeds;
(f) design of any part of works by the employer;
(g) forces of nature that an experienced contractor could not have foreseen (*Sub-Clause 17.3*).

7.18.4 *Consequences of employer's risks*

For causes falling under Clause 17.3, the contractor shall give notice to the employer and the contractor shall be entitled to an extension of time and compensation thereof (*Sub-Clause 17.4*).

7.18.5 *Intellectual and industrial property rights*

All claims under this clause shall be covered by claims within 28 days (*Sub-Clause 17.5*). The employer shall indemnify the contractor for infringements under certain conditions specified in this clause and the contractor will do so under conditions so specified in this clause.

7.18.6 *Limitation of liability*

Each party is responsible for their losses under the contract except those under Clause 16.4 (payment on termination) and Clause 17.1 (indemnities) (*Sub-Clause 17.6*). Contractors' liability under the contract shall be limited to that mentioned in particular conditions and if not so stated shall be limited to the accepted contract values excluding those under Sub-Clauses 4.19 (electricity, water and gas), 4.20 (employer's equipment and free-issue materials), 17.1 (indemnities) and 17.5 (intellectual and industrial property rights).

7.19 Clause 18: Insurance

7.19.1 *General requirements for insurances*

Where the contractor is the insuring party, each insurance shall be affected with insurers and in terms agreed by the employer (*Sub-Clause 18.1*). The contractor shall effect insurance as specified in the contract and approved by the employer within a period stated in particular conditions. The employer insuring against loss or damage shall effect insurance as specified in particular conditions. Each policy insuring against loss or damage shall provide for payments in currencies required to rectify defects. 'The contractor shall be entitled to place all insurances relating to the Contract (including, but not limited to the insurance referred to in Clause 18) with insurers from any eligible source country'.

7.19.2 *Insurance for works and contractor's equipment*

The contractor shall insure 'all works, plant, materials and contractor's documents for not less than the full reinstatement cost including the costs of demolition, removal of debris and professional fees and profit' and effect a policy for the value of the contract from the date specified and evidence submitted to the employer as per Clause 18.1 until the date of the taking-over certificate (*Sub-Clause 18.2*). The contractor shall further take out insurance for loss or damage during the defects maintenance period and this shall remain valid until the issuing of the performance certificate. The contractor shall insure all equipment during transit until it arrives on site and kept in the employer's control and for an amount equal to its replacement value. Insurance under this clause shall be effected and maintained by the contractor in the joint names of those who shall be jointly entitled to receive payments from the insurance company. This clause (18.2) makes it clear that unless otherwise stated in the particular conditions, insurances shall cover all loss and damage from any cause not listed in Sub-Clause 17.3 (employer's risks).

7.19.3 *Insurance against injury to persons and damage to property*

The contractor shall insure for his liability against any loss or damage to property, injury or death of a person arising out of a contractor's performance and occurring until the issuing of the performance certificate and for an amount stated in the Appendix to Tender (*Sub-Clause 18.3*), except that insured under Clauses 18.2 (insurance for works and contractor's equipment) and 18.4 (insurance for contractor's personnel). There is no limit to the number of occurrences. Clause 18.3 shall not apply if there is no mention of a limit in the Appendix to Tender.

Unless stated otherwise, in particular conditions, the insurance under this clause: (a) shall be effected and maintained by the contractor; (b) shall be in joint names with the employer; (c) shall cover loss or damage to employer's property (except that insured under Sub-Clause 18.2) arising out of the contractor's performance of the contract; (d) may exclude liability to the extent arising from: (i) the employer's right to have the permanent works executed, (ii) damage which is an unavoidable result of the contractor's obligation to execute the works and remedy defects and (iii) a clause listed in 17.3 (employer's risks).

7.19.4 *Insurance for contractor's personnel*

Similar to works and contractor's equipment, contractor's personnel shall be insured by the contractor for injury, sickness, disease or death (*Sub-Clause 18.4*). The employee and the engineer shall also be indemnified under this policy. For a subcontractor's employee, the insurance may be effected by the subcontractor but the prime contractor is responsible for compliance with this clause.

7.20 Clause 19: Force majeure

7.20.1 *Definition of force majeure*

Sub-Clause 19.1 defines force majeure as an exceptional event or circumstance which is beyond the control of the contractor and which could not have been reasonably provided against prior to entering into a contract or after having arisen, could not have been reasonably avoided and cannot be substantially attributed to the other party. Force majeure may include war, whether declared or not, or invasion, or an action of a foreign enemy, terrorism, revolution or civil war, riots, strikes or lockout by persons other than the contractor's or subcontractors personnel, and munitions of war or explosives radiation other than those caused by the contractor's use of them. It also covers natural catastrophes, such as earthquakes, hurricanes, typhoons and volcanic activity.

7.20.2 *Notice of force majeure*

The contractor shall give 14 days' notice from the date of such occurrence, with details of his inability to perform his obligations under the contract (*Sub-Clause 19.2*). The force majeure shall not apply to the obligations of each party to make payments under the contract.

7.20.3 *Duty to minimise delay*

Each party shall endeavour to minimise delay in the performance of the contract affected by force majeure (*Sub-Clause 19.3*). The employer shall give notice when it ceases to be affected by force majeure.

7.20.4 *Consequences of force majeure*

The contractor shall give notice of incurring delays and costs due to force majeure and if accepted by the engineer as falling under Sub-Clause 19.1 (definition of force majeure) (*Sub-Clause 19.4*), shall give an extension of time (under 8.4 – extension of time for completion) and assess payment to compensate such costs and proceed to determine it under Sub-Clause 3.5 (determinations).

7.20.5 *Force majeure affecting subcontractor*

Any additional entitlement or broader force majeure events entitling the subcontractor to compensation shall not excuse the prime contractor from any non-performance or relief under this clause (*Sub-Clause 19.5*).

7.20.6 *Optional termination, payment and release*

If non-performance due to force majeure continues beyond 84 days (or multiple periods totalling to 140 days), either party can give 7 days' notice to terminate the contract in which case the contractor shall proceed in accordance with Sub-Clause 16.3 (cessation of works) (*Sub-Clause 19.6*). In such a situation, a further process of determining the value of the work done shall commence and will include: (a) value of completed item of work for which the price is stated in the contract; (b) cost of plant and material delivered or have left vendors for delivery; (c) any other cost/liability reasonably incurred by the contractor for completing the works; (d) cost of removal of temporary works and contractor's equipment; and (e) cost of repatriation of the contractor's staff and labour employed wholly in connection with the work.

In such an event, the engineer shall determine the payments due to the contractor and issue a payment certificate as specified in this vide 19.6(a)–(e) given above.

7.20.7 *Release from performance*

If, in the event of circumstances beyond the control of the parties, it is difficult to perform obligations under the contract by any party, then the parties can be released from such performance, upon notice by the other party and the sum payable to the contractor shall be the same as due under Sub-Clause 19.6 given above (*Sub-Clause 19.7*).

7.21 Clause 20: Claims, disputes and arbitration

7.21.1 Contractor's claims: For an extension of time or for any claim arising out of the contract, the contractor shall give notice within 28 days' from the date of awareness of such a claim, failing which the contractor shall not be entitled to such an extension or payment of compensation (*Sub-Clause 20.1*).

The contractor, however, can submit any other notice applicable under the contract with all particulars of records which the engineer shall be entitled to inspect at any time and claim within 42 days of such an event occurring and the engineer shall treat the claim as interim and the contractor shall send further interim claims giving the accumulated delay and shall submit a final claim within 28 days after the end of the effects resulting from the event.

Within 42 days of receiving such claims and supporting details, the engineer shall determine the claim. Each payment certificate shall include such amounts of the claims as reasonably substantiated and the engineer shall proceed to determine claims which have been sufficiently substantiated for an extension of time and payment of extra costs.

7.21.2 Appointment of the dispute adjudication board

Disputes shall be adjudicated by a dispute adjudication board under Clause 20.4 and parties shall jointly appoint a DAB by a date specified in the Appendix to Tender (*Sub-Clause 20.2*). The DAB shall have one or three members, as stated in the Appendix to Tender. If not stated or members disagree, the DAB shall have three members.

In the case of a three-member DAB, one member each shall be appointed by the employer and the contractor. Both parties in consultation with these two members shall appoint the third member who will act as the chairman of the DAB. However, if a list of potential members is included in the contract, these names will be selected from this list.

The terms of remuneration to the members shall be mutually agreed by the parties at the time of agreeing to the terms of appointment and each party shall pay one half of the remuneration. Parties may agree to appoint in advance a replacement of such members in an emergency. The appointment of any existing member of the DAB may be terminated by mutual agreement of both parties. If not agreed by both parties, the appointment of the DAB may be terminated as per Sub-Clause 14.12 (discharge).

7.21.3 Failure to agree on the composition of the dispute board

In case of failure to appoint a DAB due to any disagreements, the appointing entity or official named in the particular conditions shall appoint member/members of the DAB (*Sub-Clause 20.3*). Such appointment shall be final and each party shall pay half the remuneration so fixed by the appointing entity or official. Each party shall be responsible for paying one-half of the remuneration of the appointing entity or official.

7.21.4 Obtaining dispute board's decision

The parties may refer any dispute under the contract in writing to the chairman of the DAB giving reference to this clause, i.e. Sub-Clause 20.4, with copies to the other party and the engineer. The DAB will proceed to refer to the parties to make available all information to settle the dispute. The DAB shall be deemed to be not acting as arbitrator. Within 84 days of receipt of reference, the DAB shall give their reasoned decision under this clause. If agreed by both parties, the DAB's decision shall be implemented. Also, if the DAB fails to give a decision within the above-stipulated period, then within 28 days of the expiry of the period, notice of dissatisfaction can be issued.

The DAB's decision is final and binding if no notice of dissatisfaction is issued by either party. If a notice of dissatisfaction is issued in time, the parties shall attempt to settle the dispute amicably before the commencement of arbitration.

7.21.5 *Amicable settlement*

Prior to commencement of arbitration, both parties shall make efforts to settle the dispute amicably (*Sub-Clause 20.5*). Arbitration can only commence on or after the 56th day of the notice of dissatisfaction is given.

7.21.6 *Arbitration*

Unless settled amicably, disputes shall be settled by international arbitration as per the rules of arbitration of the International Chamber of Commerce (ICC) (*Sub-Clause 20.6*). Arbitration shall be conducted by three arbitrators appointed in accordance with the ICC rules.

The place of arbitration shall be the neutral location specified in the Contract Data; and the arbitration shall be conducted in the language for communications defined in sub-clause 1.4 (law and language).

The arbitrators shall have full powers to open up/review/revise any decision given by the engineer or the DAB relevant to the dispute. The engineer can be called upon as a witness to give evidence. Neither party shall be limited in the proceedings before the arbitrators to the evidence or arguments previously placed before the DAB to obtain its decision, or to the reasons for dissatisfaction given in its notice of dissatisfaction. Any decision of the DAB shall be admissible in evidence in arbitration. Arbitration can commence prior or after completion of works. The obligation of all parties shall not be altered by reason of any arbitration being conducted during the progress of works.

As per Sub-Clause 20.6(b), if the contract is with domestic contractors, arbitration with proceedings can be conducted in accordance with the laws of the employer's country.

7.21.7 *Failure to comply with dispute board's decision*

In case any party fails to implement the decision of the DAB when the same has become final and binding, then such failure to implement shall be referred to arbitration under Sub-Clause 20.6 (arbitration) (*Sub-Clause 20.7*). In such an event, Clause 20.4 (obtaining DAB's decision) and Sub-Clause 20.5 (amicable settlement) shall not apply to this reference.

7.21.8 *Expiry of dispute board's appointment*

If no DAB is in place, whether for reasons of expiry of the DAB's appointment or otherwise, Sub-Clauses 20.4 and 20.5 shall not apply and the dispute may be referred directly to arbitration under Sub-Clause 20.6 (arbitration) (*Sub-Clause 20.8*).

7.22 Changes in FIDIC 2017

There are some changes to the contract structure and some relocated clauses in FIDIC 2017. There are currently 21 clauses of the General Conditions of Contract in FIDIC's 2017 contract updates, instead of the 20 clauses in the FIDIC 1999 contracts.[4] The changes in short are: (i) limitation of liability is no longer in clause 17 but has been moved to become the last sub-clause of clause 1; (ii) former "force majeure" is rechristened as "exceptional event". It is given in clause no. 18; (iii) insurance has been moved to clause no. 19; (iv) clause no. 20 is divided into two parts to deal first with "claims" (new clause 20, also covering employer's claims) and then only with "disputes" (new clause 21).

7.23 Conclusion

A construction contract is an agreement between an employer and a contractor to construct a built facility within the prevailing legal system of the country. The employer grants access and possession of the site and makes arrangements for payments and the contractor promises to complete the works as per the contract. The engineer acts as the impartial representative of the employer and performs certain determination/certifier functions under the contract. The FIDIC Red Book is a contract form which provides conditions of contract for construction works where the design is carried out by the employer. Different forms of FIDIC contracts reflect different procurement approaches. The multilateral development bank harmonised edition of FIDIC conditions of contract was elucidated in this chapter.

Notes

1 *See* http://fidic.org/about-fidic.
2   *See* http://fidic.org/bookshop/about-bookshop/which-fidic-contract-should-i-use.
3 Adapted from (Sub-Clause 13.8) FIDIC Conditions of Contract 2006.
4 *See* https://fidic.org/sites/default/files/press%20release_rainbow%20suite_2018_03.pdf Accessed on 12 March 2010.

Reference

FIDIC. (2006, March) *FIDIC Conditions of Contract for Construction: For Building and Engineering Works Designed by the Employer*, Multilateral Development Bank Harmonised Edition, Fédération Internationale des Ingénieurs-Conseils. Retrieved from: https://fidic.org/sites/default/files/cons_md b_gc_v2_unprotected.pdf Accessed on 2 June 2017.

8   Construction disputes and claims

8.1 Introduction

Construction claims are common in construction projects around the world. Construction claims mostly arise when construction contracts are ambiguous. Such claims are an important source of dispute between contracting parties in construction projects. Generally, construction projects involve a number of business parties, extend over a long period of time and are uncertain, complex and risky. Moreover, construction projects require extremely complex designs, detailed plans and specifications, several inter-related activities/work packages, high-risk construction methods, effective management, skilful supervision and close coordination. In a construction project, a multitude of stakeholders and activities to be completed within a specific time period with the specified quality and within budget, make it dispute-prone. Disputes left unresolved in construction projects will become formal claims. Several variations will bring about a claim for additional payment, if accepted. If the parties are unable to agree to a claim, a dispute arises. This chapter discusses the types of construction claims, claims between parties, management of claims, claim documentation, claim presentation and negotiation and settlement of claims.

Most construction claims are not anticipated until they are received or presented from one party to the other. It is essential to present or defend a claim completely and precisely. The best claims defence will always be the avoidance of claim-provoking conditions through proper contract formation and administration. Therefore, claims are widespread in construction projects and lead to time and cost overruns. In addition, construction contracts are an extremely lengthy, complex set of documents, which are often not well understood by the parties. This often leads to differing interpretations by different parties and thus disagreements or disputes arise regarding contractual obligations or expectations. A comprehensive understanding of the contractual terms and clauses is required to minimise or avoid construction claims and disputes. Claims arise when one party believes that the other party has not met its contractual obligations and that they deserve monetary and/or time compensation. As the project progresses, clients may request change orders, which could cause contractors' claims related to variations (Abdul-Malak, El-Saadiand Abou-Zeid, 2002). Claims can be made for problems such as delays, changes, unforeseen circumstances, insufficient information and conflicts (Figure 8.1).

Extensive transformations have taken place in the magnitude and the methods of construction. The construction of infrastructural development projects is known for its long gestation period, and uncertainties and risks are expected to arise over such long periods. The enormous cost of construction and the application of more and more mechanisation can pose unforeseen problems from when the existing contract forms were evolved. Sometimes,

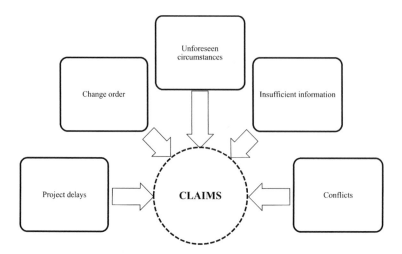

Figure 8.1 Reasons for claims.

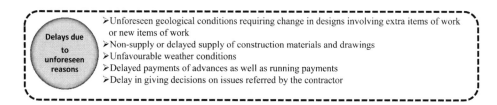

Figure 8.2 Delays due to unforeseen reasons.

contractors are not properly prepared to execute the works and are penalised in terms of the contract. At the same time, delays occur due to unforeseen geological conditions requiring a change in design involving extra items of work or new items of work, non-supply or delayed supply of construction materials and drawings, unfavourable weather conditions, delayed payments of advances as well as running payments and delays in decision-making on issues referred by the contractor (Figure 8.2). In the meantime, the unit costs escalate and the contractor is unable to execute the works within the contract price. Some of these issues should be properly covered in the contract documents so that the work proceeds smoothly without any hold-ups and the chances of disputes are minimised, if not avoided. Some of these issues are discussed further.

8.2 Claims definition

Prior to further discussions on claims, the term claim must be defined as it is used in the construction industry. In construction contracts, two main parties, even though there are many parties, present claims against each other: (i) contractor – the contractor may claim additional time for performance and/or additional compensation from the client; and (ii) client – the client may claim relief from the contractor in terms of a reduction in the contract price and/or an acceleration or delay in the contractor's performance(Gilbreath, 1992). Claims can be defined as a mechanism that allows the parties in the contract to be compensated for loss(es) or expense(s) in terms of time and cost incurred that are caused by the action, inaction, negligence or incompetency of other party(ies).

The International Federation of Consulting Engineers (FIDIC) conditions of contract define a claim to be a request or assertion by one contracting party to the other party for an entitlement or relief under any clause of the conditions of contract or otherwise in connection with, or arising out of, the contract or the execution of the works. In many instances, a contractor's claim is a legitimate request for additional compensation, in terms of cost and/or time, due to a change in the terms of the contract. Identifying and managing construction project risk factors can reduce the frequency and severity of claims.[1] Thus, a claim is a legitimate request for additional compensation (cost and/or time) due to a change in the terms of the contract. The majority of claims are settled through negotiation, adherence to the terms of the contract or some mutually agreeable adjustment in time and cost of performance between the owner and the contractor(s) (Gilbreath, 1992).

The government is the largest contracting body, and as the custodian of public funds and the welfare of the people, it is expected to incorporate fair terms and conditions in the contract and provide a clean field of activity for the contracting parties – a model employer of contractors to fit in with the needs of society. The purpose of a contract is to accomplish the desired objective as per the stipulations laid down in the agreement to be signed by the client and the contractor.

8.2.1 *Unfair contracts*

Unfortunately, contracts entered into with the government are not known for their spirit of give and take. Most of the contracts are one-sided, and a contractor can, at most, try to make the best of the bargain. Due to the vast experience gained by the government over the years, contract documents contain carefully and skilfully drafted terms and conditions, safeguarding the interest of the government against all possible eventualities and breaches which the government or its representative – the operator of the contract – may commit. These clauses are now tailored to safeguard the government/client's interest due to the various modifications and changes in light of decisions of arbitrators or courts which might have, in the past, impaired the interests of the government.

8.3 Types of construction claims

There are different types of claims. Adrian (1993) categorised claims based on: (i) the cost of the claim (i.e. small, medium and large claims); (ii) who initiates the claim (i.e. owner-related delay claims and uncontrollable event and delay claims); (iii) public and private claims (i.e. in publicly and/or privately funded projects); (iv) the type of project (i.e. claims on hospital projects, claims on school buildings); (v) geographical area (i.e. flood or landslide claims and the like, resulting from the geographical location of projects which are susceptible to adverse weather conditions and/or natural disasters or acts of God). Furthermore, claims can be contractual claims, extra contractual claims, quantum meruit claims, ex-gratia claims and counter claims. Contractual claims are claims arising out of the express provision of a particular contract. 'Common law claims' or extra contractual claims are for damages for breach of contract under common law. Quantum meruit (means 'what one has earned') claims provide remedy for a person who has performed work under the instruction of the client but no price has been agreed or where a new contract has replaced the original contract and payment is claimed for work done under the replaced contract. An ex-gratia claim or 'sympathy claim' or 'out of good will claim' is one where no legal remedy is available to the contractor but arises out of hardship.

Marsh (2003) categorises claims under three headings: (i) ex gratia; (ii) from excessive ordering of variations; and (iii) default by the client in his obligations under the contract. Ex-gratia claims are claims with no legal or contractual base but the contractor considers the client has a moral obligation to meet the claim and pay compensation. Even though the client has no obligation to meet such a claim, he may be prepared to pay on the grounds of natural justice. Suppose there were serious and unforeseeable difficulties in the performance of a contract, such as exorbitant oil price hikes which caused the contractor to become bankrupt. Ex-gratia claims are allowed in such situations to complete the project in a timely manner. Change orders result in additional time or costs and result in a claim. Under all contracts there are some obligations the client must fulfil, for instance, timely land acquisition; obtaining environmental and all other necessary approvals before the start of work; providing detailed drawings and specifications for advanced planning and execution; making the site available and supplying information and facilities which if not provided on time and to specification will result in the contractor incurring additional costs. Claims on recovery of those costs relate to different variations. Common categories of claims identified by Kumaraswamy (1997a) are variations due to site conditions, client changes and design errors, unforeseen ground conditions and ambiguities in contract documents.

Variation or a change in the quality and/or quantity of an item of work (time, material, plants or working method) which is contained and specifically indicated in the contract document can considerably impact on the cost and time of the project and lead to claims. Another type or cause of claim is additional works because it will be accompanied by extra costs which could result in the contractor incurring extra expenses, thereby amounting to a claim. Other common causes of claims are related to: (i) late handing over of site; (ii) extension of time; (iii) excusable events (caused by client/client representative such as postponement of work); (iv) non-excusable events (caused by contractor such as project delays caused by contractor's incompetence); and (v) external events (caused by neither the contractor nor the client, but events such as natural calamities).

A study by Enshassi, Mohamed Choudhry and Mohamed El-Ghandour (2008) provides the following claims factors caused by owners:

1. residents' interferences during project implementation caused by a delay in the contractor's activities;
2. unexpected increase in material prices;
3. site possession with obstacles (license, land occupation, etc.);
4. material rejection because of unacceptable quality or specifications;
5. changes in material type and specification during construction;
6. continuous verbal instructions to contractor;
7. cardinal changes in the quantity plus or minus;
8. owner's financial difficulties because of delayed release of funds from donors;
9. delay in progress payments of the contractor;
10. changes to currency value (index value);
11. lack of support from the owner to his supervision team;
12. owner's slow decision-making;

13. owner's direct interference in a project without any coordination and ignoring his supervision team;
14. supervision team lacking authority and showing weakness in decision-making;
15. supervision team requiring the contractor to supply materials of a higher standard than that specified in the contract;
16. owner's poor control and monitoring of his supervision team;
17. adversarial relations between the contractor, the owner and the supervision team;
18. poor judgement of the supervision team in estimating time and resources;
19. low-quality assurance and control in the project;
20. issue of change in site location or conditions;
21. owner uncooperative with the contractor regarding work activities and following up with the supervision team;
22. project termination or suspension of some main activities during project implementation; and
23. lack of experience of the supervision team in project supervision.

Insufficient predesign investigations (feasibility studies, project economic and financial analysis, social impact studies, sustainability studies and environmental impact assessment [EIA], mapping and geotechnical investigations, etc.), faulty design concept, defective bid document, unrealistic specifications and errors in drawings often become the reasons for claims arising from contracts. Ineffective subcontractors, incompetent supervision, lack of safety in construction and inferior materials control are some of the major issues to be addressed at the time of project execution to minimise claims.

8.3.1 Problem areas of dispute

In his book, Levy (2009) lists 13 areas that most often lead to disputes and claims: (i) plans and specifications that contain errors, omissions and ambiguities; (ii) plans that have not been properly coordinated; (iii) incomplete or inaccurate responses or non-responses to questions or problems presented by one party in the contract to another party; (iv) inadequate administration of an owner's, architect's or contractor's responsibilities; (v) unwillingness to comply with the intent of the drawings by the owner, architect or contractor; (vi) site conditions that differ materially from those represented in the contract documents; (vii) unforeseen subsurface conditions; (viii) a change in conditions; (ix) discrepancies in the plans and/or specifications; (x) breaches of contract by any party; (xi) disruptions to the normal pace of construction by disputed change orders, acceleration of the work or lack of decisiveness by the architect, owner or contractor; (xii) delay to the work caused by any party to the contract; and (xiii) inadequate financial strength on the part of the owner, the contractor or any of the subcontractors or vendors.

Problem areas of dispute as given by Chan and Suen (2005) are: payments, variations, extension of time, quality of works, project scope definition, risk allocation, technical specifications, management, unrealistic client expectations, availability of information, unclear contractual terms, unfamiliarity with local conditions, difference in the way of doing things, poor communication, adversarial approach in handling conflicts, lack of team spirit, previous working relationships, lack of knowledge of local legal system, conflicts of law and jurisdictional problems.

Kumaraswamy (1997b) provides a list of common root causes and proximate causes for disputes and claims, as shown in Figure 8.3.

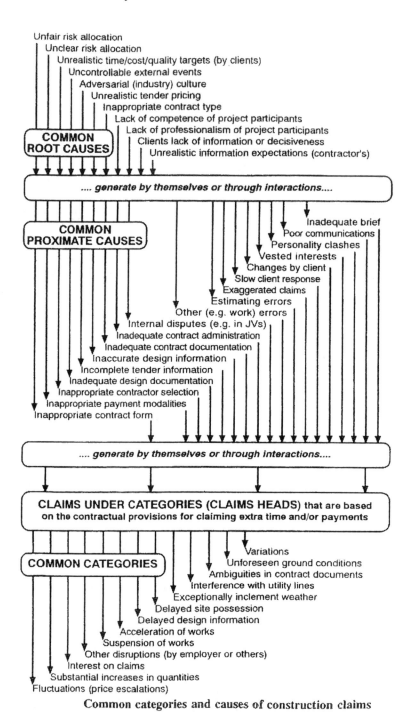

Common categories and causes of construction claims

Figure 8.3 Common root causes and the proximate causes of disputes and claims.

Source: Kumaraswamy, M.M. (1997) Common categories and causes of construction claims. *Construction Law Journal*, 13(1), pp. 21–34.

8.4 Claims between parties

8.4.1 Clients' claims against contractor

Clients have the right to claim to recover from issues that may economically harm the client. The client/owner raises claims against his contractors for defective work, late completion of the work, termination or breach of contract, variations in the specification of materials, property damage, defective work, etc. The client can raise claims for one or more of these reasons.

8.4.1.1 Defective work

Clients/owners dissatisfied with their contractor's output/product/project may claim damages that include the cost of repair, replacement or removal of the defective work. Often, the work does not meet the contractual specifications or is not fit for its intended purpose.

Contractors are responsible for the specified quality of work. Contractors seldom follow quality specifications. Designers also have a role in this due to lack of adequate specifications at the design stage. In addition, change orders from the client also frequently play a role in variations in quality.

8.4.1.2 Contractor's late completion of the work

The owner may claim damages when a delay is caused by the contractor (Gilbreath, 1992).[2] Typical damages claimed by the owner in this regard are loss of use of the facility and the increased cost of other delayed works.

8.4.1.3 Termination or breach of contract

Another type of claim one encounters, though seldom, is that arising from termination or breach of contract. This generally occurs when a contractor: (i) fails to complete the work; or (ii) for some reason leaves the job site. In this situation, owners usually demand to be compensated for the increased cost, over and above that paid to the contractor, for completing the work through other means. Contractors also claim damages when they feel they have been unjustly removed from a project or otherwise prevented from completing their contractual work. Both these situations arise when the contract, in effect, has been terminated by one party.

8.4.1.4 Specification of materials

Discrepancies may result due to differences in the interpretation of the contractual material specifications. Omissions of contract specifications are common, with the detrimental result that decisions have to be waited for or directions given by the owner's engineer at the job site. Consequently, the project gets delayed and/or the scope of the job increases. Once the issue is material specifications, multiple complications arise as every owner who desires to save money starts raising claims.

8.4.2 Contractors' claims against owner

Most of the claims in the construction industry are those raised by contractors against owners. In their study, Bakhary, Adnan and Ibrahim (2015)[3] found (i) design changes introduced at the post-tender stage; (ii) project implemented in unduly short time periods with inadequate site investigation, design work, tender and contract documentation; and (iii) inadequate definition and/or specification of the precise scope of the contract works as the three main reasons for construction claims given by contractors. The 16 reasons for claims in the construction industry that these authors introduced in their study are:

1.   Design changes introduced at the post-tender award stage.
2.   Project implemented in a short time period with inadequate site investigation, design works, tender and contract documents.
3.   Inadequate definition and/or specification of the precise scope of the contract works.
4.   Incomplete and/or uncoordinated design.
5.   Changes in employer's/user's requirements arising during the post-tender award stage.
6.   Lack of clarity in the employer's requirements and/or inadequacy of the design brief.
7.   Acceptance of unclear/imprecise tender offers without proper clarifications, negotiations or recording of changes.
8.   Contracting environment relatively more competitive with a larger number of players and narrow profit margins.
9.   Parties generally more aware of their rights and relatively more litigious/claim conscious.
10.   Changes arising out of statutory, local authority sources.
11.   Employer's/contract administrator's failure and/or neglect to meet the relevant contract obligations.
12.   Deterioration in the general standard of professionalism of designers, contract administrators and contractors.
13.   More use of claims' consultants.
14.   Sudden swings in economic and market conditions.
15.   A shift in the philosophy of project implementation in an adversarial/confrontational direction.
16.   Negative effects of political factors.

In the literature, frequently referred claims raised by contractors against owners are: (i) late or defective owner-furnished information, generally in the form of drawings or specifications and owner-furnished material or equipment; (ii) changes in the regulatory requirements; (iii) changed or unknown site conditions; (iv) restrictions in work method, including delay or acceleration of contractor's performance; (v) ambiguous contracts or contract interpretations; and (vi) changes in economic and market conditions.

Three simple rules can be promulgated to avoid making claims: (i) know exactly what the contract requires; (ii) do what the contract requires, but without interference; and (iii) do not carry out anything else, without proper documentation.

8.5 Management of claim

Management of a claim is the process of implementing and coordinating resources in order to improve the claim process, from identifying and analysing to preparing and presenting it, then attempting to resolve it (Kululanga, Kuotcha, McCaffer and Edum-Fotwe, 2001; Bakhary, Adnan and Ibrahim, 2014). Claim management consists of: (i) claim identification;

(ii) claim quantification; (iii) claim planning; (iv) claim execution; and (v) claim resolution (Jebel, 2008). Claims may be raised depending on: (i) the structure of the contract; (ii) the terms of the project; and (iii) what events happened during the project; therefore, predicting and dealing with these is not allowed in the contract (Rubaie, 2002). Depending on the decision of the opposite party against which the claim is made, the claim may be settled agreeably or it may take the form of a dispute.

Claims not settled amicably may take the form of a dispute. This generally happens because the claim was not substantiated by the facts/evidence. Subsequent to acceptance of the claim by the contractor as extra work, the contractor informs the engineer within the time frame stipulated and clarifies the contractor's intention to provide extra rates for the same. If the claim is not allowed, an adjudication process is carried out as per the provisions set out in the contract. The general claims management process followed are claim identification, claim notification, claim documentation, claim presentation, claim negotiation and claim resolution (Figure 8.4).

8.5.1 *Claim identification*

The identification of a claim situation is the primary stage of the entire claims management process. Generally, claim situations arise out of subtle variations in field conditions, from job site delays or as a result of differences in contract interpretation. In these and in all other instances, a claim situation must be recognised and identified as soon as it occurs.[4]

Previously, civil engineering works were less complicated and less expensive. Relationships between the owner, architect or engineer and the contractor, as well as between the contractor and his subcontractors, were much closer and less formal. Therefore, the scope for claims was less. However, the current market conditions have changed. The state of the economy and high interest rates have made owners much more anxious to hold down original capital outlay, and avoid additional costs.

Revisions are made to standard contract forms intended to transfer increased liability to the contractors and subcontractors, as well as require more onerous warranties. All these revisions enhance the possibility for disputes and claims. Consequently, it is important to recognise the situation and handle it in a realistic, positive and sensible manner.

8.5.2 *Causes of dispute*

Several studies have been conducted on different types and causes of construction disputes. Daoud and Azzam (1999) listed the major causes of disputes in construction contracts as: (i) alterations; (ii) lack of understanding; (iii) variations in legislation and guidelines; (iv) poor communication; and (v) the impact of local culture. Kumaraswamy (1997a) studied common causes and disputes and categorised them into two types: proximate causes and root causes

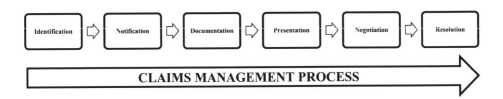

Figure 8.4 General claims management process.

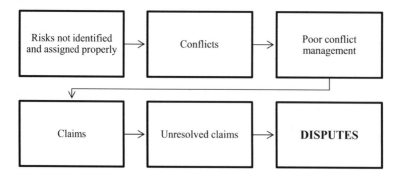

Figure 8.5 Construction disputes path.

(ranging from unfair risk allocation to unrealistic expectations of contractor). According to his study, common causes of claims are inaccurate design information, inadequate design information, slow client response to decisions, poor communication and unrealistic time targets. The study by Bristow and Vasilopoulos (1995) reveals that there are five major causes for disputes in construction: (i) impracticable expectations by the parties; (ii) indistinct contract documents; (iii) poor communication among project contributors; (iv) lack of cooperation among contributors; and (v) failure of contributors to instantly handle changes and projected conditions. In their study, Hadikusumo and Tobgay (2015) identified and categorised claims as: (i) differing site condition; (ii) delays of project participants; (iii) changes in design and specification; (iv) force majeure; and (v) omissions/ambiguous contract provisions. If a risk management plan is unable to control or manage risks and there is no cooperation among contributors, conflicts arise. Poor conflict management leads to claims. If a claim is not clearly resolved, it may lead to a dispute. The construction disputes path is given in Figure 8.5.

Construction is an extremely unpredictable business. Every project is unique and extremely unpredictable, possessing its own particular conditions and challenges with changes such as cost, design and material which can lead to disputes. All projects are confronted with situations waiting to incite disputes even under the best conditions. Generally, the causal factors contributing to construction disputes are from the side of the employer, contractor, consultant and the contract itself. Typical sources of disputes and claims are worth noting. Figure 8.6 provides a classification of the major causes of construction disputes as per contributor and cause of dispute.

Any clause in the contract possibly is capable of turning out to be the origin of a claim. However, the contract law and contracting system prevailing in the world along with experienced consultants and design firms have, in fact, reduced the occurrence of claims. Generally, …Generally, claims may be identified as falling into one of four main groups: (i) changed conditions, conditions different from those presented in the contract documents; (ii) additional work; (iii) delays strictly beyond the contractor's control; and (iv) contract time extension due to changed conditions, or owner-caused delays.[5]

8.6 Claim notification

The identification of a claim must be followed by notification. Once the claim is ready and submitted by the contractor, the next step will be a claim notification. The contractor is subsequently required to inform the engineer within the stipulated time frame and

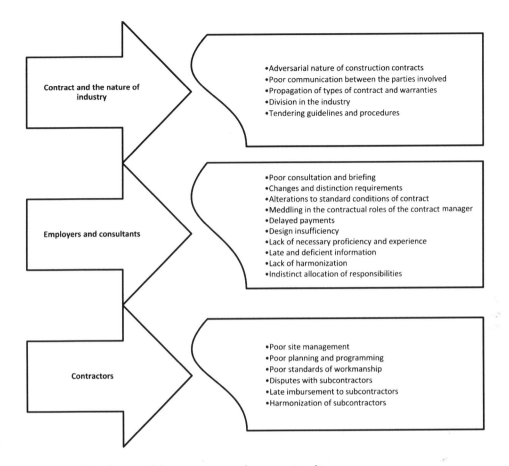

Figure 8.6 Classification of the major causes of construction disputes.

clarify his intention to claim extra rates for the same. The client or representative must be formally notified of a claim or a change order if the contractor intends to seek an equitable adjustment for additional time or costs. The claim is supported by necessary documents such as letters, vouchers and drawings. Such a notification allows both parties to verify conditions, to assemble facts and to resolve disputes while the issues are fresh in their minds (Mitchell, 1998). The client's representative or his office performs a schedule analysis to compute the time impact and break down the cost into different cost components to estimate the cost impact. FIDIC Sub-Clause 20.1 explains the notice period and other details (refer Chapter 7) for projects under FIDIC conditions of contract.

8.6.1 *Claim documentation*

The coverage of documentation required for a particular construction project will depend on the type of contract. The contractor should keep a record of all changes and regularly convey it to the owner as stated in the contract. This will allow both sides to foresee claims. Many records are documented as required: (i) by law; (ii) by the terms of the contract[6]; (iii) to monitor, control and stay up to date with the ongoing work; (iv) as data for estimating future work; and (v) for preserving the contractor's rights under the contract. Above all, it confirms the quantity of work done and is an indication of the progress of the ongoing project work. Most of these records pertain to the contractor's job. All verbal orders and memos given should be written down and personal diaries should be maintained to record all the solid facts and figures because diaries are valid records in court. Photographs and videos also need to be kept.

8.6.2 Claim presentation

The entire supporting documents for the claim, for instance, drawings, specifications, written instructions, cost breakdowns, measurement records, etc., will be compiled and submitted to the client for assessment. While preparing the claim for presentation, the party must ensure that the relevant position of the contract law where the remedy is provided is precisely stated.

Supporting evidence will depend on the type of claim. However, supporting evidence such as site records, correspondence, record of site meetings, site diaries, programmes and progress schedules, payments claimed and made, etc., can greatly contribute to proving a claim. In circumstances where a claim is lost, for direct loss and/or expense relating to a variation, it will be necessary that it is supported by: (i) a copy of an instruction for variation validly issued and acted upon; (ii) the date of issue and receipt of instruction relating to variation; (iii) a brief on the nature of variation; (iv) the time at which it was necessary to carry out the work pursuant to the variation instructions; (v) justification that carrying out the variation instructions certainly affected the progress of the work; (vi) the extent of the work that was affected by implementing the variation instructions; (vii) proof that the contractor made a proper and timely application to the engineer in respect of the direct loss and/or expense; and (viii) substantial supporting particulars and proof of direct loss and/or expenses suffered or incurred. Ground of claim like reference of the relevant contract clauses which permits the contractor for the claim and where the contract clauses do not offer any remedy, reference of the contract law section where the claim can be legally justifiable and admissible should be provided.

8.6.3 Negotiation and settlement of claims

Contract claims can be determined either during construction or at the end of the construction project by negotiation. The settlement of claims requires that both parties are willing to compromise on the claims. Humans are considered to have an innate ability for negotiation. Weak points should not be emphasised when negotiating a settlement of claims since the main purpose of a negotiation is to achieve a settlement. Well planned and documented negotiations can reach a settlement of claims.

A claim needs to be resolved mutually through negotiation or through the intervention of a dispute resolution body. Work stoppage due to a claim should be avoided; however, if this is inevitable, attempts should be made settle the claim amicably.

Notes

1 Adapted from World Bank. (2018, September)*Contract Management Practice*. The World Bank, Washington, DC. Retrieved from: http://pubdocs.worldbank.org/en/277011537214902995/ CONTRACT-MANAGEMENT-GUIDANCE-September-19-2018-Final.pdf Accessed on 26 May 2019.

2 Gilbreath, Robert D. (1992) *Managing Construction Contracts: Operational Controls for Commercial Risks*, 2nd Edition. John Wiley & Sons, New York, p. 204.

3 Bakhary, N.A., Hamimah Adnan and Azmi Ibrahim. (2015) A study of construction claim management problems in Malaysia. *Procedia Economics and Finance*, *23*, pp. 63–70.

4 Mitchell, R.S. (1998) *Construction Contract Claims, Changes and Dispute Resolution*. American Society of Civil Engineers, New York, p.3.

5 *See* http://www.maxwideman.com/papers/construction/dispute.htm.

6 *See* http://www.maxwideman.com/papers/construction/record.htm.

References

Abdul-Malak, M.A.U., El-Saadi, M.M. and Abou-Zeid, M.G. (2002) Process model for administrating construction claims. *Journal of Management and in Engineering*, *18*(2), pp. 84–94. Retrieved from: https://ascelibrary.org/doi/pdf/10.1061/%28ASCE%290742-597X%282002%2918%3A2%2884%29 Accessed on 26 May 2019.

Adrian, J.J.(1993) *Construction Claims – A Quantitative Approach*. Stipes Publishing L.L.C., Champaign, IL.

Bakhary, N., Adnan, H. and Ibrahim, A. (2014) A study of construction claim management problems in Malaysia. *Global Journal of Arts Education*, *4*(2), pp.67–75.

Bakhary, N.A., Adnan, Hamimah and Ibrahim, Azmi (2015) A study of construction claim management problems in Malaysia. *Procedia Economics and Finance*, *23*, pp. 63–70.

Bristow, D. and Vasilopoulos, R. (1995) The new CCDC 2: Facilitating dispute resolution of construction projects. *Construction Law Journal*, *11*, pp. 95–117.

Chan, E.H.W. and Suen, H.C.H. (2005) Dispute resolution management for international construction projects in China. *Management Decision*, *43*(4), pp.589–602.

Daoud, O.E.K. and Azzam, O.M. (1999) Sources of disputes in construction contracts in the Middle East. *Technology, Law and Insurance*, *4*(1), pp. 87–93.

Enshassi, A.A., Choudhry, Rafiq Mohamed and El-Ghandour, Said Mohamed (2008) Owners' perception towards causes of claims on construction projects in Palestine. *Arab Gulf Journal of Scientific Research*, *26*(3), pp. 133–144. Retrieved from: https://iugspace.iugaza.edu.ps/bitstream/handle/20.500.12358/26459/Enshassi%2C%20Adnan_131.pdf?sequence=1&isAllowed=y Accessed on 27 May 2019.

Gilbreath, Robert D. (1992) *Managing Construction Contracts: Operational Controls for Commercial Risks*, 2nd edition. John Wiley and Sons, New York.

Hadikusumo, B.H.W. and Tobgay, S. (2015) Construction claim types and causes for a large-scale hydropower project in Bhutan. *Journal of Construction in Developing Countries*, *20*(1), pp. 49–63.

Ameli, Jebel and M. (2008) The plan of claims management in civil construction projects (case study). In: *4th International Project Management Conference*. Civilica Publication, Tehran, Iran.

Kululanga, G.K., Kuotcha, W., McCaffer, R. and Edum-Fotwe, F.(2001) Construction contractors' claim process framework. *ASCE Journal of Construction Engineering and Management*, *127*(4), pp. 309–314.

Kumaraswamy, M.M. (1997a) Conflicts, claims and disputes in construction. *Engineering, Construction and Architectural Management*, *4*(2), pp. 95–111.

Kumaraswamy, M.M. (1997b) Common categories and causes of construction claims. *Construction Law Journal*, *13*(1), pp. 21–34.

Levy, S.M. (2009) *Construction Process Planning and Management: An Owner's Guide to Successful Projects*, Chapter 10, p. 254.Elsevier, MA. Retrieved from: https://ebookcentral.proquest.com/lib/nicmar-ebooks/reader.action?docID=453061&ppg=262 Accessed on 27 May 2019.

Marsh, Peter (2003) Managing variations, claims and disputes. In: J. Rodney Turner (Ed), *Contracting for Project Management*, p. 130.Routledge, London. Retrieved from: https://ebookcentral.proquest.com/lib/nicmar-ebooks/reader.action?docID=429863&ppg=6 Accessed on 27 May 2019.

Mitchell, R.S. (1998) *Construction Contract Claims, Changes and Dispute Resolution*, p. 3.American Society of Civil Engineers, New York.

Rubaie, F. (2002) *Contract Law*. Behnami Publication, Tehran, Iran.

9 Arbitration, conciliation and dispute resolution

9.1 Introduction

Construction contracts are unique in nature because of the existence of 'unforeseen' circumstances that can impact on a project in terms of cost, time and quality, and contain due provisions on how such 'unforeseen' issues are dealt with in the contract conditions. An agreement which is enforceable by law is considered to be a contract. The Indian Contract Act of 1872 contains the law relating to contracts.

Construction projects are continuous in nature and are generally spread over a number of years. However, for several reasons, projects do not always progress as desired and delays cost money for both the client and the contractor (see Chapter 8). If a delay is an excusable delay (caused by acts of God, labour strikes, etc.), the contractor or client is permitted a time extension. Monetary compensation is not allowed in such situations. In the case of liquidated damages in the contract, the client will need to extend the project's date of completion albeit no costs for extension can be considered. Concurrent delay occurs when two or more delays are created within the same time frame and they cancel each other out, and neither party can recover damages (Levy, 2009). Delays can result in disputes and claims.

Disputes can be resolved either through litigation or through an alternative dispute resolution (ADR) mechanism, which includes mediation, conciliation and arbitration. When the parties to the contract differ on a specific interpretation or the documents on which they both agreed lead to disputes, such disputes are generally resolved in court. Litigation is time-consuming and expensive. Thus, the construction industry has devised a faster means of settling disputes, i.e. arbitration. Arbitration is a faster way of resolving disputes as it is less time-consuming and, at the same time, most participants in the process are players in the construction industry who are acquainted with how the industry operates.

9.2 The Arbitration and Conciliation Act, 1996

The Arbitration and Conciliation Act, 1996 is the major legislation regulating domestic arbitration, international commercial arbitration and the enforcement of foreign arbitral awards. It repealed the three statutory provisions for arbitration: (i) the Arbitration Act, 1940; (ii) the Arbitration (Protocol and Convention) Act, 1937; and (iii) the Foreign Awards (Recognition and Enforcement) Act, 1961. The Arbitration and Conciliation Act, 1996 provides a clear framework for both international commercial arbitration and domestic arbitration. It provides for laws relating to domestic arbitration, international commercial arbitration, the enforcement of foreign arbitral awards, conciliation and the UN model law to make Indian law accord with the law adopted by the United Nations Commission on

International Trade Law (UNCITRAL). In order to strengthen Indian arbitration and promote alternative dispute resolution, The Arbitration and Conciliation (Amendment) Act, 2015 bill was presented before the Lok Sabha, to amend the 1996 Act.

The 1996 Act has two important parts: Part 1 concerns arbitration conducted in India and the enforcement of awards thereunder; and Part 2 provides for the enforcement of foreign awards to which the New York Convention or the Geneva Convention applies. Parts 3 and 4 discuss conciliation and supplementary provisions, respectively.

9.3 Domestic arbitration

Domestic arbitration is an ADR method wherein disputes between parties are settled through the intervention of a third person and without having recourse to a court of law. The ADR is a mechanism by which a dispute is referred to a nominated person. The nominated person decides the issue in a 'quasi-judicial' manner after hearing both sides. Usually, the disputing parties refer their case to an arbitral tribunal. The decision of the tribunal is known as an 'award'. One of the major advantages of this system is when a dispute concerns a technical matter; the parties can select an arbitrator who possesses the appropriate qualifications or skills in the trade.

9.4 International commercial arbitration

International commercial arbitration is 'an arbitration relating to disputes arising out of legal relationships', whether contractual or not (Arbitration and Conciliation Act, 1996). It is considered commercial under the law in India and where at least one of the parties is:

- an individual who is a national of, or habitually resident in, any country other than India; or
- a body corporate which is incorporated in any country other than India; or
- a company or an association or a body of individuals whose central management and control is exercised in any country other than India; or
- a government of a foreign country.

9.5 Claim

Throughout the execution of a project, numerous problems can occur which cannot be resolved between project participants. Usually, such problems involve a contractor requesting: (i) either a time extension; or (ii) reimbursement of an extra cost; or (iii) sometimes both. Such requests are referred to as a claim. If the client agrees to the contractor's claim and grants him an extension of time or reimbursement of an additional cost, or both, the matter is resolved. However, if the owner does not agree to the claim raised by the contractor and there are dissimilarities in their interpretations, the issue takes the form of a dispute. In this chapter, we will consider various methods of dispute settlement. However, it is useful to first clarify a few basic terms.

9.6 Principal causes of disputes and claims

According to Levy (2009), 13 areas most often lead to disputes and claims (refer to Chapter 8). Mohsin (2012) identifies three major kinds of claims: (i) common law claims;

(ii) ex gratia claims; and (iii) contractual claims. Common law claims arise from causes which are outside the express terms of a contract; for instance, if the client or architect were negligent or in some way hindered the progress of the works, resulting in loss to the contractor. Ex gratia claims have no legal basis but are claims that the contractor considers the client has a moral duty to meet (refer to Chapter 8 for details). Contractual claims arise from the express terms of a contract and are, by far, the most frequent kind of claim. They may relate to any or all of the following: (a) fluctuations; (b) variations; (c) extensions of time; and (d) loss and/or expense due to matters affecting the regular progress of the works (Ramus, Birchall and Griffith, 2006). Other types of claims are: different site conditions claims; difference in pricing and measuring claims; acceleration claims which occur when a contractor's work is expedited to complete a particular work activity earlier than planned; damage claims (actual damages where the owner is entitled to the cost of completing the job or of remedying those defects that are remediable, liquidated damages and delay damages) (Ivor, 1997). Some other types of claims are: (a) when there are changes including additions, alterations, variations and deletions, disputes arise; (b) failure to complete the project within time and budget gives rise to disputes; (c) claims arising out of an unknown subsurface, called differing site condition; (d) insufficient information from the client side or a delay in payment from the client side with the possibilities of further delay give rise to disputes.

9.6.1 Disputes and their effects on the project

Generally, disputes occur in a project owing to a delay in the project. They take place when one party raises a demand that is denied by the other. One of the parties to the contract may feel aggrieved that its rights and privileges have been dishonoured or that dues payable have not been paid or short paid or paid late. The affected party may approach the other party asking for compensation, which is declined. Likewise, the client may conclude that the quality, speed and volume of deliverables are not as per agreement and may appeal to the other party to set things right, which is not done. All such situations and others similar to these may lead to disputes.

9.6.2 Differing site conditions

The most risky latent condition in construction is the unknown subsurface. Once construction work starts, the contractor may find that the composition of the soil and many other things do not correspond to the contract documents and may differ materially from the contract documents. Contractors base their site work estimates on information provided by the owner's engineer. The government has a '15 per cent rule', i.e. if the actual quantity of the material encountered exceeds 15 per cent, a changed condition will be established. For that, the contractor should present: (a) the time, date and condition when these differing conditions were observed; (b) the project superintendent's entry of the event in the daily log or daily report; (c) photos; (d) a statement from the field personnel involved in the discovery, such as the excavating contractor, the contractor's foreman or the equipment operator; and (e) a statement explaining the operation that was taking place at the time and the exact location to make the claim.[1]

9.6.2.1 Type I and type II claims

The two types of differing site condition claims are Type I and Type II claims. A Type I differing site conditions claim usually refers to the subsurface or latent physical conditions at the site that differ materially from those shown in the contract, whereas Type II claims arise when the contractor encounters unknown or unusual physical conditions that differ materially from those ordinarily encountered. For a Type I claim, the contractor must prove: (a) the contract documents include the subsurface or latent conditions that form the basis of the claim; (b) the contractor's interpretation of the contract documents is reasonable; (c) the contractor relied on these interpretations when he prepared the estimate of the work; (d) the subsurface or latent conditions actually encountered were materially different from those represented in the contract documents; (e) the actual conditions discovered were reasonably unforeseen; (f) the costs included in the claim are solely representative of the costs to correct the materially differing conditions.[2]

A Type II differing site conditions claim occurs when the actual physical condition encountered at the site differs materially from conditions normally encountered. In the case of a Type II claim, the client has not specified any substantive representations or indications in the contract documents relative to the expected site conditions (Long, Lane and Kelly 2018). In support of a Type II claim, the contractor must prove: (a) the usual conditions the contractor would have encountered based on the information included in the contract documents; (b) the actual conditions encountered; (c) the physical conditions encountered differed materially from the known or usual conditions; (d) the encountered conditions created an increase in the cost of the work.

9.7 Dispute settlement methods in the construction industry

Due to the construction industry's important role and the contribution that it makes to the growth and development of the economy, the complexities of construction project execution make conflicts and/or disputes inevitable. Disputes require resolution and various methods of dispute resolution are available, such as conciliation, mediation and mini-trials, to assist parties in reaching a settlement without having to resort to arbitration or litigation. These methods are largely informal and their principal features are that they are cheap and non-binding. In his study, Chong (2012) explains the stages of dispute resolution and their features. According to Chong, there are six stages of dispute resolution and they always begin with a grievance. Chong's six stages of dispute resolution are shown in Figure 9.1.

Following the introduction of the Arbitration and Conciliation Act, 1996, the role of the alternative dispute resolution systems in the country is increasing. The ADR refers to any means of settling disputes outside a courtroom. It includes arbitration, mediation, negotiation and conciliation, which together are known as ADR because they are usually considered an alternative to litigation. Arbitration is a speedy and effective remedy to resolve disputes between parties by arbitrators with expertise in technical, commercial or like fields, selected by parties' own choice to the extent possible, or otherwise, with the involvement of the courts (India Juris, 2004). The 1996 Act was established to provide an effective and expeditious dispute resolution framework, which would inspire confidence in the Indian

Grievance → Negotiation → Mediation → Adjudication → Arbitration → Litigation

Figure 9.1 The stages of dispute resolution.

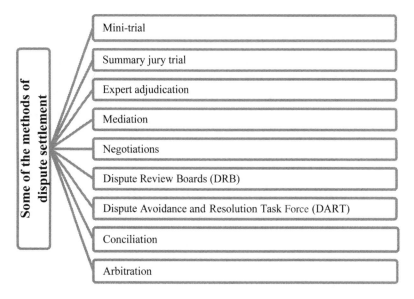

Figure 9.2 Some of the methods of dispute settlement.

dispute resolution system, attract foreign investment and restore international investors' confidence in the reliability of the Indian legal system to suggest an expeditious dispute resolution mechanism (Krishna, Oinam and Kaushik, 2009, October). Some dispute settlement methods are discussed below (Figure 9.2).

9.7.1 Mini-trial negotiations

A mini-trial is not a trial. There is neither a judge nor lengthy procedures. It may be seen as a form of evaluative mediation (Brown and Marriott, 1982; Green, 1982; Wilkinson, 1990). The decision-making is fast and is made by authorised representatives from top management with managerial and often technical skills, sitting across the table from a neutral advisor. In effect, the mini-trial is a structured form of negotiated settlement. All the parties take part in a mini-trial voluntarily, and any party can quit the procedure at any time (Edelman, Carr and Creighton, 1989). Both parties lay bare the facts to measure the strength of each side in the presence of the neutral advisor who only evaluates the merits of various issues and gives his opinion, which is not binding but is persuasive. A mini-trial is successful once there is a mutual agreement.

9.7.2 Summary jury trial

This alternative is popular in the United States and is similar to a mini-trial but has more observers to act as a jury to help the neutral advisor, the judge, who gives his recommendations. The 'summary jury trial' has been publicised as an effective means to facilitate case settlement (Alfini, 1989). This system accommodates more experts, some on law and some on technical matters, and is well-suited as a pre-trial evaluation on a large complex project. It is frequently used to resolve certain categories of civil cases, especially low-value tort cases

(Croley, 2008). The summary jury trial, which normally takes 1 day to complete, permits the parties to submit their dispute to an empanelled jury, which renders a non-binding verdict (Lambros, 1986).

9.7.3 Expert adjudication

Adjudication is an accelerated and cost-effective means of dispute resolution. Unlike other means of resolving disputes, adjudication involves a third-party intermediary imposing a binding decision on the parties in dispute, and is final unless and until it is reviewed by either arbitration or litigation. Expert adjudication is the adjudication by quasi-judicial bodies comprising technical and legal experts with a provision for appeal to a multidisciplinary appellate body and is becoming an increasingly preferred mode of dispute resolution in public-private partnerships. Expert adjudication involves a less thorough investigation than the normal form of arbitration. Moreover, the parties who consent to accept the opinion of an expert adjudicator may be subject to the risk that the expert adjudicator is wrong. Under the system of expert adjudication, disputes are referred to one or more experts in the field. The expert hears, examines and understands the respective viewpoints. The expert/adjudicator then gives his expert opinion, which parties may find very difficult to ignore. A comparison of arbitration and expert adjudication is given in Table 9.1.

9.7.4 Mediation

Mediation is by far the most popular of ADRs and requires an expert with intuitive diplomacy to resolve discrepancies using skill and persuasion. Mediation is a form of ADR in which a third party, a mediator, assists the parties in negotiating a settlement. General conditions require mediation before taking a dispute to arbitration. Mediation is a structured voluntary process in which the mediator assists each party in the dispute to reach a negotiated settlement of their differences. The mediator listens to both sides, initially in each other's presence, serving as an intermediary between the parties to the dispute. Subsequently, the mediator discusses his settlement strategies with each party to the exclusion of the other, giving his expert viewpoint with full knowledge of the subject matter and the relevant law. The mediator continuously discusses with the parties in search of an acceptable resolution and ultimately encourages both parties to reach a settlement and, if not, considerably reduces the areas in dispute.

9.7.5 Negotiations

Negotiation is a dispute resolution method that is used to settle disputes in construction contracts when they arise during construction. In negotiations, both sides are represented by experts who are familiar with the strengths and weaknesses of the respective case. Through discussion, both legal and technical experts convince the parties that litigation can do no better than the settlement concluded on negotiation. Early negotiation may prove successful if attempted before the parties have had a chance to formulate strong positions regarding the dispute (Mohsin, 2012).

9.7.6 Dispute review boards (DRB)

The dispute review board was commenced in the United States in 1952 during the construction of the Central Artery/Tunnel in Boston (Harmon, 2009). The DRB system was

Table 9.1 Arbitration vs. expert adjudication

Sl. no.		Arbitration	Expert adjudication
1.	Source of powers	The power of the arbitrator comes from the contract and the arbitration ordinance	The powers of the expert are prescribed exclusively by the agreement between the appointing parties
2.	No dispute	In the absence of a formulated dispute, there can be no arbitration since judicial or arbitral duties do not arise until there is a dispute	In a determination by an expert, there need not necessarily be a dispute. The expert is not determining a dispute, he is deciding what to do in all the circumstances
3.	Nature of the inquiry	1. Arbitration is adversarial in nature 2. Arbitrator must act on the materials provided by the parties and unless specially empowered by them, cannot conduct investigation of his own into the matters in dispute 3. Arbitrator has to conduct the arbitration according to an agreed set of rules	1. The inquiry conducted by the expert adjudicator is inquisitorial in nature 2. The expert adjudicator can make his or her own investigation 3. Expert adjudication is not subject to any rules of procedures
4.	Challenging the decision	Decision of the arbitrator cannot be criticised or challenged for mistakes	Decision of the expert adjudicator can no longer be criticised or challenged for mistakes, unless the expert exceeds the scope of his or her jurisdiction or is guilty of fraud or collusion
5.	Immunity from suit	As arbitrators' functions are of a judicial nature, like a judge, the law has granted an arbitrator immunity from suit	Expert adjudicators are liable in tort to a party suffering loss if they perform their appointed task negligently (Sutcliffe vs. Thackrah [1974] AC72)
6.	Enforceability of award	Arbitrator's award can be enforced as a judgment of the court	Expert adjudicator's award cannot be enforced as a judgment of the court

Source: Plunkett, D. (1995) Expert adjudicators: Who are they? Hong Kong Lawyer, November 1995. In Lau, Kin-Ho Lewis (1996). The role of alternative dispute resolution methods in the construction industry and the application of these methods in Hong Kong. (Doctoral Dissertation), Department of Surveying, University of Hong Kong, Hong Kong, p. 15. Retrieved from: https://hub.hku.hk/R44qNl1U9S/eyJhbGciOiJIUzI1NiIsInR5cCI6Ik pXVCJ9.eyJuYW1lIjoiNTc5ZGYwOTAxZDUyIiwiZW1haWwiOiI3ODlhZmM4NzFiNGZkZGExZTk3NTc0Y 2RmZTA5N2M0YTI0MTZmY2ZmMTciLCJoYW5kbGUiOiIxMDcyMi8zNzAyNSIsInNlcSI6IjEiLCJpYXQiO jE1OTE5Mzk3MTgsImV4cCI6MTU5MjAyNjExOH0.4ZomyJ0jUu5xzn1iJPIN-ZqdgCiQyEGqosfipIM8YHk/Fu llText.pdf Accessed on 11 June 2019.

first launched in India in 1994 in World Bank–financed projects valuing US$50 million or more. The DRB is a concurrent mechanism for resolution of disputes before recourse to arbitration or litigation. The DRB is constituted at the very beginning of a construction project. It consists of independent and impartial professionals. The DRB: (i) tracks progress of construction; (ii) encourages dispute avoidance; and (iii) assists in resolving disputes that may arise during the execution of the project. It is a standing body of individual(s) who remain attached to a project for its entirety and are independent of any institution. The DRB conducts brief status meetings and site visits, has discussions with the employer and the contractor's representatives and becomes familiar with the project procedures, progress and potential disputes (ICAI, 2016, April).

A DRB is a three-member panel, mutually chosen by the contractor and the client, which is present throughout the course of the contract, and whose responsibility is to hear disputes contemporaneously with their occurrence (Harmon, 2009). Two members, chosen by the contractor and owner, should review the plans and specifications as necessary to understand the work as well as the contract conditions and select a third member who will supplement their expertise and experience. The third member will act as the chairperson of the board.

The Indian Council of Arbitration suggests DRBs for all major construction contracts over Rs. 50 crore so that disputes can be resolved concurrently (Figure 9.3). It again suggests a one-member DRB for projects with contract values between Rs. 50 crore and 200 crore and a three-member DRB for projects above Rs. 200 crore. The DRB hears both sides and makes recommendations to the authority named on the contract to give interim decisions, which in the final stage are hard to challenge. This method can be used for an ongoing project or as a pre-arbitration ADR.

The contractor shall give to the engineer, within 28 days after the contractor has become aware, or should have become aware, of the event or circumstance, notice for extension of time for completion and/or any additional payment, with contemporary records as may be necessary to substantiate the claim. Within 42 days after receiving a claim or any further particulars supporting a previous claim of the contractor, the engineer is expressly required to respond with approval, or with disapproval and detailed comments (Sub-Clause 20.1 – contractor's claims). The engineer shall proceed in accordance with the International Federation of Consulting Engineers (FIDIC) multilateral development banks (MDB) Sub-Clause 3.5 (determinations) to agree or determine (i) the extension (if any) of the time for completion (before or after its expiry) in accordance with FIDIC (MDB) Sub-Clause 8.4 (extension of time for completion), and/or (ii) the additional payment (if any) to which the contractor is entitled under the contract.

9.7.7 *Dispute avoidance and resolution task force (DART)*

The dispute avoidance and resolution task force was formed during an American Arbitration Association's conference (April 1991) as an industry-wide coalition to promote awareness, understanding and the use of private dispute prevention and resolution methods and to

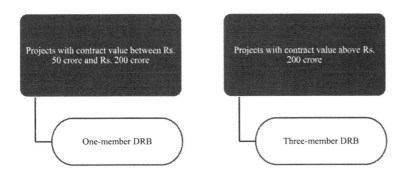

Figure 9.3 DRB for projects with size of contract.

encourage the use of these methods as standard practice for the construction industry. The DART stands for/represents all sections of the industry, i.e. public and private owners, architects, engineers, contractors, subcontractors, sureties, insurers and lenders (József, 1998). The conference advised the creation of Engineer-officers whose job is 'Claim Avoidance'.

9.7.8 Conciliation

Conciliation differs from mediation in that a neutral party evaluates the dispute and then issues a proposal for resolving the dispute that is presented to the parties for approval or rejection. Conciliation is a non-adjudicatory and non-adversarial ADR mechanism relating to a settlement process in which a neutral third party, i.e. a conciliator, facilitates and steers the disputant parties to reach a reasonable and acceptable settlement. The conciliator assists parties, privately and collectively, to identify the issues in a dispute and to develop proposals for satisfactory settlement of the dispute. 'The conciliation not only mitigates the risk of differences developing into disputes, it improves communication and creates an atmosphere of trust and cooperation which is conducive to constructive negotiation' (Moore, 1986).

Conciliation is a method of third-party intervention adopted when bipartite negotiations fail to resolve a dispute. The conciliator is not empowered to decide any dispute. Conciliation is a 'voluntary non-binding process'. A conciliator does not give any judgement as such, but persuades parties to come to a settlement. Under this system of dispute resolution, the parties present their cases and after a series of joint and separate discussions and clarifications, the conciliator develops options for settlement. These options are given to the parties in dispute, one after the other, to ascertain the response from each party until the case is either settled or it becomes apparent that a settlement will not be reached and conciliation can then be ended (Figure 9.4). Its interest-based direct negotiations rather than positional bargaining is perceived as a fairer procedure.

9.8 Arbitration as a dispute settlement mechanism

The only field in which the courts in India have recognised ADR is the field of arbitration.[3] For numerous reasons, disputes can arise between a contractor and an employer/engineer throughout the performance of a contract. In such situations, disputes should be resolved amicably. There are various ways to resolve disputes, such as mediation, as discussed above. In both the arbitration and litigation contexts, mediation is an early step in the resolution of disputes. Figure 9.5 illustrates the differences between mediation and arbitration as ADR mechanisms, as documented by Lipsky and Ronald Seeber (1997) based on their study among legal counsels of large US corporations.

Arbitration is a popular method of dispute resolution because it is binding. Arbitration is based on an agreement among parties to refer a dispute to a neutral third party, an arbitrator, who has the authority to make a binding award. It is a method of dispute resolution that can be agreed by both parties, which may be binding and which can have the support of the law. It involves the disputants laying their case before an independent, impartial and expert arbitrator who, from his knowledge of the law and an examination of the facts of the case, makes a decision which, if the parties so agree, will be binding (Stewart, 1998).

Arbitration has a long history in India. The Indian Contract Act of 1872 permits the settlement of disputes by arbitration under Section 28. The legislations relevant to arbitration started with the Arbitration Act, 1899 which was rooted in the English Arbitration Act, 1899. Afterwards, various statutes governing dispute resolution, such as the Code of

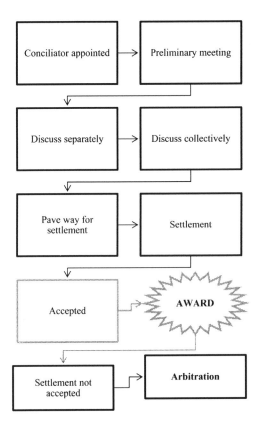

Figure 9.4 Conciliation procedure.

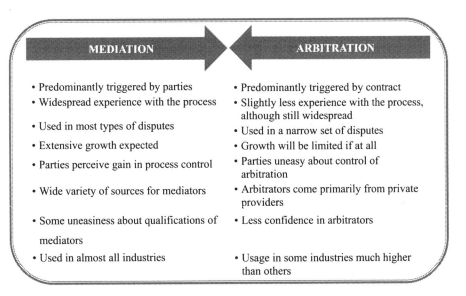

Figure 9.5 Important differences between mediation and arbitration.

Civil Procedure, 1908, the Arbitration Act, 1940 and the Specific Relief Act, 1963 were made. The Arbitration Act, 1940 was superseded by the Arbitration and Conciliation Act, 1996. The enactment of the Arbitration and Conciliation Act, 1996 was to incorporate the UNCITRAL model laws as adopted by the United Nations Commission on International Trade Law on 21 June 1985. The Act provides statutory recognition to conciliation as a distinct mode of dispute settlement.

Conciliation is a process of amicable dispute settlement by parties with the assistance of a conciliator. The difference between arbitration and conciliation is as follows:

1. under arbitration the award is the decision of the third party or the arbitral tribunal; and
2. in the case of conciliation, the decision of the parties is arrived at through the mediation of the conciliator.

There are several advantages to ADR including that it is less expensive and time-consuming. The advantages of ADR as given in 'The Law Commission's 222nd Report on Need for Dispensation of Justice through ADR' are given in Figure 9.6.

9.8.1 Types of arbitration proceedings

Arbitral proceedings are categorised into two types: (i) ad hoc arbitration; and (ii) institutional arbitration.[4] In ad hoc arbitration, the parties determine the conduct of the arbitration proceedings without an arbitral institution. The arbitration proceedings are agreed on and arranged by the parties themselves.

Section 11 of the 1996 Act aids the ad hoc arbitration process: (i) if the parties are unable to agree on who will be the arbitrator; or (ii) one of the parties is reluctant to cooperate in appointing an arbitrator. Under such circumstances, the Chief Justice of a high court or the supreme court or their designate will appoint an arbitrator.[5] For international commercial arbitration, the Chief Justice of India or his designate and the Chief Justice of a high court or his designate for domestic arbitration will appoint an arbitrator.

9.8.1.1 Institutional arbitration process

Arbitration administered by an arbitral institution is known as 'institutional arbitration'. The parties specify an institution for the resolution of an arbitral dispute between them, such

- Less expensive
- Less time-consuming
- Free from technicalities as in the case of conducting cases in law Courts
- Parties are free to discuss their differences of opinion without any fear of disclosure of this fact before any law courts.
- Parties have the feeling that there is no losing or winning side between them but at the same time their grievance is redressed and their relationship is restored.

Figure 9.6 The advantages of ADR.

as the Indian Council of Arbitration or the International Centre for Alternative Dispute Resolution. International institutions include the International Court of Arbitration, the London Court of International Arbitration and the American Arbitration Association. Both these Indian and international arbitration institutions have ideal rules expressly formulated for conducting arbitration.

9.8.1.1.1 ADVANTAGES OF INSTITUTIONAL ARBITRATION IN COMPARISON WITH AD HOC ARBITRATION

Institutional arbitration is professionally administered, highly reputed and widely recognised. Consequently, institutional arbitration is considered superior to non-administered arbitration, i.e. ad hoc arbitration. The ad hoc arbitration refers to the arbitration agreed to and arranged by the parties themselves, without recourse to an arbitration institution. Under institutional arbitration, the institution maintains a panel of arbitrators together with their profiles and the parties choose from the panel. Arbitration is agreed to and arranged by the parties themselves without recourse to an arbitration institution under ad hoc arbitration. Figure 9.7 illustrates the advantages of institutional arbitration in comparison with ad hoc arbitration.[6].

9.8.2 Essential characteristics of arbitration

Arbitration is a mechanism for the settlement of disputes. It is consensual in nature, i.e. arbitration is a matter of contract and a party cannot be required to submit to arbitration any

	Institutional Arbitration	Ad hoc Arbitration
Arbitration Procedures	There is no need to worry about formulating rules or spend time on making rules.	Arbitration procedures will have to be agreed to by the parties and the arbitrator. This needs cooperation between the parties. When a dispute is in existence, it is difficult to expect such cooperation. In institutional arbitration, the rules are already there.
Infrastructure Facilities	The arbitral institution will have infrastructure facilities for conduct of arbitration; they will have trained secretarial and administrative staff. There will also be library facilities. There will be professionalism in conducting arbitration. The costs of arbitration also are cheaper in institutional arbitration.	Infrastructure facilities for conducting arbitration is a problem, so there is temptation to hire facilities of expensive hotels. In the process, arbitration costs increase. Getting trained staff is difficult. Library facilities are another problem.
Arbitrators	The institution will maintain a panel of arbitrators along with their profiles. The parties can choose from the panel. It also provides for specialized arbitrators.	These advantages are not available.
Experienced Committee	Many arbitral institutions have an experienced committee to scrutinize the arbitral awards. Before the award is finalized and given to the parties, it is scrutinized by the experienced panel. So the possibility of the court setting aside the award is minimum.	Not available in ad hoc arbitration. Hence, there is higher risk of court interference.
Fee	The arbitrator's fee is fixed by the arbitral institution. The parties know beforehand what the cost of arbitration will be.	The arbitrator's fee is negotiated and agreed to. The Indian experience shows that it is quite expensive.
Rules	The arbitrators are governed by the rules of the institution and they may be removed from the panel for not conducting the arbitration properly	There is no such fear.
Capability of Arbitrator	The arbitrator becomes incapable of continuing as arbitrator in institutional arbitration, it will not take much time to find substitutes. When a substitute is found, the procedure for arbitration remains the same. The proceedings can continue from where they were stopped	These facilities are not available in ad hoc arbitration
Confidentiality	As the secretarial and administrative staff is subject to the discipline of the institution, it is easy to maintain confidentiality of the proceedings.	It is difficult to expect professionalism from the secretarial staff.

Figure 9.7 The advantages of institutional arbitration in comparison with ad hoc arbitration.

dispute which it has not agreed to submit. Arbitration is a private procedure and arbitrators are normally chosen by the parties or arbitral institutions on their behalf. It is an impartial procedure allowing each party to present its case. The arbitration procedure leads to a final and binding determination of the rights and obligations of the parties.

9.8.3 Objectives of the Arbitration and Conciliation Act, 1996

The Arbitration and Conciliation Act, 1996 is derived from the United Nations Commission on International Trade Law model law on commercial arbitration. The main objectives of the Arbitration and Conciliation Act, 1996 was to: (i) contain both international and domestic arbitration and conciliation; (ii) make provisions for an arbitral procedure which is fair, efficient and capable of meeting the needs of the arbitration; (iii) minimise the supervisory role of courts in the arbitral process and thus ensure minimal judicial intervention; (iv) ensure that an arbitral tribunal functions within the framework of the Act; (v) empower an arbitral tribunal to use mediation and conciliation to encourage settlement of disputes; (vi) provide that a settlement reached by the parties as a result of conciliation proceedings will have the same status and affect as an arbitral award; (vii) provide that the arbitral tribunal gives reasons for its arbitral award; and (viii) provide that every arbitral award is enforced in the same manner as if it were a decree of the court.

9.8.3.1 Highlights of the Arbitration and Conciliation (Amendment) Act 2015

The amended act of the Arbitration and Conciliation Act, 1996, i.e. the Arbitration and Conciliation (Amendment) Act, 2015 came into effect on 1 January 2016. The core reforms in the new Act are: high courts have exclusive jurisdiction over matters of international commercial arbitration; discourages the court from accepting an application for interim relief following the constitution of the arbitral tribunal; courts are prevented from blindly referring parties to arbitration, without ensuring that the parties will be able to get the desired relief through arbitration proceedings; does not allow the setting aside of an international commercial award under the guise of public policy, on the grounds of a patent illegality appearing on the face of the award; the arbitrator is required to disclose the existence of any relationship or interest of any kind which is likely to give rise to justifiable doubts, in writing; the Supreme Court of India or any authority chosen by the supreme court has the power to appoint a sole or third arbitrator, where the parties or co-arbitrators are unable to arrive at an agreement as to such appointment, or the applicable arbitral institution fails to make this appointment; the arbitral tribunal must render the award within a period of 12 months from its constitution, which may be extended by a further period of six months; provides for a fast-track procedure for conducting arbitral proceedings, subject to the parties' agreement; and detailed provisions related to costs.

9.8.3.2 Process of arbitration

At the time of entering into a contract, the parties agree that in the case of a conflict, they would seek to resolve the matter through an arbitrator. Generally, the name of the probable arbitrator, agreed upon by both parties, is given in the contract itself. The first step, if a dispute arises, is the issuing of an 'arbitration notice' by any of the parties. The subsequent steps are: the response by the other party; and after the appointment of an arbitrator, a decision on the rules and procedures, the place of arbitration and language. The arbitration

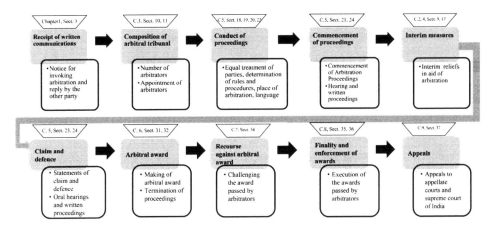

Figure 9.8 Stages of arbitration proceedings as per the Arbitration and Conciliation Act, 1996.

proceedings begin with formal hearings and written proceedings. The arbitrator, if required, issues interim reliefs and then a final award which is binding on both parties. The complex part arises if either of the parties, dissatisfied with the award, challenges it before the court. It can come before the appellate court or the supreme court depending upon the matter. The stages of arbitration proceedings as per the Arbitration and Conciliation Act are given in Figure 9.8.

9.8.4 Arbitration agreement

The provision of arbitration can be made at the time of entering the contract itself, so that if any dispute arises in the future, the dispute can be referred to an arbitrator as per the agreement. A dispute can also be referred to arbitration after the dispute has arisen. Consequently, before any party initiates arbitration proceedings, it is essential that there is an arbitration agreement setting out the terms of arbitration. Generally, a condition of arbitration is given in construction contracts providing comprehensive terms for the appointment of arbitrators and proceedings. If such a condition does not exist, a special arbitration agreement is required to be signed by both parties before they can take recourse to arbitration.

9.8.5 Advantages of arbitration

Arbitration has several advantages over law courts (Figure 9.9). Disputes taken to arbitration can be resolved faster than a lawsuit. Arbitration has all the legal advantages of a court, one can get the entitled benefit as per the law and, at the same time, the whole case is heard, discussed and presented in a more informal and professional manner without reservations. Consequently, arbitration takes less time than proceedings in court and ends up being more economical than a case that goes to trial. Arbitration involves a panel of arbitrators who are either attorneys who practice in a certain area or other industry professionals who are familiar with that area. The decision of the arbitrator is usually simpler, shorter and based on common sense and practical wisdom. The court strictly observes the rules of evidence and

The parties can choose their own: (i) professionally qualified and capable arbitrator; (ii) arbitrator based upon the technical complexity of the claim; and (iii) place of hearing and timings convenient to them.

The arbitrators can: (i) ensure and accelerate the whole process with mutual consent of the parties; (ii) engage expert to unfold or focus on certain intricate points for clarifications; (iii) facilitate negotiated settlement between the parties even amidst the arbitration; and (iv) grant interest without any inhibition from the date of cause.

The arbitration has all legal advantages of a court. The arbitral award shall be final and binding on the parties and persons claiming under them respectively.

The parties can: (i) mutually negotiate, settle and inform the arbitrators during the arbitration process; and (ii) seek correction of any arithmetical or clerical mistakes in an award within 30 days of receipt of award.

The award is: (i) final and binding on the parties, unless challenged; and (ii) to be made within four months from the day entering into reference.

Figure 9.9 The advantages of arbitration.

the civil procedure code in hearing and disposing of a case and gives reason for its decisions, which is not the case in arbitration.

9.8.6 *Process of arbitration*

As arbitration is a private system of dispute resolution, it has to be agreed on between the parties. Generally, the appointment of an arbitrator falls under one of three methods: (i) an arbitrator or arbitrators are nominated by the parties in the arbitration agreement; (ii) an arbitrator or arbitrators are appointed by the parties; (iii) an arbitrator or arbitrators are appointed by a third party designated by name or holder for the time being of any appointment. By and large, in construction, contract conditions contain arbitration agreements. Once an arbitration agreement is fixed, either in the contract or later, the parties must use arbitration (Whitfield, 2012). As per the arbitration agreement, an arbitrator is appointed for each party.

Arbitrators, by inviting the parties in writing, fix the place and time and prepare a schedule that is acceptable to the parties, for submitting the claim by claimants or the reply by respondents, including counterclaims, if any. As per Section 23 (1) of the Arbitration Act, 1996, a claim shall be submitted with authentic supporting material as evidence by each party to the arbitrator and the other party. In his statement of claim, the claimant must set out the facts of his case and the remedy he is seeking so that the other party will know what he has to meet. Once the issues in dispute are clearly accepted by both parties and the arbitrator, the arbitrator decides the date and place of hearing which he confirms in writing to both parties. At this stage, the arbitrator conducts the hearing continuously as may be required. The arbitrator must conduct the hearing in accordance with the arbitration agreement. A presentation should be given by the claimant during the hearing of the dispute and the claim. Both parties should be present at the time of presentation. An absent party, who is willing to cooperate in arbitration, will be given a chance by fixing a convenient date. If there is disagreement from any party, after due written chances, the arbitrator has privilege to proceed with the process of an ex-parte decision. The award could be challenged if proper opportunity is not given. The arbitrator gives the reasoned award after duly conducting the arbitration process. Procedurally, the award must be made by the arbitrator himself, signed by him and attested by a witness. The arbitrator is not bound to give reasons for his decision unless otherwise specifically provided in the arbitration agreement. The award is binding on both parties unless it is challenged by the other party.

9.8.6.1 *Default of a party*

Section 25 of the Arbitration Act, 1996 provides that unless otherwise agreed by the parties, where, without showing sufficient cause: (i) the claimant fails to communicate his statement of claim in accordance with Section 23 (1), the arbitral tribunal shall terminate the proceedings; (ii) the respondent fails to communicate his statement of defence in accordance with Section 23 (1), the arbitral tribunal shall continue the proceedings without treating that failure in itself as an admission of the allegations by the claimant and shall have the discretion to treat the right of the respondent to file such statement of defence as having been forfeited; (iii) a party fails to appear at an oral hearing or to produce documentary evidence, the arbitral tribunal may continue the proceedings and make the arbitral award on the evidence before it.

9.9 Conclusion

In sum, arbitration is a domestic forum of dispute resolution. The causes of disputes could be due to the contractor or the client. The arbitrator resolves the disputes between the client and the contractor and his verdict is legally enforceable. Evidence, submitted to the arbitrator by the employer or the contractor, is analysed and then the verdict is given in favour of any of the parties. Clause 73 of the Arbitration and Conciliation Act, 1996 [26 of 1996] as amended by the Arbitration and Conciliation (Amendment) Act, 2015 [3 of 2016] 'settlement agreement' says 'When it appears to the conciliator that there exist elements of a settlement which may be acceptable to the parties, he shall formulate the terms of a possible settlement and submit them to the parties for their observations. After receiving the observations of the parties, the conciliator may reformulate the terms of a possible settlement in the light of such observations'.

Notes

1 For details, refer to Chapter 10 of Levy, S.M. (2009).
2 Ibid., p. 262.
3 *See* http://lawcommissionofindia.nic.in/reports/report222.pdf p. 18 paragraph 1.27 Accessed on 22 June 2019.
4 The Law Commission's 222nd Report on Need for Dispensation of Justice through ADR (alternative dispute resolution) elaborated on the concept of ad hoc and institutional arbitration. Retrieved from: http://lawcommissionofindia.nic.in/reports/report222.pdf Accessed on 23 June 2019.
5 As per the Arbitration and Conciliation (Amendment) Act, 2015, the Supreme Court of India or any authority chosen by the supreme court has the power to appoint a sole or third arbitrator, where the parties or co-arbitrators are unable to arrive at an agreement as to such appointment, or the applicable arbitral institution fails to make this appointment.
6 Op. cit., Law Commission's 222nd Report for details, pp. 26–28.

References

Alfini, J.J. (1989) Summary jury trials in state and federal courts: A comparative analysis of the perceptions of participating lawyers. *Journal on Dispute Resolution*, 4(2), pp. 213–234.

Brown, H. and Marriott, A. (1982) *ADR Principles and Practice*. Sweet and Maxwell, London.

Chong, Heap-Yih (2012) Selection of dispute resolution methods: Factor analysis approach. *Engineering, Construction and Architectural Management*, 19(4), pp. 430–431.

Croley, S. (2008) Summary jury trials in Charleston County, South Carolina. *Loyola Law Review*, 1585, 41 Loy. Retrieved from: https://digitalcommons.lmu.edu/cgi/viewcontent.cgi?article=2640 &context=llr Accessed on 8 June 2019.

Green, E. (1982) *Mini-Trial Handbook*. Center for Public Resources, New York.

Hajdú, József (1998) The methods of alternative dispute resolution (ADR) in the sphere of labour law (the case of USA, Australia, South Africa and Hungary). Acta Juridica et Politica Tomus LIV. Fasciculus 8, Szeged. Retrieved from: http://acta.bibl.u-szeged.hu/6976/1/juridpol_054_fasc_008 _001-078.pdf Accessed on 18 June 2019.

Harmon, Kathleen M. (2009) Case study as to the effectiveness of dispute review boards on the central artery/tunnel project. *Journal of Legal Affairs and Dispute Resolution in Engineering and Construction (ASCE)*, 1(1), pp. 18–31. Retrieved from: https://ascelibrary.org/doi/full/10.1061/%28ASCE%2 91943-4162%282009%291%3A1%2818%29 Accessed on 14 June 2019.

ICAI. (2016, April) *Standard Operating Procedures for Dispute Boards in India*. Indian Council of Arbitration, New Delhi. Retrieved from: http://www.icaindia.co.in/DB-ica/Final-SOP.pdf Accessed on 14 June 2019.

Juris, India (2004) *Arbitration Laws in India*, p. 6. International Corporate Legal Consultants, New Delhi.

Lambros, Thomas D. (1986) The summary jury trial – An alternative method of resolving disputes. *Judicature*, 69, pp. 286–290.

Lester, Edelman, Carr, Frank and Creighton, James L. (1989) *The Mini Trial*, p. 1. US Army of Engineers, Fort Belvoir, Virginia. Retrieved from: https://apps.dtic.mil/dtic/tr/fulltext/u2/a224260. pdf Accessed on 8 June 2019.

Levy, S.M. (2009) *Construction Process Planning and Management: An Owner's Guide to Successful Projects*, Chapter 10. p. 254. Elsevier, MA. Retrieved from: https://ebookcentral.proquest.com/lib/ nicmar-ebooks/reader.action?docID=453061&ppg=262 Accessed on 27 May 2019.

Lipsky, David B. and Seeber, Ronald (1997) The use of ADR in US corporations: Executive summary. *Cornell University School of Industrial and Labor Relations*. Retrieved from: https://mdcourts.gov/site s/default/files/import/macro/pdfs/reports/cornellstudy1997execsummary.pdf Accessed on 14 June 2019.

Long, Richard J., Lane, Robert J. and Kelly, James E. (2018) *Differing site Conditions*, p. 21. Long International Inc., Littleton, CO. Retrieved from: http://www.long-intl.com/articles/Long_Intl_D iffering_Site_Conditions.pdf Accessed on 28 June 2019.

Mohsin, Mohammed Al (2012) Claim analysis of construction projects in Oman. *International Journal on Advanced Science Engineering Information Technology*, 2(2), pp. 73–78.

Moore, C.W. (1986) *The Mediation Process: Practical Strategies for Resolving Conflict*. Jossey-Bass, San Francisco, CA. In Preez, Olive du. (2004) Conciliation: A founding element in claims management. *Procedia – Social and Behavioral Sciences*, 119, p. 117. Retrieved from: https://www.sciencedirect .com/science/article/pii/S1877042814021065 Accessed on 20 June 2019.

Plunkett, D. (1995) "Expert Adjudicators Who are they?" *Hong Kong Lawyer*, November 1995.

Ramus, Jack, Birchall, Simon and Griffith, Phil (2006) *Contract Practice for Surveyors*, 4th edition. Elsevier Linacre House, Jordan Hill. Cited by Mohsin, Mohammed Al. (2002) Claim analysis of construction projects in Oman. *International Journal on Advanced Science Engineering Information Technology*, 2(2), pp. 73–78. Retrieved from: http://citeseerx.ist.psu.edu/viewdoc/download?doi= 10.1.1.1000.7540&rep=rep1&type=pdf Accessed on 7 June 2019.

Sarma, Krishna, Oinam, Momota and Kaushik, Angshuman (2009, October) *Development and Practice of Arbitration in India: Has it Evolved as an Effective Legal Institution?* Working Paper, No. 103, p. 4. Center on Democracy, Development and The Rule of Law, Freeman Spogli Institute for International Studies, Stanford University, Stanford.

Seeley, Ivor (1997) *Quantity Surveying Practice*, 2nd edition. McMillan Press Ltd., London.

Stewart, Susan (1998) *Conflict Resolution: A Foundation Guide*, pp. 16–17. Waterside Press, Winchester, England, Retrieved from: https://search.proquest.com/legacydocview/EBC/3416322?accountid=34791. Retrieved from: NICMAR-ebooks Accessed on 20 June 2019.

Whitfield, Jeffery (2012) *Conflict in Construction*, p. 126. John Wiley & Sons, Incorporated. ProQuest Ebook Central. Retrieved from: http://ebookcentral.proquest.com/lib/nicmar-ebooks/detail.action?docID=967282 Accessed on 25 June 2019.

Wilkinson John, H. (1990) A primer on mini-trials. In: H. Wilkinson John (Ed), *Donovan Leisure Newton and Irvine ADR Practice Book*, pp. 171–180. Wiley, New York.

10 International construction project exports

10.1 Introduction

Export is the most common method of entry into international markets. Exports expands the markets, brings economies of scale due to increased production, more foreign exchange and employment opportunities. India began construction project exports in the 1970s. The export of engineering goods on deferred payment terms and the execution of turnkey projects and civil construction contracts abroad are collectively referred to as 'project exports' (RBI, 2003). The Memorandum of Instructions on Project Exports and Service Exports (PEM), generally known as the Project Export Manual, contains instructions for exporters engaged in project exports. The current chapter provides details of the global construction market and project exports from India. The complexities of international contracting, factors in bidding for international projects, the overseas contract tendering process, the clearance of project export proposals, requirements relating to completed projects, the steps to be taken by exporters on completion of contracts and preparation of the final report are explained in this chapter, which includes a case study.

10.2 Global construction market and project exports from India

The international construction market has been growing steadily for nearly a decade, as per the data given by the *Engineering News-Record* (ENR) on the top 400 contractors. International contract revenues reached US$405 billion in 2018 from US$290.6 billion in 2009. The majority of contractors come from Canada (28.7%) and Europe (27.0 per cent) (Figure 10.1), with the maximum revenue earned by Canada (Figure 10.2). 'Global construction services exports are growing at a Compound Annual Growth Rate (CAGR) of 9.4 per cent between 2005 and 2018 and India's share in it has nearly tripled over the same period' (Table 10.1) (Exim Bank, 2019). Even though India has tremendous potential with technological capabilities and skilled and comparatively cost-effective human capital, India could not capture much of the share of the international construction market. The contracts secured by the Project Exports Promotion Council of India's member companies in 2013–14 are given in Table 10.2. An examination of the bidding conversion ratio[1] shows that project exporters could have won almost half the bids (Table 10.3). India's project exports focus on the regions of Sub-Saharan Africa, the Middle East and North Africa. India's major competitors in project exports are China, South Korea, Turkey, Spain, Italy, France, Germany and the United States. Some of the reasons for India's low share of the international construction market are discussed below.

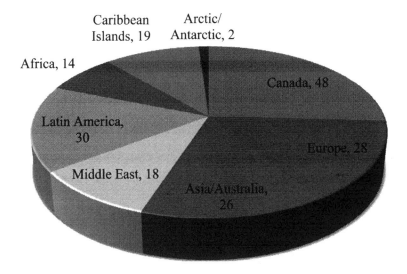

Figure 10.1 Number of firms in each region.

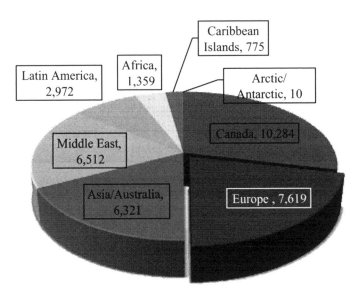

Figure 10.2 Revenue from each region in US$ millions.

10.2.1 *Factors affecting project exports from India*

Major factors affecting construction project exports from India are: (i) low equity base; (ii) lack of experience in certain specialised works which inhibits Indian exporters from securing large-size projects funded by multilateral agencies; (iii) lack of consortium approach; (iv) inferior materials and equipment; (v) inferior quality of output; (vi) no price advantage to

Table 10.1 Global construction services exports and India's share

Year	2005	2009	2010	2011	2012	2013	2014	2015	2016	2017	2018
ENR											
International contract revenue in US$ billions[a]	–	290.6	259.4	282.1	309.4	324.2	331.9	344.1	366.4	374	405
Exim bank											
Global construction services exports (US$ billion)[b]	34	89.8	87.9	96.4	100.4	107.4	118.5	103.7	99.1	104.6	109.4
India's share (%)[b]	1	0.9	0.6	0.9	0.9	1.1	1.4	1.4	2.1	2.2	2.9

Source: [a]ENR, May 27/June 3, 2019. [b]IMF, Exim bank research. In Exim Bank. (2019), Project exports from India: Strategy for reenergizing and reorienting, Occasional Paper No. 193, *Export-Import Bank of India*, p. 36.

Table 10.2 Project exports from India

	2013–14	2014–15	2015–16	2016–17	2017–18 (Apr–May)
US$ million	4,436.19	5,493.04	5,014.9	8,146.7	801.53
Rs. crore	26,197.95	33,581.38	33,056.78	54,024.18	5,341

Source: Project exports promotion council of India. Retrieved from: https://www.projectexports.com/allpagecms/view/project-exports-performance Accessed on 11 February 2020.

Table 10.3 Trends in bidding for overseas projects by Indian exporters

Year	No. of overseas projects bid for	No. of overseas projects won	Overall bid conversion ratio (%)	Average of bid conversion ratio (%)
2014–15	167	37	22.2	37
2015–16	178	39	21.9	15.3
2016–17	268	74	27.6	46.6
2017–18	292	75	25.7	38.5
2018–19	242	96	39.7	45.1

Source: Exim Bank. (2019) Project exports from India: Strategy for reenergizing and reorienting, Occasional Paper No. 193, *Export-Import Bank of India*, Mumbai, India, p. 35.

Note
[a] Sample size is 13 companies.

cheap labour due to the increasing level of mechanisation and government regulations in certain countries prohibiting the import of manpower; (vii) foreign competitors get cheap finance with less guarantee costs compared to Indian project exporters; (viii) non-competitive prices; (ix) deviation from specifications; (x) procedural non-compliance such as timely submission of bid bonds with confirmation from local banks and non-submission of complete bid documentation requirements; (xi) lower rating received by Indian bidders vis-à-vis foreign bidders in prequalification; (xii) long delivery schedules; (xiii) cancellation and re-tendering of bids; (xiv) clients preference for known and established exporters; (xv) influence of local agent; and (xvi) political relations with the host country. In addition, the bulk of infrastructural work needed for development in Middle East countries, one of the major markets, is nearly at an end. Since these countries gained sufficient knowledge in the design and execution of large projects, they have started awarding major projects to their local contractors. Another reason is that the current types of projects require more expertise which many contractors from India are yet to develop.

10.3 Complexities of international contracting

What makes international business strategies different from domestic business strategies is the differences in the socioeconomic and political environment. Competing in the global arena exposes a company to different cultures, languages and economic, legal and political systems. Major economic risks include exchange rate variation, interest rate variation and inflation. Other risks are technical failure of the process and risk of delay in completion. Financial risk is the major risk faced by global competitors as it relates to foreign exchange volatility. Until 1973, markets operated with fixed exchange rates. Subsequently, the rates were allowed to float, at times causing sudden and extreme changes in the market. Currency shifts have become one of the major factors in determining a company's ability to compete worldwide. In 1990, the Brazilian government imposed strict economic policies to control inflation. This effectively paralysed the market for multinationals in that country. The Gulf crisis resulted in economic loss to many multinationals. A list of risks in international construction project exports is provided in Figure 10.3. The identification, assessment and valuation of risks are often difficult in the case of international contracts. Specific problems in the international construction contracting business are explained in the current section.

10.3.1 Political and legal differences

Political risks largely cover any action that might be taken by host governments that impairs project economics. These would include expropriation; inconvertibility; the imposition of new taxes or the withdrawal of previously agreed subsidies; the implementation of tariffs or other barriers affecting the project's ability to source needed equipment or supplies on international markets; and unilateral changes to, or arbitration of, key contract provisions. The failure of local authorities to grant adequate or timely price increases for, say, infrastructure projects, could be just as devastating to project economics as are the more familiar instances of political risk. Political risks also include severe limitations placed on the repatriation of profits to the home country or artificial exchange rates or high withholding taxes imposed on repatriated funds or temporary freezing of the funds of a subsidiary company by a foreign control authority. 'Complex legal hoops, prolonged submission procedures, and changes in the legal policies of the government also cause difficulties for owners, and contractors in the project implementation process as well as in payment and final settlement' (Nguyen and Nguyen, 2020).

The political environment of a country affects its government's actions and regulative framework which can cause variations in contract conditions and consequent amendments to the scope of a contract. Haendel (1979) defines political risk as 'Government interference, through specific acts or events, with the conduct of business or in terms of overall government policy towards foreign investors'. Ashley and Bonner (1987) defined political risk as 'the occurrence of politically motivated events that affect the multinational contractor's ability to operate effectively in the host country'. As per Butler and Joaquin (1998a) 'political risk is the risk that a sovereign host government will unexpectedly change the

International environment
- international commercial dispute or trade war
- international economic instability
- economic sanctions or embargoes
- extent of involvement in international organizations
- extent of interaction between host and home countries
- good neighbor relations with surrounding countries
- extent of regional and international cooperation
- degree of economic dependence
- level of debt outstanding
- unfavorable trade balance
- level of protectionism
- degree of free trade

Industry-specific factors
- level of industry competition
- congruence with national economic goals
- extent of natural resource seeking
- level of industry maturity
- rate of return in the industry
- role of industry in national economy
- availability of alternative suppliers
- degree of industry concentration

Project-specific factors
- project desirability to the host country
- level of public opposition to the project
- size of project
- project duration
- prioritized location of the project
- technical and managerial complexity
- sufficient external funding of project
- advantageous conditions of contract

Firm-specific factors
- organization culture differences
- strong relationship with governments
- good relations with power groups
- degree of acceptance of the firm
- degree of localization
- size of affiliate firm or firm
- leverage ratio of the affiliate firm
- ownership share of the affiliate firm
- firm's degree of internationalization
- contribution of the firm to the local economy
- involvement of local business interests
- experiential knowledge of political risks
- level of technology and technology transfer
- dependence on the local market
- level of firm diversification
- extent of firm's market dominance
- misconduct of contractors

Host country
- degree of government stability
- extent of popular support for the government
- degree of consensus in policy making
- level of democracy
- recent or impending independence
- forthcoming elections
- restraints to retaining power
- level of governmental control in economy
- factional conflict
- consistency in reform progress
- policy uncertainty
- different policies in local and central government
- poor public decision-making process
- degree of red tape
- poor enforcement mechanisms
- vague laws and regulations
- ineffective legal system
- adverse legal rulings
- judicial unpredictability
- independence of the judiciary and the executive
- distribution of income and wealth
- cultural differences
- religious and ethnic tensions
- racism and xenophobia
- unfavorable attitude toward foreign businesses
- negative media reports
- urbanization pace
- population density
- communication and language barriers
- reliability and creditworthiness of entities
- gross domestic product (GDP) growth rates
- per capita income
- currency instability
- inflation
- unemployment rate
- fiscal and monetary expansion
- exchange rate volatility
- debt in default or rescheduled
- credit ratings
- access to capital markets

Figure 10.3 Risks in international construction project exports.

"rules of the game" under which businesses operate'. Socio-political instability has both direct and indirect negative consequences on overseas construction projects (Deng and Low, 2014). The risk factors in international construction compiled by Deng and Low are given in Figure 10.3. A host government's policy changes are a major source of political risk (Butler and Joaquin, 1998b). Discrimination against foreign investors has often been triggered by deteriorating political relations between governments (Casson and Lopes, 2013).

The political risk associated with a construction project can be defined as the returns a project investment could suffer as a consequence of government actions which deny the rights of a construction contractor. Political risks are associated with government actions which deny or restrict the right of an investor/owner: (i) to use or benefit from his/her assets; or (ii) which reduce the value of the firm. Political risks include war, revolutions, government seizure of property and actions to restrict the movement of profits or other revenues from within a country.[2] Political risks include hostilities with a neighbouring country or region; fragmented political structure; fractionalisation by language, ethnic and regional groups; restraints to retaining power; imposition of augmented taxes and tariffs relating to the project; mindset, including expropriation or nationalism, corruption and dishonesty; social conditions (population density and distribution of wealth); imposition of foreign exchange controls limiting the transfer of funds outside the host country; limiting availability of foreign exchange; failure of government departments to grant necessary permits; social conflicts such as demonstrations, strikes and street violence; instability because of non-constitutional changes; nationalisation; changes in law such as imposing new safety, health/environmental standards and regulations and changes in labour laws; attitude of opposition group; probability of opposition group taking over; attitude towards foreign investment; quality of government management; anti-private sector influence; relationship with the company's home government; relationship with neighbouring countries; and religion. Some of these factors are discussed in this section.

The political and legal frameworks of overseas markets are different from those of the home country market. Complexities generally increase when a company starts doing business in more countries. Moreover, the political and legal framework is not the same in all states or regions of many home markets. For instance, the political and legal environment is not exactly the same in all states in India. The reasons for political risk are considered as the likelihood of unfavourable outcomes produced by political events such as wars, regime changes, revolutions, political violence, riots, insurrections, terrorist attacks, coups, etc., and government activities such as expropriations, unfair compensations, foreign exchange restrictions, illegal interferences, changes in laws, corruption, poor enforcement of contracts and labour restrictions (Aliber, 1975; Zhuang, Ritchie and Zhang, 1998).

A government's actions can change the outcome of a private business venture's decisions. Robock (1971) classified political risk into macro and micro risk categories. Macro risks are those politically motivated events that impact foreign enterprises in a general sense while micro risks are factors impacting a specific firm or business sector. 'Two of the more extreme examples of these risks are expropriation (taking over of property from its owner for public use or benefit) and nationalization'.

According to Lloyd (1974), political actions affect business decisions through political and social stability and controls on the flow of private capital. Factors influencing political and social stability are: (i) strong internal factions (religious/racial/language/tribal or economic); (ii) social unrest and disorder; (iii) recent or impending independence; (iv) new international alliances and relations with neighbouring countries; (v) forthcoming elections; (vi) extreme programmes; (vii) vested interests of local business groups; and (viii) proximity to armed conflict. According to Lloyd, governments under economic pressure adopt many strategies to control the flow of private capital: (i) unattractive exchange control and currency regulations; (ii) restrictions on the registration of foreign companies; (iii) restrictions on foreign management; (iv) restrictions on local borrowing; (v) expropriation and nationalisation; (vi) unsatisfactory tax laws and regulations; (vii) restrictions on imports and exports; (viii) limitation on the types or areas of activity; and (xi) lack of clarity of the local corporation laws. Exchange control regulations can adversely affect business through controls on: (i) repatriation of profit; (ii) remittance of dividends earned; and (iii) foreign exchange available for the necessary imports.

Lloyd's study further states that governments or governmental agencies can control the performance of overseas firms in numerous ways: (i) confiscation (i.e. loss of assets without compensation); (ii) expropriation with compensation (i.e., loss of freedom to operate); (iii) operational restrictions through control of market share, product, characteristics and employment policies; (iv) loss of transfer freedom for goods, personnel or finance; (v) breaches, or unilateral revisions, in contracts and agreements, discrimination through taxes and compulsory subcontracting; (vi) damage to property from riots, revolutions and wars.

10.3.1.1 Expropriation, nationalisation and confiscation

Expropriation arises when a government seizes the assets of a business or corporation. 'The power of the government to take private property for public use without the consent of the owners through a process is called expropriation' (Ndjovu, 2015). Expropriation refers to the forced disinvestment of the equity ownership of a foreign direct investor (Kobrin, 1980, 1984; Minor, 1994). The government of the host country monopolises its coercive power to define and enforce property rights. The law prevailing internationally does not define the compensation for expropriation. According to the United Nations Conference on Trade and Development (UNCTAD, 2012): states have a sovereign right under international law to take property held by nationals or aliens through nationalization or expropriation for economic, political, social or other reasons. In order to be lawful, the exercise of this sovereign right requires, under international law, that the following conditions be met: (a) Property has to be taken for a public purpose; (b) On a non-discriminatory basis; (c) In accordance with due process of law; (d) Accompanied by compensation.

Expropriation includes changes in the contract terms and contract revocation under which a government ends the contractual agreement at a given phase of the project and assumes ownership of whatever investment the foreign contractor has made in it. Contractors working in equipment-intensive projects are particularly exposed to this kind of risk, because they often invest millions of dollars in equipment. Similarly, contractors participating in the operation of a project as a way of receiving their payment are vulnerable to sudden and unexpected expropriation.

Twenty-first-century expropriations show the continuing relevance of these kinds of political risk (Jensen et al., 2012). For instance, the Namibian government issued expropriation orders to 18 farmers in 2005. In 2006, Venezuela seized oil fields from France's Total and Italy's Eni. Bolivia nationalised its natural gas industry in 2005.[3]

Blaszczyk's (2008) study on DuPont's Iranian project describes how they tried to reduce the political risk. The political and religious revolution of 1979 led to alterations in the strategies of multinational corporations (MNCs) from other countries operating in Iran. DuPont deliberately formed a joint venture (JV) with the Behshahr Industrial Group of Tehran to defend the firm from hostile intervention. However, the DuPont plant was eventually nationalised. Consequently, DuPont filed suit in the United States against Polyacryl Iran Corporation (PIC) and the government of Iran and in 1984 obtained US$42 million in compensation.

10.3.1.2 War and riot

Firms involved in international construction must be alert to probable civil disorder. Riots or war-like circumstances might induce the firm to discontinue works at the site resulting in increased costs and interrupting works elsewhere. The contractor can bargain for contract terms which deal specifically with civil unrest. Clause 19 of the International Federation of Consulting Engineers (FIDIC) conditions of contract (2006) entitles a contractor to receive extra costs. Sub-Clause 19.1 (definition of force majeure) states that force majeure may include war, whether declared or not or invasion, or an action of foreign enemy, terrorism, revolution or civil war, riots, strikes or lockout by persons other than contractor's or sub-contractors personnel, and munitions of war or explosives radiation other than those caused by contractor's use of them. It does cover natural catastrophes, such as earthquake, hurricane, typhoon or volcanic activity. In such situations, the contractor shall give notice of suffering delay and incurring costs due to force majeure and if accepted by the engineer, as falling under Sub-Clause 19.1 (definition of force majeure), shall be given an extension of time.

10.3.2 Political risk management

Traditional management options used to mitigate potential political risk include the following approaches (Figure 10.4).

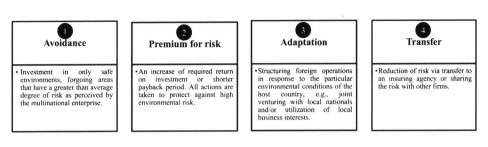

Figure 10.4 Some of the traditional management options used to mitigate potential political risk.

10.3.2.1 Strategies for political risk management

A review of the literature reveals that several studies (Bonner, 1981; Ashley and Bonner, 1987; Liu, Zhao and Li, 2016; Chapman, 2001; Chang, Deng, Zuo and Yuan, 2018; Deng and Low, 2013; Pallant, 2010) on international construction projects' political risk management identified many possible strategies. A compilation of strategies from those studies shows that making a higher tender offer; conducting market research; buying risk insurance; adopting optimal contracts; implementing a localisation strategy; avoiding misconduct; adopting closed management of the construction site; supporting environmental protection; abiding by the traditional local culture; making contingency plans; obtaining the corresponding guarantee; implementing an emergency plan; forming joint ventures with local contractors; sending staff on training programmes; settling disputes through renegotiation; choosing suitable projects; building proper relations with host governments; creating links with local businesses; maintaining good relations with powerful groups; changing operation strategies; controlling core and critical technology; choosing a suitable entry mode; employing capable local partners; building a reputation; and allocating extra funds and maintaining good relations with the public can be possible strategies for international construction projects' political risk management.

10.3.2.2 Political risk insurance

Political risk insurance is an instrument for businesses to mitigate and manage risks occurring from the adverse actions or inactions of governments and their agencies. It helps international construction contractors and lenders mitigate risk through insurance against unfavourable government actions or war, civil strife, terrorism, currency transfer restriction and inconvertibility, expropriation and breach of contract. Political risks covered by some of the organisations are presented in Table 10.4.

10.3.2.3 Political risks covered

Any company investing abroad should take full advantage of the subsidised forms of insurance that provide cover at a reasonable cost against the many kinds of political and country risks that are encountered in politically unstable countries abroad.

Political risk can be prevented to a certain degree, by inviting the participation of the local contractors as subcontractors and also the local workforce on the project. This is a classical hedge against government interference in a foreign investment project. Another way of hedging the political risk is to raise funds for the project from the local capital market. Many governments will allow foreign investment locally only if a substantial portion of the funds invested in the venture are brought in from abroad.

Table 10.4 Political risks covered by some of the organisations

Agency	Currency inconvertibility/ transfer restrictions	Confiscation, expropriation, nationalisation	Political violence/war	Default on obligations (loans, arbitral claims, contractual, etc.)	Terrorism	Other risks covered
OECD						
EFIC (Australia)	Yes	Yes	Yes	Not found	Not found	Cover can also be provided for other political events such as selective discrimination and arbitral award default
OEKB (Austria)	Yes	Yes	Yes	Not found	–	–
ONDD (Belgium)	Yes	Yes	Yes	Yes	Not found	–
EDC (Canada)	Yes	Yes	Yes	Yes	Not found	–
COFACE (France)	Yes	Yes	Yes	–	Not found	Changes in host country legislation; denial of justice in countries with which France has no bilateral investment agreement
PWC (Germany)	Yes	Yes	Yes	Yes	Yes (included in war risks)	–
SACE (Italy)	Yes	Yes	Yes	Yes	Yes (including sabotage)	Embargo; force majeure including natural disasters; exchange rate fluctuation due to laws adopted by the host country
ATRADIUS DSB (Netherlands)	Yes	Yes	Yes	Yes	Not found	Some commercial risks also covered; force majeure including natural disasters; default on local authorities' obligations
NEXI (Japan)	Yes	Yes	Yes	Yes	Not found	Force majeure
KEIC (Korea)	Yes	Yes	Yes	Yes	Not found	–
Türk Eximbank (Turkey)	–	–	–	–	Not found	–
ECGD (UK)	Yes	Yes	Yes	Yes	Not found	–
OPIC (US)	Yes	Yes	Yes	Yes	Yes (as a stand-alone policy)	Coverage of project-specific risks
Non-OECD						

ECGC (India)	Yes, for works of a capital nature abroad only	Yes	Yes	Not found	Not found	For construction works abroad: exchange rate fluctuation, failure of the employer to pay the amounts due
ECIC (South Africa)	Yes	Yes	Yes	Yes, for works of a capital nature abroad only	Not found	For works of a capital nature abroad: insolvency
SINOSURE (China)	Yes	Yes	Yes	Yes	Not found	–

Source: Gordon, Kathryn. (2008) Investment guarantees and political risk insurance: Institutions, incentives and development. *OECD Investment Policy Perspectives 2008*, ISBN 978-92-64-05683-1.

Notes

[a] The risks included in the table are those found on the website of the political risk insurers. In some cases, the risks shown are available only for certain sectors, projects, activities or asset types. In other cases, it can only be purchased on a stand-alone and/or bespoke (tailor-made contract) basis.

[b] Includes politically motivated violence: revolutions, rebellions, civil disturbances, war, etc.

10.3.3 *Cultural differences and international construction*

The construction industry is a project-based industry; therefore, one needs to have a greater understanding of the factors that contribute to project-level cultural issues and their impact on a project. Cultural diversity is one of the complex dilemmas in international construction. Culture is originated and regulated by society where the human being is born and brought up. Many sociologists have defined culture. Kroeber and Kluckhohn (1952) define culture as: Culture consists of patterns, explicit and implicit, of and for behavior acquired and transmitted by symbols, constituting the distinctive achievement of human groups, including their embodiment in artifacts; the essential core of culture consists of traditional (i.e., historically derived and selected) ideas and especially their attached values; culture systems may, on the one hand, be considered as products of action, on the other hand as conditioning elements of further action. Wegley (1997) defines culture as: Each human society has a body of norms governing behavior and other knowledge to which an individual is socialized, or encultur-ated, beginning at birth. … Human culture in the technological sense includes the insignificant and mundane behavior traits of everyday life, such as food habits, as well as the refined arts of a society. 'Culture is comparable to the programming of an individual's mind'. Culture determines the attitudes, codes of conduct and expectations that guide people's behaviour and the way that they interpret messages (Loosemore and Al Muslmani, 1999).

Hofstede's (1983) study is the most widely referenced study internationally on culture. His study identified 'four cultural dimensions' that rate culture, i.e. (i) individualism versus collectivism; (ii) power distance; (iii) uncertainty avoidance; and (iv) masculinity versus femininity. Individualism versus collectivism is generally used to describe general dissimilarities between Western views of self and Eastern viewpoints on the concept of self. Western individualist cultures have a tendency to conceptualise the self as a comparatively independent and autonomous entity whereas Eastern collectivist cultures view the self not as detached from the surrounding context (Wang, 2003). Power distance deals with the degree of power each person exerts or can exert over other persons and the power in this context indicates the extent to which a person is capable of influencing other people's ideas and behaviours. Uncertainty avoidance assesses the extent to which members of a society 'feel either uncomfortable or comfortable in unstructured situations. Unstructured situations are novel, unknown, surprising, and different from usual. The basic problem involved is the degree to which a society tries to control the uncontrollable' (Hofstede, 2001). Masculinity versus femininity is related to the division of emotional roles between women and men. Hofstede later added two more dimensions, i.e. long-term versus short-term orientation and indulgence versus restraint. The former speaks about 'the choice of focus for people's efforts: the future or the present and past' and the latter 'the gratification versus control of basic human desires related to enjoying life' (Hofstede and Minkov, 2010). His studies show that each country has its own unique cultural dimension.

Awareness of the cultural factors of a host country can aid project planning. The efficiency of project execution can be improved if project authorities can incorporate cultural factors. For instance, fasting for periods of time is a religious practice in many Arab countries, similar to Christians in Western countries (Alkharmany, 2017). Therefore, the productivity of the workers also depends on cultural factors such as religion. Proper planning can ensure that any delays in executing the work due to a reduction in productivity are eliminated during such occasions. Many researchers used 'Hofstede six dimension model' to arrive at a culture index of countries.

An expatriate needs to adjust his/her behaviour to the norms and rules of the host nation (Konanahalli et al., 2012). Failure to adjust to the culture of the host nation will result in incomplete tenure of job (Harzing, 1995), poor performance and poor job satisfaction

(Naumann, 1993). Conversely, successful adjustment will bring about the integration of some elements of the expatriate's culture with that of the host country (Kim and Ruben, 1988). Therefore, well-adjusted expatriates can perform successfully. Black and Stephens (1989), Black (1988) and Black, Mendenhall and Oddou (1991) identified different aspects of adjustment: (i) interaction adjustment; (ii) work adjustment; and (iii) general adjustment. The interaction adjustment as propagated by Black and Stephens deals with the comfort levels when dealing or interacting with host country nationals at work and in non-work situations. The work adjustment involves adapting oneself to the new job tasks, roles and environment. The general adjustment deals with overall adjustment to living in a foreign land and adjusting to its culture and comprises factors such as housing conditions, healthcare, cost of living, etc.[4]

Many studies have found that behaviours, cultural profiles and the way people organise their time differ in different countries. Persons with a polychronic attitude do two or more things simultaneously whereas a monochronic person does one thing at a time. It appears that North American and Northern and Central European people are monochronic and Mediterranean, South American, African and Asian people are polychronic in nature. 'These two extremes in behavior with regard to time can have important implications in projects when monochronic and polychronic people work together' (Duranti and Di Prata, 2009).

The study of Pheng and Leong (2000) shows that strong personal relationships, a high social conscience more towards collectivism, a focus on social status and a polychronic culture prevail in China. There are also studies on project culture. A study by Zuo, Zillante and Coffey (2009) attempted to find out Chinese contractors' perceptions of the impact of project culture on the performance of construction projects and discovered a significant contribution of project culture to project outcomes. Another study by Anderson (2003), conducted in Norway, shows that a stronger task orientation improves the chances of staying within the project budget.

Zwikael, Shimizu and Globerson (2005) studied the differences in project management style between the Japanese and the Israelis and found significant cultural differences between the two countries. Israeli project managers are more focused on performing "scope" and "time" management processes was given more focus by Israelis whereas "formal communications and cost management" are more frequently used by Japanese project managers. It is interesting to note that although Israeli project managers pay more attention to time planning, their projects result in higher schedule overruns, as compared to projects performed in Japan. Japanese managers make more use of the communication planning process since teamwork is highly regarded in Japan, and this cannot be practiced without an effective communication system.

International contractors are expected to gain knowledge of the history, culture, meeting and greeting corporate culture, acceptable dress codes, customs, etc., of the countries where they are planning to take up projects because it varies from country to country (Dimatteo, 2017).

10.3.4 *Economic environment*

Often, the economic environment determines the survival and success of a business, especially exports of projects. The economic factors prevailing in a host country have immense influence on a project's execution and its success. These economic factors include gross domestic product (GDP) and its growth; per capita income; inflation; markets for goods and services; availability of capital in the host country; foreign exchange reserves of the host country; foreign trade and balance payments; and the economic policies of the host country. Fiscal and monetary policies, foreign investment policies, export import policies, government debt as a percentage of GDP and five-year plans have a direct impact on the success of any business. Fiscal and monetary policies provide details on government policy regarding public expenditure, taxation, supply of credit to a business, etc. Policies regulating the inflow of foreign investment in various sectors are specified in foreign investment policies. Similarly, barriers and controls on exports and imports are given in export import policies. High rates of inflation, by and large, result in constraints on international contractors as they augment the different costs of a project, for instance, the purchase of construction materials and machinery and the payment of wages and salaries to employees. Therefore, understanding the economic environment of different countries will help to predict the economic trends and events that affect the future performance of the project business.

10.3.4.1 Exchange rate fluctuations

Exchange rate fluctuations affect the operating profits of international contractors. If country 'A' has higher inflation, then the currency of country 'A' will depreciate in comparison to the currency of another country, say country 'B'. Unstable exchange rates influence the competitiveness of international contractors when procuring materials and labour from different countries. If the deutsche mark strengthens 4% against the dollar and the German inflation rate is 1%, a U.S. exporter to a German market served primarily by German producers would see its dollar price rise 5%. If, however, the inflation rate in the United States is 4%, or 3% higher than the German inflation rate, the operating margin of the U.S. producer will rise by only one percentage point (Lessard and John Lightstone, 1986). The change in relative competitiveness depends on changes in the real exchange rate.

10.3.5 Other risks

Exchange rate fluctuations are part of operating risks. Other operating risks include tax increases, imposition of import restrictions, rent seeking through taxes of or perceived excess profits, local ownership and prohibition on the repatriation of profits. Other risks include the unwillingness of a host country to make a good faith effort to meet its financial obligations; creeping expropriation (changes in contract terms, etc.); variation in working habits affecting the productivity of the labour force; lack of experience in specific types of projects; local laws and customs placing constraints on the contractor's activities; delays in progress payments due to bureaucratic procedures or the inability of the owner to meet his obligations; delays in labour mobilisation; delays resulting from differences and variations in the standards of materials ordered from international markets; delays in materials delivery due to transportation problems; materials shortages; owner inefficiency and delays in port.

The following is a list of other business risks normally faced by construction contracting firms in their overseas projects:

   i.   Inaccurate estimate of costs of the project by the contractor.
   ii. Using heavy front-loading to finish other jobs.
   iii. Inaccurate cash flow projections.
   iv. Inadequate time scheduling and delays.
   v.   Failure to predict changes in interest rates in association with future interest payments as in the case of Euro credits.

vi. Unwillingness of a host country to make a good faith effort to meet its financial obligations.

vii. Variation in working habits affecting the productivity of the labour force.

viii. Lack of experience in specific types of projects, including the use of front-end technology in foreign areas.

ix. Assumption of the role of the developer by the contractor which assumes important financial or other responsibilities.

x. Local laws and customs placing constraints on the contractor's activities.

xi. Contractor liability for the design of the project, especially for its structural reliability in addition to construction risks.

Delays in

a. progress payments due to bureaucratic procedure or the inability of the owner to meet his obligations;

b. labour mobilisation;

c. materials delivery due to transportation problems, materials shortages, owner inefficiency, port, etc.;

d. delays resulting from differences and variations in the standards of materials ordered from international markets.

Although the above-listed factors do not cover all business risks facing a firm in the international project export scenario, they are indicative of the number of diverse circumstances that can affect its financial position.

Contractors not acquainted with a new country and its construction environment can cause complications, including accuracy in assessing the viability of the project; client's incompetent supervisors; alterations in project scope; contractors trying to maximise profit by overlooking project specifications; using inferior materials for more profit because of a fixed price contract; inadequate capacity or experience of the staff on the project; lack of consistent support from subcontractors; inappropriate allocation of resources to the project; and poor communication between the client and the stakeholders in the project implementation process.

Many problems start from the bidding itself. The lack of transparency in bidding (breakdown of packages, projects not yet approved but with proceeded bidding), conflicting or incomplete contracts, improper contracts signing, unfair allocation of responsibilities and risks, inappropriate contract quotes related to the bidding process and project evaluation. The lack of transparency and fraud in bidding leads to the selection of incompetent and inexperienced contractors who cannot ensure the design and construction work in accordance with specified requirements, causing problems during construction. In addition, a negative situation in bidding and collusion in bidding increases total investment cost. Lack of experience in contract drafting or loose terms in the contract can cause controversies during project implementation. Especially if potential risks (fluctuation, global instability, etc.) are not anticipated and are tightly bound into the contract, the project's progress will be prolonged and will often lead to excess contract estimates.[5]

10.4 Bidding for international projects

10.4.1 *Factors in bid decision*

The decision to bid is based on the owner's credit worthiness, the availability of funds, the reputation of the funding agency, a detailed financial examination of the specific project and the project's overall consistency with the long-term corporate goals.

10.4.2 *Domestic bidder price preference*

If so indicated in the particular instructions to applicants (PITA), a 7½ per cent margin of price preference for qualifying domestic bidders shall apply in a bid evaluation as per the World Bank. Based on information submitted by bidders and available at the time of notification, the client will inform prequalified bidders of their eligibility to qualify for the domestic bidder price preference (subject to subsequent confirmation at bid evaluation), in accordance with the following.

A domestic bidder is the bidder who meets the following criteria:

i.   for an individual firm:
    a.   is registered in the country of the borrower;
    b.   has more than 50 per cent ownership by nationals of the country of the borrower;
    c.   does not subcontract more than 10 per cent of the contract price, excluding provisional sums, to foreign contractors.
ii. for a joint venture between domestic firms:
    a.   individual member firms shall satisfy (i)(a) and (i)(b) above;
    b.   the joint venture shall be registered in the country of the borrower;
    c.   the joint venture shall not subcontract more than 10 per cent of the contract price, excluding provisional sums, to foreign firms.

10.5 Prequalification

Tenders for international projects are advertised in newspapers worldwide. Competitive bidding on an international basis is the most common method for choosing among interested contractors, particularly for public projects. Prequalification is helpful for the owner because he is not pressed to accept a low bid from an unreliable contractor. It is also good for the contractor because he will not spend time and money preparing his bid unless he qualifies. In some cases, the owner decides to prequalify all bidders on the basis of documents establishing the firm's expertise and capacity.

10.5.1 *Screening of contractors on their project execution capability*

To ensure the project execution capabilities of contractors, a working group examines the experience, competence and capability of the applicant's prime contracting firm and its main subcontractors in the execution of large-value projects. In order to ensure the capabilities of contractors, the working group considers proposals from contractors meeting the criteria on turnover, net worth and work experience by the screening committee as elaborated in Figure 10.5. The criteria for selection vary for different categories of contractors, i.e. prime contractor, subcontractor to foreign prime contractor and subcontractor to Indian prime contractor.

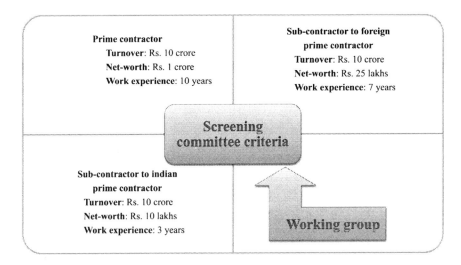

Figure 10.5 Screening committee criteria.

10.6 Project identification process

This is the most important stage in deciding whether to participate in a tender. At this stage, the following information is required.

10.6.1 Tender information

This is related to the type, nature, size, location, source of funding, currency of payment, etc., of a project. This information is available from the following sources/publications:

   i. MEED (Middle East Economic Digest) for projects coming up in Middle East;
  ii. AED (Africa Economic Digest) for projects coming up in Africa;
 iii. Development Business, a United Nations publication giving information about projects aided by the World Bank, African Development Bank, Asian Development Bank, etc.;
 iv. Project Export Promotion Council of India (PEPCI) Newsletters;
  v. Engineering Export Promotion Council (EEPC) Newsletter;
 vi. Indian Mission Abroad; and
 vii. Internet.

10.6.2 Economic information

In deciding on a country to do business with, important information about the country is required, for example:

  i. Economy of the country
 ii. Basic economic parameters:
    • Per capita income
    • Direction of trade

- Balance of trade
- Balance of payments

iii. Political situation of the country

iv. Economic and political history

v. Relationship with India

vi. Foreign exchange restrictions/currency of country

vii. Rules and regulations for foreign companies

viii. Import–export regulations

ix. Is the country free from excessive external debts?

x. Future plans of the country

xi. At least for the duration of the proposed project whether the foreign exchange resources, present or anticipated, would be adequate to cover the payments in foreign exchange due to the project exporter – are the facilities for converting the local currency into convertible currencies smooth and effective?

xii. Is the rate of inflation high?

xiii. Particulars of the development plans of the country – outlining the scope for foreign contractors sector-wise.

xiv. Do the local banks provide loans to foreign contractors?

xv. What is the rate of interest applicable to foreign contractors?

xvi. Extent of project exporters country's involvement in the host country's economy in trade, manufacturing, technical collaboration, etc., and the experience of the exporters companies already operating in that country.

10.6.3 *Local contractors*

The structure of a major international construction project is usually quite complex, often involving various different international companies. Many a times employer may be only from host country and the prime contractor, nominated subcontractors, suppliers, consultants, financiers and labourers may be from different countries. Therefore, jurisdictional risks can be very significant in international projects. Tying-up with local contractors and employing the natives can reduce the probability of being treated with discrimination and opposition by the host country government, political parties and local personnel. The following is important in this context.

a. Whether any local contractors are available and if so, how active are they? Is it possible, if necessary, to join with them on a mutually acceptable basis so as to improve the Indian exporter's image/effectiveness in that country?

b. Is there any law stipulating that prequalification (PQ) applications/tenders should be submitted through a local partner that should be a member of a consortium or a joint venture?

c. In the case of foreign contractors, does any nation(s) enjoy any favourable treatment?

10.6.4 *Securing further contracts*

The prospects for securing further contracts in a country are important criteria in getting permissions for executing projects abroad. This is relevant to the operations of the exporter in as much as the equipment and other resources available on completion of the project under reference may be made use of if more projects come later.

10.6.5 *Other information*

i. Until the exporter's application for prequalification is accepted, what sort of representation should Indian exporters have in that country? Should it be in the form of posting a resident representative or through a local agent or through a representative of financial institutions present there?

ii. What is the experience in the country regarding payment delays to contractors? Are they reasonably prompt and is there scope for remedial action in case of delays or non-payment?

iii. What is the experience of the country in the release of various bank guarantees, which might be given by the project exporters?

iv. Are these bonds returned promptly after the event takes place or are they duly cancelled or are they kept by the authorities on some pretext or other?

v. What is the position regarding obtaining visas for expatriates required for execution of the project?

vi. Is there any regulation making it mandatory to employ a certain minimum number of local labourers?

vii. Do the local regulations provide for:
    a. easy repatriation of expatriates' salaries?
    b. repatriation of the permissible percentage provided in the contract of the project receipts?

viii. What are the formalities needed for winding up the establishment on completion of the project?

ix. Are these formalities very stiff, requiring the continued presence of some people for a long period?

This information is available in the profiles (development plan of the country) of each country and is maintained by Exim Bank and the Chamber of Commerce or can be obtained on request from Indian Missions abroad. Although all the above information is important in deciding to do business in a country, the most important of these is probably to know the future plans of the country as expressed by its own planning institutions. Exim Bank/PEPCI/ Indian Missions in foreign countries maintain and update these details, which companies can request for their own use.

10.7 Procedure after receipt of tender documents

As soon as a corporate office receives the tender documents, they are studied in detail and a detailed note is prepared which contains the following information about the tender:

i. Scope of work and nature of job.
ii. Source and extent of funding.
iii. Whether turnkey, lump sum or item rates tender.
iv. Details about clients/consultant for the project.
v. Currency of payment.
vi. Payment terms.
vii. Whether mobilisation advance payable or not.
viii. Retention money, if any.
ix. Requirement of bank guarantees for:

- Bid bond
- Performance guarantee
- Mobilisation advance
- Retention money
- Secured advance
- Maintenance period

x. Bid bond amount and its validity.
xi. Time of completion.
xii. Mobilisation period.
xiii. Maintenance period.
xiv. Liquidated damages payable in case of delay.
xv. Any specific requirements with respect to the deployment of:
- Manpower
- Plant and Machinery
- Materials

xvi. Provision for payment of taxes and duties.
xvii. If the import of manpower, material, plant and machinery is allowed from outside the country.
xviii. Arbitration clause.
xix. Date and place of submission offer.

In order to price the project, a site survey also needs to be conducted at the overseas project site by a group nominated by the company. Once it is decided to participate in the tender, the tendering process starts.

10.8 Tendering process

Often, limited time is available for submission of tenders (about 45 days, even 90 days are given) and generally the bidder runs against time to complete the tender workings. In the case of turnkey contracts, project design is done either in-house or by engaging outside consultants. The items and materials required for the works are identified. Some specialised items specified by clients are only available in third countries (neither in India nor in the country of the project) and documents are required to be sent to various outside parties for their offers.

10.8.1 Site survey

A team composed of an engineer and a finance representative is sent to the country of the project for about 5–7 days to collect the following local information .

- geographical information;
- site of works (location, approachability, etc.);
- local rules and regulations;
- banking details and cost of finance;
- availability of:
  - materials, materials cost, transportation cost, etc.
  - labour, labour rates and visa problems
  - equipment, cost/hiring cost
  - markets and transportation cost
  - communication facilities and cost

- subcontractors
- power and cost
- water and cost
- cost of fuel;
- Cost of setting up offices;
- assessment of mobilisation;
- assessment of completion time;
- disposal of plant and machinery after work;
- interpreters and their availability.

The team collects all the local information including rates, makes contact with local parties, clients and the Indian Mission and collects as the information required to frame the offer. The team keeps in touch with Exim Bank's head office and passes on any important or urgent information during their stay. However, in the case of countries where the contracting firm has offices, the firm does not send representatives; instead, details will be collected through the contractor's overseas office.

10.8.2 Tender working

After the team returns or on receipt of the site survey report, the tender is worked out. A detailed analysis of each item of work is done based on the rates, information received from overseas suppliers, etc. The project overheads are calculated and after adding profits, etc., the tender value is worked out.

10.8.3 Preparing cost estimates

Pricing is the most important part of tendering and there is potent risk of leakage. After studying the tender documents, various items of work in the bill of quantities (BOQ) are broken down in to 'A', 'B' and 'C' categories. 'A' category items are those items in the BOQ which are vital and will have a tremendous impact on the overall tender cost, consequently they need to be analysed in great detail. Generally 'A' category items contribute 60 per cent of the total cost of the project. 'B' category items are those items that contribute about 30 per cent of the work and 'C' category items may be generally be classified as minor items and contribute the balance cost of the work. This helps decide which items get more time and are analysed in detail.

In the analysis of items, the optimum numbers and types of machinery/equipment along with other resources required for carrying out the work are calculated. The equipment can be hired locally or can be utilised from an existing project or purchased new or reconditioned.

Purchasing new machinery involves a large capital investment and care should be taken to minimise this too, as one should try to maximise the use of the existing plant and equipment. At this stage, efforts are also made to maximise exports from India without compromising on competitiveness. However, a judicious balance of what is to be purchased vis-à-vis existing equipment to be used needs to be competently addressed, without compromising the progress on site or being out priced.

On the basis of the number of effective hours (calculated on the basis of machine efficiency, operator efficiency and type of work to be executed) an piece of equipment has been used, depreciation is generally calculated and is ultimately charged to the item of work or a group of items. Care should be taken not to enhance/decrease the depreciation figure as the same will have adverse effect in calculating the cash profits.

The availability of the materials required to execute the project is an essential feature of pricing, since materials may constitute up to 60 per cent of a project's cost. Thus, materials should be procured from the best manufacturers/dealers to the specifications of the tender. A most important feature, apart from the specifications, is supply/delivery of materials in time because all international projects have to be completed within the specified period and delay in completion of a project will impose a heavy penalty in the form of liquidated damages.

The next step in pricing is to find out the manpower requirements. Generally, staff required for executing the work is charged in the items, and staff required for supervision is charged in the overheads. All these calculations added together, such as the cost of machinery (depreciation spares; repair and maintenance; petroleum, oil, and lubricants; duties etc.), the cost of materials including wastage and duties, the cost of manpower and the cost of subcontracted items required, give the basic cost of the project.

After calculating the basic cost of the project, the next step is to calculate the overheads. The overheads generally include costs towards cover provided by the Export Credit Guarantee Corporation (ECGC), supervision cost, corporate tax, insurances, profit, pre-tender expenses, contingencies, agency commission, bank charges, financing and escalation. Before calculating the final sale price of the bid, it is very important to correctly calculate the foreign and local currency required for the project.

10.9 Overseas contract tendering

A company can start the tendering procedure after the tender value has been worked out and the tender workings are ready. Once the proposal is ready, the contracting company's bankers transmit their counter-guarantee for a bid bond to their correspondent bank in the country of the project. On the basis of this counter-guarantee, the local bank issues the bid bond which is enclosed along with the tender. This bid bond from the local bank is to be issued and collected by the tenderer's representative for submission of tender. For new countries where the contracting firm is not operating, the firm will have to send their representative for submission of tender. However, in the case of a tender in countries where the firm's units are located, the tender for submission is generally sent by post and the firm's local representative collects the bid bond from the local bank in the host country along with the tender.

A favourable reference from a previous client can create more confidence. The reference could be on quality as well as quantity. If sufficient foreign references do not exist, impressive Indian references can be given, taking care of course, to mention the climate and difficult features, etc. encountered locally, to create confidence for parallel situations. Therefore, try to include a favourable reference from a previous client while submitting the tender.

10.10 Post-tender follow-up

As much information as possible should be obtained on a competitor's bid and this starts right from the opening of bids, if it is a public opening. All the technical and commercial advantages of a bid should be protected and qualified in economic terms. Post-tender follow-up is generally done through direct contact with overseas clients by fax, e-mail or through Indian Missions, an agent or the firm's local office.

10.11 Constraints

In international contracting, tendering is the biggest constraint on Indian contractors. The major constraints are presented below.

10.11.1 *Delay in getting information*

There is always a time lag in getting information on overseas project's invitation of tender. It may take up to 10 days. Once the decision has been made to tender for the work, the mission/agent/office is requested to send the documents, which can take a minimum of 7 days even if they are sent by courier or any other means such as e-mail.

10.11.2 *Time lag in collection of information*

To prepare a tender and collect local information, Indian contractors need to send a team to carry out a site survey in the foreign country. Taking into consideration foreign exchange relations and visa requirements, a person needs a minimum of 7–10 days before he can depart. The team returns after 7 days with all the site information. In short, about 1 month is already lost by the time reliable data from the site has been collected.

10.11.3 *Processing time*

To work out a tender, e-mails are sent to overseas parties to get their offer for the specialised materials and equipment required for the project. With the communication facilities available in India, it can take a few days to get a reply from these parties as they also require time to frame their offers. With the tender closing date approaching, the Indian contractors have no option but to work day and night, running against time to complete the offer. Normally, the maximum time required for to prepare a tender is 10–12 weeks and the minimum time is 8–9 weeks.

10.11.4 *Delay due to delayed bid bond*

After preparing the tender, the contractor's bankers will have very little time to transmit their counter-guarantee for the bid bond to the foreign country. So, should a problem or error occur in transmission, there is little or no time to correct it. If the contractor is able to correct the mistake, then getting the bid bond on time may be difficult, resulting in the offer being rejected.

10.11.5 *Delay in getting bank guarantees*

Companies have problems obtaining bank guarantees as the value is large and there are restrictions due to prudential norms. Therefore, two or three banks have to be approached for guarantees. This is an expensive and time-consuming process.

10.12 Clearance of project export proposals

The Government of India has set up a working group through which all export projects must be routed. Therefore, once the contracting firm/exporter has been awarded the project/contract, the exporter should submit an application in the prescribed form to the working group through the bank of the exporter within 15 days of entering into contract. Exporters who have secured orders by undertaking supply contracts on deferred payment terms or those who have secured turnkey/civil construction contracts abroad require the approval/assistance of different institutions such as the Reserve Bank of India (ECD), Exim Bank and ECGC Ltd. besides their own bank as per the Memorandum of Instructions on Project Exports and Service Exports. Without Exim Bank's working group approval, it is not possible to execute overseas construction projects. No agency such as bankers, the Reserve Bank of India (RBI) or ECGC Ltd. can extend any facility to a contractor without the working group's approval.

The working group approves export contracts involving cash or deferred payment and/or issues guarantees where the value of the contract is more than US$100 million (as per Project Exports Manual 2003). Contracts valued up to US$100 million can be cleared directly by the bankers of the project exporting firm with the agreement of Exim Bank, if bid proposals are in conformity with the Reserve Bank of India's guidelines. Currently, post bid clearance is given. The objective of the working group is to maximise exports from India.

The working group is composed of the following members (Figure 10.6):

   i. Exim Bank of India
  ii. Reserve Bank India's Exchange Control Department
 iii. ECGC Ltd.
 iv. Representatives of the Government of India
  v. Project exporter
 vi. Main sub-suppliers
vii. Subcontractors
viii. Other associates
 ix. Concerned authorised dealers

Figure 10.6 Members of the working group.

10.12.1 *Permission required from India*

The facilities/permissions required from India by any contractor for executing an overseas project are:

- Bank guarantees for a bid bond, performance, mobilisation advance, etc.
- Permission for borrowings in foreign currency.
- Permission for opening of bank accounts overseas.
- Permission for opening of site and central office in the country of the project.
- ECGC construction works policy.

Therefore, the essential requirements are opening a bank account (in a third country and in the country where the bid is to be submitted), site offices, liaison offices, guarantees such as a bid bond, performance guarantee, retention money guarantee, mobilisation advance guarantees, sanctioning of overdraft limit, pre-shipment credits and letter of credit limit.

A bid bond is an essential feature of today's tenders. Timely and clear advice through the issuing bank and a fool-proof arrangement for the collection of bid bonds at the site are necessary to avoid disqualification. Photocopies of bid bonds, the issuing bank's letter and counter signatures of competent authorities such as the foreign mission sometimes help if the bid bond does not arrive in time. Unless the wording of the bid bond is specified by the tender, it is wise to give only an overall bid bond in consultation with a banker who will advise on a fair, yet safe, wording.

To enable the working group to consider the proposal, a complete copy of the proposal, regarding the tender, is required to be submitted to each member of the working group at least 3 weeks before the closing date of tender. This proposal should be in the prescribed pro forma and contain complete details about materials, machinery and labour requirements, their source of procurement, cost breakdown, expenses on overheads, cash flow for the project and a host of other information about the project, company and clients.

As per the rules, copies of the Exim Bank proposal are to be submitted to all members of the working group and the Ministry of Finance through the contracting firm's commercial bankers along with their intention to provide bank guarantees for the project. However, to save time, an advance copy of the proposal should be sent to all working group members and the recommendations of the contractor's bankers follow the same.

On receipt of the proposal, Exim Bank goes through the proposal details and invites the party for preliminary discussion. Once Exim Bank and other members are satisfied with the proposal and clarifications, a working group meeting to consider the proposal is held within 7 days of receipt of the application. Following this meeting, during which questions are raised regarding the project, the financial conditions and the capabilities of the firm to execute the project in a timely manner, the proposal is cleared and the commercial bankers are asked to issue the guarantees. If Exim Bank is not satisfied with the company's ability to execute the project or it is not satisfied regarding any of the terms and conditions of the contract or it has any doubts about the client's ability to pay or about the country of the project, it rejects the proposal.

Once Exim Bank clears the proposal, the contracting company's bankers transmit their counter-guarantee for a bid bond to their correspondent bank in the country of the project. On the basis of this counter-guarantee, the local bank issues the bid bond which is to be enclosed along with the tender. This bid bond from the local bank is to be issued and collected by the tenderer's representative visiting the country for submission of the tender.

For new countries in which the contracting firm does not operate, the firm will have to send their representative for submission of the tender. However, in countries where the firm's units are located, the tender for submission is generally sent by post and the firm's local representative collects the bid bond from the local bank in the host country along with the tender.

A favourable reference from a previous client can create more confidence than anything else. The reference could be on quality as well as quantity. If sufficient foreign references do not exist, impressive Indian references can be given, taking care of course, to mention the climate and difficult features etc., encountered locally, to create confidence for parallel situations. Therefore, try to include a favourable reference from a previous client when submitting the tender.

10.12.2 Procedure of post-award clearance

10.12.2.1 For smaller projects

If a project value is under US$200 million, all facilities required for executing the project can be extended by the authorised dealer with the concurrence of Exim Bank. For such projects, within 15 days of entering into contract, the exporter should submit to his bankers a DPX1 application form (for turnkey and deferred payment supply contracts) or a PEX1 application form (for civil construction contracts) in six copies along with six copies of the contract. Authorised dealers should deal with all applications made by exporters in connection with project exports.

10.12.2.2 Details to be furnished by civil construction contractors for post-award approval

There are two types of application forms: DPX1 and PEX1. The DPX1 is for turnkey and deferred payment supply contracts and the PEX1 is for civil construction contracts. The contractor should submit six copies of the completed PEX1 application form and six copies of the contract to an authorised dealer within 15 days after being awarded the overseas construction contract. The PEX1 contains two parts, i.e. Part A and Part B. Part A of PEX1 contains general details of the exporting firm; broad particulars of other construction contracts executed abroad during the previous 3 years; a brief descriptive account of the firm covering the technical, financial and management aspects and whether the applicant is on the approved list of the Project Export Promotion Council of India; details of the overseas contracts with which the firm is associated (country, nature of the contract, value, present position); details of the overseas buyer; the status of the exporter – whether a prime exporter or a consortium member; details of the construction contract and its value (in foreign currency, equivalent Indian rupees and exchange rate); provisions in the contract in respect of penalties/liquidated damages and ceiling; period of delay for which minimum penalty becomes applicable; price escalation; force majeure; arbitration and laws governing the contract; payment terms (advance payment, down payment, progress payment, retention money, deferred payment terms); currency of payment; security (nature of security to be furnished by the buyer for down payment, progress payment and deferred receivable and interest); foreign exchange outgo; third country imports; construction equipment required for execution of the contract (details of equipment to be exported from India or from a third country or to be transferred from other projects); arrangement of procurement made; time schedule; mobilisation plan for workforce and materials; proposed programme of work with time schedule; programme evaluation and review technique (PERT) and bar chart;

manpower requirements and their cost; place/s where site/liaison office/s will be required and its estimated expenditure; amount and nature of overseas expenditure involved (currency-wise); details of overseas agent; details of foreign currency/local borrowings; estimated foreign exchange repatriation to India; profitability estimates; project financiers; and the procedure for the certification of bills and time span at each stage for passing the bills and receiving payments. Part B of PEX1 contains facilities required by exporters such as fund-based (pre-shipment credit, deferred credit, credit against export incentives, etc.) and non-funded (advance payment guarantee, performance guarantee, retention money guarantee, guarantee for borrowings abroad, other guarantees and letters of credit [L/Cs], etc.) facilities. Other facilities required are site office/s overseas, liaison office/s overseas, overseas bank account/s (currency-wise), initial remittance/transfer from other projects if any, required, subject to repatriation to India, risks cover (type of cover from ECGC – political/comprehensive/specific services contracts policy, etc.). All these will have to be furnished by the project exporting firm as per PEX1.

10.12.2.3 For larger projects

If the project value exceeds US$100 million, the project exporting firm will have to submit an application through its banker (authorised dealer) to the Exim Bank.

The authorised dealer should immediately forward copies of the application together with copies of the contract and the banker's comments in forms DPX2/PEX2, as the case may be, forwarded to the office of the Reserve Bank of India (Exchange Control Department) within whose jurisdiction the head office of the exporter is situated, and also to ECGC, Mumbai, and Exim Bank.

10.12.3 Authorised dealer's (banker's) form: PEX2

Comments on the application are to be furnished by the authorised dealer (bankers) in form PEX2 to the Exim Bank/working group by the financing/participating bank/s in respect of the following items:

1. Management of the applicant company/firm.
2. Applicant's export performance during the previous 3 years.
3. Applicant's financial position (based on a study of the balance sheets, profit and loss accounts of the applicant for the preceding 3 years and other available data) and dealings with banks.
4. Terms of construction job payment.
5. Capacity of the applicant to fulfil his obligations under the contract.
6. Estimates of cost and profitability furnished by the applicant.
7. Extent to which the bank is willing to provide various facilities to the applicant and the main terms on which they are proposed to be extended. Details of the present credit facilities, if any, sanctioned to the applicant.
8. Extent of Exim Bank's participation required, giving reasons why the bank will not be able to finance the entire transaction.
9. Credit report on the employer in the private sector.

In addition to the above, the bank should recommend the proposal for approval and agree to extend the facilities sought by the exporter in form PEX2.

10.12.4 *Working group meeting*

Exim Bank will convene a meeting of the working group within 1 week of receipt of the application to consider the final terms and conditions of the contract and to grant a package post-award clearance for the contract.

10.12.5 *Follow-up of turnkey/construction projects*

Exporters and all their Indian subcontractors executing turnkey contracts or civil construction contracts abroad should furnish progress reports in form DPX3 on a half-yearly basis (June and December) to the concerned approving authority, viz., authorised dealer/Exim Bank/WG as the case may be, and the concerned regional office of the RBI through their bankers within 1 month from the date of expiry of the relative half-year. A copy of the report may be sent to ECGC/Exim Bank in all cases where their risk/guarantee cover participation has been obtained. The final report in DPX3 should clearly indicate the completion of the project and full compliance with the requirements relating to completed projects.

10.13 Requirements relating to completed projects

10.13.1 *Steps to be taken by exporters on completion of the contract*

Close foreign currency accounts and transfer the balances to India.

- Wind up the site and liaison offices opened abroad.
- Cancel the guarantees for performance of the contract and other guarantees issued and return to exporters.
- Liquidate all overseas borrowings/overdrafts obtained, if any, and cancel counter-guarantees.
- Make suitable provision for the payment of taxes, customs and other statutory obligations in the country of the project.
- Dispose of the equipment, machinery, vehicles, etc., purchased abroad or arrange for their import into India. (In case the machinery etc., is to be used for another overseas project, the market value [not less than the book value] should be recovered from the project to which the equipment/machinery has been transferred.)
- Recover funds, if any, transferred to other overseas project/s and repatriate them to India.

10.14 Final report

- A report giving a full account of the various steps taken should be sent by the exporter through his bankers to the concerned AD/Exim Bank.
- The report should also be sent to Exim Bank/ECGC because of their participation in risk sharing.
- Where the project export proposal was approved at the level of the working group, the report may be sent to Exim Bank and ECGC.

10.14.1 *Documents to be forwarded along with final report*

- A completion or final handing over certificate.
- A certificate from the overseas bank regarding closure of the account held with it.
- A statement of remittances made to India.

- Bank certificates for the repatriation of funds to India.
- Tax clearance certificate/no tax liability certificate for the overseas project.
- Bills of entry for the reimportation of machinery.
- Statements of income and expenditure and profit and loss account of the project duly certified by a chartered accountant/project manager.

10.15 Case study[6]

This is a case study on a large-value international construction project conforming to international standards and specifications executed successfully by an Indian construction contracting company.

The ABC project details

- Client: Middle-East (a government authority)
- Consultant: European company
- Contractor: XYZ Company, an Indian company
- Subcontractors: Various Indian companies
- Project price : US$4500 million
- Construction period: 30 months

Project scope

- Route length: 38.30 km
- Track length: 95.67 km
- Gauge: Standard 1435 mm
- Designed speed: 250 kmph[7] (pass train) and 140 kmph (freight train)
- Track: 96 km
- Sleepers: 140,000
- Ballast: 0.25 mm^3
- Stations: 2
- Air-conditioned workshop: 29,000 m^2
- Roads: 46 km
- Buildings: 402 (87,000 m^2)
- Bridges: Major – 4; minor – 97
- Sub-ballast: 4.10 mm^3
- Earth work: 4.75 mm^3

Funding

Sources of funding for this project were from government resources. It was not funded through financial institutions such as the Asian Development Bank or the World Band. There was no assurance of funding for the project period; however, an oil rich country was a positive factor.

The XYZ Company decided to execute an overseas project in addition to domestic projects to increase their revenue and profits. Accordingly, they applied for registration with the Project Export Promotion Council of India (PEPCI), an organisation promoted by the Ministry of Commerce and Industry, Government of India. After examining the company's track record in the execution of projects, adherence to time and cost schedules, profitability on earlier projects, net worth, total assets, etc., the PEPCI granted registration.

Subsequently, the company started searching for international civil construction projects coming up in different parts of the world. To find information on a project, they referred to *Development Business* (a publication of the United Nations), PEPCI's newsletter, the Engineering Export Promotion Council's newsletter, Middle East Economic Digest, etc. They located a few potential projects from these publications. After discussions, the senior management decided to bid for the ABC project.

Later, a group of engineers from the XYZ Company were tasked with collecting information about the project site (site survey) and studying the economic, socio-cultural and political risk factors of the country where the project was coming up and about the neighbouring countries. Factors such as political continuity, attitude towards foreign investors and profit, nationalisation/expropriation, enforceability of contracts, government incentives, inflation, economic growth, bureaucratic delays, communication and transportation, professional services other than construction, social conflicts, quality of government management, legal framework, current account balance, international reserves, foreign debt, budget performance, etc., were also studied. Another task given to this group was to study all the contract clauses and the pricing of the project and the possibility of repeat projects from the same country and to identify from which country or countries they could import the materials, manpower and equipment required for the project. After identifying the countries from where they could import the inputs required for the project, they had to study the economic aspects, especially Exim rules and regulations and details of the currency of the country. After studying the above factors, the group of engineers submitted a report to the company.

Political impact and law and order of the host country

Political impact on the project was satisfactory due to a stable government. The law and order situation in the host country was good.

Political relations and labour laws

Political relations with India were excellent and a large number of Indian companies had been awarded large-value infrastructure contracts. Labour laws were also favourable. The import of Indian and other country's labour was permitted which could lead to increasing productivity.

Contract clauses, contract price and currency

Contract clauses contained favourable conditions for payments. Contract prices had good rates compared to those prevailing in other countries in the Middle East. There was a strong local currency and surplus funds were repatriable.

Taxation, import restrictions

No personal taxation or corporate tax existed with the host country and India. Anything pertaining to the project could be imported into the country of the project. Another peculiarity of the country was there was no corruption.

International political situation and country classification

The international political situation was not good; however, it did not materially affect construction and a large number of international and Indian companies were currently working on infrastructure projects in the country. ECGC Ltd. cleared the undertaking of projects in the country.

International competitive bidding

The company was the lowest bidder. With almost a zero risk situation, XYZ Company decided to participate without any joint venture and prequalified for the project. In association with Indian private sector companies as the company's nominated associates, the company participated in the bidding process and gave a responsive, competitive and compelling bid. Ultimately, the company was awarded the project. The contract was secured by international competitive bidding. The contract was finalised without a reduction in price; however, conditions were negotiated and clarification was obtained on some of the weaknesses of the company on their performance capability as brought out by their competitors. One of the major issues was that the equity in the company was very low. In answer to the question of how the company proposed to organise financial support for the project, the Indian Exim Bank, the State Bank of India and ECGC had agreed to provide all funded and non-funded facilities to the project with a counter-guarantee from the government.

Another major issue raised was the company's lack of experience in executing projects to speed standards of 250 Kilometres per hour (Kmph). The maximum speed of Indian railways is 100–120 Kmph. How could the company provide assurances regarding its capability to execute projects to speed standards of 250 Kilometres per hour? The answer given was that this capability depended on the track structure and parameter and the rolling stock. The rolling stock was not part of the contract. Regarding the track structure, similar track structure had been adopted on the entire high-speed track. The methodology to be adopted had been explained in the technical bid. After verbally convincing and creating confidence, the work was awarded to the XYZ Company.

The company promised the client that they would complete the project on time, within budget, to international standards, to the satisfaction of the international consultants and to the client's full satisfaction. XYZ Company was further challenged that it would be awarded the next contract on a negotiated basis, based on its performance in executing this project. For this project, the company formed an association with large private sector companies as their nominated associates working as a cohesive team, some of which were already working in the country near the project site.

Post-award activities of the company

Within 15 days of being awarded the contract, the company completed the PEX1 (application form for civil construction contracts). The project exporters and all their Indian subcontractors furnished details by completing the PEX1. The PEX1 is the application form submitted to the working group for permission to execute the project. The PEX1 requires details of the foreign contractor (where applicable) and the main subcontractors for services; the total value of the machinery, construction equipment and materials required for the execution of the contract; and estimates of the cost of construction and profitability, etc. The PEX1 was completed and submitted to the company's (exporter's) bank. As per the rules, the project exporter submitted six copies of the application and six copies of the contract to the authorised dealer, the company's bank. After receiving the application, the bank completed the banker's form (PEX2), as per the Project Export Manual, and forwarded it to Exim Bank and members of the working group, i.e. Exim Bank, ECGC Ltd., the Reserve Bank of India's Exchange Control Department (ECD).

Within 7 days of receipt of the application, Exim Bank arranged a meeting of the members of the working group. The working group was composed of Exim Bank, RBI (ECD), ECGC Ltd., representatives of the Ministry of Finance, the Government of India, the prime contractor and the prime contractor's bank, the subcontractor and the subcontractor's bank. The working group examined all the contract clauses and terms and conditions of the project and approved the project.

The project exporter, as per the Project Export Manual, must open, hold and maintain separate foreign currency accounts for each project under execution abroad. Also, the company must open temporary site offices and make payments for third-country purchases (machinery/equipment/materials), etc. for the project. The working group gave permission to open foreign currency accounts and temporary site offices, pay the agency commission and avail of temporary overseas borrowings. Following the working group's permission, the authorised dealer of the company agreed to furnish all the necessary guarantees required in connection with the execution of the project abroad,. After receiving permission from the working group to execute the project, the company started project execution.

Acceptance of deferred payments

During project execution, due to financial constraints, the client had to stop the budget provisions on all projects and duly informed the contractors and the government. They gave a choice to close the project or accept deferred payments. As the project had passed the halfway mark, the financial implications of both alternatives were worked out. The company recommended continuing the project on deferred payment of the US dollar component with need-based financing by Exim Bank. Both governments negotiated the contract for deferred payment of the US dollar component between Exim Bank and the company and the project was successfully completed with good relationships maintained. Subsequent payments (by the client) of deferred payment with interest provided the company with large amounts due to the devaluation of the Indian rupee.

Follow-up of project execution

The company and all its Indian subcontractors, as per the Project Export Manual 2014, furnished progress reports in form DPX2 on a half-yearly basis (June and December) to the working group and the regional office of RBI through their bankers.

Steps taken subsequent to project completion

The steps taken by the company on completion of the contract were: (i) closed the foreign currency accounts and transferred the balances to India; (ii) closed the site and liaison offices opened abroad; cancelled the guarantee for performance of the contract and other guarantees issued and returned to the exporters; fully liquidated overseas borrowings/overdrafts obtained and cancelled counter-guarantees; made suitable provision for the payment of taxes, customs and other statutory obligations in the country of the project; disposed of some of the equipment, machinery, vehicles, etc., purchased abroad and arranged to import some to India, transferring the balance to the second project and recovering the market value of the equipment transferred and repatriated to India. Subsequently, a report giving full account of the various steps taken was sent by the company through its bankers to Exim Bank and ECGC Ltd. The company forwarded the following along with the final report:

1. A completion or final handing over certificate.
2. A certificate from the overseas bank regarding closure of the account held with it.
3. A statement of remittances made to India.

4. Bank certificates for the repatriation of funds to India.
5. Tax clearance certificate/no tax liability certificate regarding the overseas project.
6. Bills of entry for the reimportation of machinery.
7. Statements of income and expenditure and the profit and loss accounts of the project duly certified by a chartered accountant/project manager.

After successful completion of the project, the client was awarded another contract worth US$1300 million in the vicinity of the company's camp area, with an increased price of 13 per cent over the earlier contract prices.

10.16 Conclusion

The international construction contracting business is very complex and risky due to a dearth of sufficient environmental information about the host country and international construction experience. Correspondingly, the post-award procedure followed in India is also different from internal projects. This chapter discussed the complexities of the international contracting business, bidding and procedures related to post-award clearance of international projects in India and steps to be taken by project exporters during and after completion of a project along with a case study.

Notes

1 Number of bid won/number of bid submitted.
2 Refer to the online glossary of the Political Risk Insurance Centre, a website sponsored by the MIGA www.pri-center.com/.
3 *See* for more examples Keeton, G. and G. White, G. (2011) Is the nationalisation of the South African mining industry a good idea? [Conference paper]. Biennial conference of the Economic Society of South Africa, 5–7 September 2011. Department of Economics and Economic History, Rhodes University.
4 Adapted from Konanahalli A. et al. (2012).
5 Nguyen, Phong Thanh and Phu-Cuong Nguyen. (2020) pp. 5237-5241.
6 The author acknowledges with thanks the late V.K.J. Rane for his valuable inputs on the project.
7 Kilometre per hour.

References

Aliber, R.A. (1975) Exchange risk, political risk and investor demands for external currency deposits. *Journal of Money, Credit and Banking*, 7(2), pp. 161–179.

Alkharmany, Abdullah. (2017) Project management: The effect of Saudi national culture on the attitudes of key stakeholders towards delay in construction projects in Saudi Arabia (Doctoral Dissertation). University of Brighton, Brighton, England. Retrieved from: https://cris.brighton.ac.uk/ws/portalfiles/portal/4781058/Final_Thesis_Alkharmany.pdf Accessed on 5 December 2019.

Anderson, E.S. (2003) Understanding your project organization's character. *Project Management Journal*, 34(4), pp. 4–11. Retrieved from: https://csbweb01.uncw.edu/people/rosenl/classes/OPS100/Understanding%20Your%20Project%20Organizations%20Character.pdf Accessed on 9 December 2019.

Ashley, David B. and Bonner, Joseph J. (1987) Political risks in international construction. *Journal of Construction Engineering and Management*, 113(3), pp. 447–467. Retrieved from: https://ascelibrary.org/doi/pdf/10.1061/%28ASCE%290733-9364%281987%29113%3A3%28447%29 Accessed on 25 November 2019.

Black, J.S. (1988) Work role transitions: A study of American expatriate managers in Japan. *Journal of International Business Studies*, 19(2), pp. 277–294.

Black, J.S., Mendenhall, M. and Oddou, G. (1991) Toward a comprehensive model of international adjustment: An integration of multiple theoretical perspectives. *Academy of Management Review*, 16(2), pp. 291–317.

Black, S.J. and Stephens, G.K. (1989) The influence of the spouse on American expatriate adjustment and intent to stay in Pacific Rim overseas assignments. *Journal of Management*, 15(4), pp. 529–544.

Blaszczyk, Regina Lee. (2008) Synthetics for the Shah: DuPont and the challenges to multinationals in 1970s Iran. *Enterprise and Society*, 9(4), pp. 670–723. Retrieved from: https://www.academia.edu/10003709/_Synthetics_for_the_Shah_DuPont_and_the_Challenges_to_Multinationals_in_1970s_Iran Accessed on 26 November 2019.

Bonner, J.J. (1981) *Political Risk Analysis System for Multinational Contractors*. Massachusetts Institute of Technology, Cambridge, MA.

Butler, K.C. and Joaquin, D.C. (1998a) A note on political risk and the required return on foreign direct investment. *Journal of International Business Studies*, 29(3), pp. 599–607. Retrieved from: http://citeseerx.ist.psu.edu/viewdoc/download?doi=10.1.1.849.5640andrep=rep1andtype=pdf Accessed on 25 November 2019.

Butler, Kirt C. and Joaquin, Domingo Castelo. (1998b) A note on political risk and the required rate of return on foreign direct investment. *Journal of International Business Studies*, 29(3), pp. 599–607.

Casson, M. and Lopes, T.D.S. (2013) Foreign direct investment in high-risk environments: An historical perspective. *Business History*, 55(3), pp. 375–404.

Chang, T., Deng, X., Zuo, J. and Yuan, J. (2018) Political risks in Central Asian countries: Factors and strategies. *Journal of Management in Engineering*, 34(2), Article ID 04017059.

Chapman, R.J. (2001) The controlling influences on effective risk identification and assessment for construction design management. *International Journal of Project Management*, 19(3), pp. 147–160.

Deng, X. and Low, S.P. (2013) Understanding the critical variables affecting the level of political risks in international construction projects. *KSCE Journal of Civil Engineering*, 17(5), pp. 895–907.

Dimatteo, Larry. (2017) *International Business Law and the Legal Environment: A Transactional Approach*. Routledge, New York.

Duranti, G. and Di Prata, O. (2009) Everything is about time: Does it have the same meaning all over the world? [Paper presentation]. In: PMI® Global Congress 2009—EMEA, Amsterdam, North Holland, The Netherlands, Project Management Institute, Newtown Square, PA. Retrieved from: https://www.pmi.org/learning/library/everything-time-monochronism-polychronism-orientation-6902 Accessed on 3 December 2019.

Exim Bank. (2019) Project exports from India: Strategy for reenergizing and reorienting. Occasional Paper No. 193, pp. 35–36, Export-Import Bank of India. Retrieved from: https://www.eximbankindia.in/Assets/Dynamic/PDF/Publication-Resources/ResearchPapers/120file.pdf Accessed on 11 February 2020.

Fisher, Glen. (1988) *Mindsets: The Role of Culture and Perception in International Relations*. Intercultural Press, Yarmouth, Maine.

Haendel, D. (1979) *Foreign Investments and the Management of Political Risk*. Westview Press, Boulder, CO.

Harzing, A.W. (1995) The persistent myth of high expatriate failure rates. *International Journal of Human Resource Management*, 6(2), pp. 457–475.

Hofstede, Geert. (1983) The cultural relativity of organizational practices and theories. *Journal of International Business Studies*, 14(2), pp. 75–89. Retrieved from: https://alingavreliuc.files.wordpress.com/2010/10/the-cultural-relativity-of-organisational-practices-and-theories.pdf Accessed on 3 December 2019.

Hofstede, G. (2001) *Culture's Consequences: Comparing Values, Behaviors, Institutions, and Organizations Across Nations*, 2nd edition. Sage, Thousand Oaks, CA.

Hofstede, G., Hofstede, Gert Jan and Minkov, Michael. (2010) *Cultures and Organizations: Software of the Mind*, 3rd edition. McGraw-Hill, New York.

Jensen, Nathan M., et al. (2012) *Politics and Foreign Direct Investment.* University of Michigan Press. ProQuestEbook Central. Retrieved from: https://ebookcentral.proquest.com/lib/nicmar-ebooks/detail.action?docID=3415091.

Keeton, G. and G. White, G. (2011) Is the nationalisation of the South African mining industry a good idea? Biennial conference of the Economic Society of South Africa, 5–7 September 2011. Department of Economics and Economic History, Rhodes University, Eastern Cape, South Africa.

Kim, Y.Y. and Ruben, B.D. (1988) Intercultural transformation: A systems theory. In: Young Yun Kim and William B. Gudykunst (Eds), *Theories in Intercultural Communication*, pp. 280–298. Sage, Newbury Park, CA.

Kobrin, S.J. (1980) Foreign enterprise and forced divestment in LDCs. *International Organization*, 34(01), pp. 65–88.

Kobrin, S.J. (1984) Expropriation as an attempt to control foreign firms in LDCs: Trends from 1960 to 1979. *International Studies Quarterly*, 28(3), pp. 329–348.

Konanahalli, A. et al. (2012) International projects and cross-cultural adjustments of British expatriates in Middle East: A qualitative investigation of influencing factors. *Australasian Journal of Construction Economics and Building*, 12(3), pp. 31–54.

Kroeber, A.L. and Kluckhohn, Clyde. (1952) *Culture: A Critical Review of Concepts and Definitions*, p. 132. The Museum Press, Cambridge, MA.

Lessard, Donald R. and Lightstone, John B. (1986, July) Volatile exchange rates can put operations at risk. *Harvard Business Review*, pp. 107–114. Retrieved from: https://hbr.org/1986/07/volatile-exchange-rates-can-put-operations-at-risk Accessed on 27 January 2020.

Liu, J., Zhao, X. and Li, Y. (2016) Exploring the factors inducing contractors' unethical behavior: Case of China. *Journal of Professional Issues in Engineering Education and Practice*, 143(3), Article ID 04016023.

Lloyd, B. (1974) The identification and assessment of political in the international risk in environment. *Long Range Planning*, 7(6), pp. 24–32. Retrieved from: https://www.sciencedirect.com/science/article/pii/0024630174901290 Accessed on 27 November 2019.

Loosemore, M. and Muslmani, H.S.Al. (1999) Inter-cultural communication in the Gulf: Inter-cultural communication. *International Journal of Project Management*, 17(2), pp. 95–100. Retrieved from: https://reader.elsevier.com/reader/sd/pii/S0263786398000301?token=B9B276A18C64BE6061DF7EA8B124B863EB023EDF39E0C2EF2DA2961642FAFBBAE0FC76891A68CBE4F7222E8CAF4BABB7 Accessed on 6 December 2019.

Minor, M.S. (1994) The demise of expropriation as an instrument of LDC policy, 1980–1992. *Journal of International Business Studies*, 25(1), pp. 177–188.

Naumann, E. (1993) Antecedents and consequences of satisfaction and commitment among expatriate managers. *Group and Organization Management*, 18(2), pp. 153–187.

Ndjovu, Cletus Eligius. (2015 December) Compulsory land acquisitions in Tanganyika: Revisiting the British colonial expropriation principles and practices. *International Journal of Scientific and Technology Research*, 4(12), pp. 10–19. Retrieved from: http://www.ijstr.org/final-print/dec2015/Compulsory-Land-Acquisitions-In-Tanganyika-Revisiting-The-British-Colonial-Expropriation-Principles-And-Practices-.pdf Accessed on 27 November 2019.

Nguyen, Phong Thanh and Nguyen, Phu-Cuong. (2020) Risk management in engineering and construction: A case study in design-build projects in Vietnam. *Engineering, Technology and Applied Science Research*, 10(1), pp. 5237–5241.

Pallant, J. (2010) *SPSS Survival Manual: A Step by Step Guide to Data Analysis Using SPSS.* Open University Press, Buckingham, UK.

Pheng, Low Sui and Leong, C.H.Y. (2000) Cross-cultural project management for international construction in China. *International Journal of Project Management*, 18(5), pp. 307–316.

RBI. (2003) Memorandum of Instructions on Project Exports and Service Exports. Reserve Bank of India.

Robock, S. (1971, July/August) Political risk: Identification and assessment. *Columbia Journal of World Business*, 6(4), pp. 6–20.

UNCTAD. (2012) *Expropriation.* UNCTAD Series on Issues in International Investment Agreements II, p. 16. United Nations, New York and Geneva. Retrieved from: https://unctad.org/en/Docs/unc taddiaeia2011d7_en.pdf Accessed on 27 November 2019.

Wang, P.Z. (2003, December) Assessing consumer vanity in Australia and China. [Conference Paper] In: *Australian and New Zealand Marketing Academy Conference*, pp. 1457–1461. ANZMAC, Adelaide, Australia. Retrieved from: https://opus.lib.uts.edu.au/bitstream/10453/2217/3/2003 000846.pdf Accessed on 3 December 2019.

Wegley, Charles. (1997) Culture. *Academic American Encyclopaedia.* 5. Deluxe Library Edition, pp. 415–417. Grolier Inc., Danbury, CN.

Xiaopeng, Deng and Low, S.P. (2014) Exploring critical variables that affect political risk level in international construction projects: Case study from Chinese contractors. *Journal of Professional Issues in Engineering Education and Practice, 140*(1), p. 04013002. Retrieved from: https://ascelibrary .org/doi/pdf/10.1061/(ASCE)EI.1943-5541.0000174 Accessed on 3 December 2019.

Zhuang, L., Ritchie, R. and Zhang, Q. (1998) Managing business risks in China. *Long Range Planning, 31*(4), pp. 606–614. Retrieved from: https://reader.elsevier.com/reader/sd/pii/S002463019880 0538?token=646B49DB09429DA2BA908110F7280FF953D2A07221A3CB4B1B4FAE2659F F7CE0F1597AFD5133BD2F171734A56628A792 Accessed on 3 December 2019.

Zuo, Jian, Zillante, George and Coffey, Vaughan. (2009) Project culture in the Chinese construction industry: Perceptions of contractors. *Australasian Journal of Construction Economics and Building, 9*(2), pp. 17–28. Retrieved from: https://digital.library.adelaide.edu.au/dspace/bitstream/2440/8419 1/2/hdl_84191.pdf Accessed on 9 December 2019.

Zwikael, O., Shimizu, K. and Globerson, S. (2005) Cultural differences in project management capabilities: A field study. *International Journal of Project Management, 23*(4), pp. 454–462. Retrieved from: https://reader.elsevier.com/reader/sd/pii/S0263786305000530?token=2DDF8CFDF0A17A 07B716F6FC2EF12EB9150367E8949C9BE4DAD6E662AEC402FCC7EC35CEE4770DD6D3BF C68A32A8B6FB Accessed on 9 December 2019.

11 Identifying, analysing and managing construction project risk

11.1 Introduction

The success of infrastructure projects is significantly influenced by the appropriate management of the risks related to a project. Risk is the probability that an event might occur which would lead to a change in the project's circumstances that were assumed while forecasting the costs and benefits of the project and would have an impact on the project's objectives. Risk management is an ongoing process that continues throughout the lifecycle of a public-private partnership (PPP) project. The general process of risk management is presented in Figure 11.1. Risk mitigation strategy is the method of developing alternatives and actions to enhance the opportunities and reduce the threats to a project's objectives. This chapter identifies, analyses and manages a construction project's risk.

Executing construction projects in a foreign country is generally regarded as a high-risk business, mainly because the contractor may be short of adequate overseas environmental information and overseas construction experience. Similar construction projects would possibly have entirely dissimilar risk characteristics in different regions. Therefore, it is often difficult for a beginner to identify new risks and complexities in a new overseas construction project environment. A further complication arises in assessing these risks and the subtle impact of any associations between them. However, ignoring these risks is irresponsible, and will result in unrealistic decisions. Conversely, identifying and assessing all new risks and their associations and interactions is a very complex, slow and expensive process. Once this type of complex scenario is experienced, identifying and controlling these vital risk factors in overseas projects becomes enormously significant. Numerous unique risks are encountered with overseas development projects.

Global Construction 2030, a report by Global Construction Perspectives and Oxford Economics UK, forecasts that the volume of construction output will grow by 85 per cent to $15.5 trillion worldwide by 2030, with three countries – China, the United States and India – leading the way and accounting for 57 per cent of all global growth.[1] The report further states that the 'construction market in India will grow almost twice as fast as China to 2030, providing a new engine of global growth in emerging markets'. According to the World Economic Forum, the investment required globally for infrastructure projects is at least US$4 trillion (or 5% of global gross domestic product [GDP]) per year until 2030 (World Economic Forum and the Boston Consulting Group, 2014). India's urban population is expected to grow to 165 million by 2030, swelling Delhi by 10.4 million people to become the world's second largest city. Global economic uncertainties have been reflected in the international construction market over the past 4 years. This is evident from the results of the *Engineering News-Record* (ENR) Top 250 International Contractors survey.

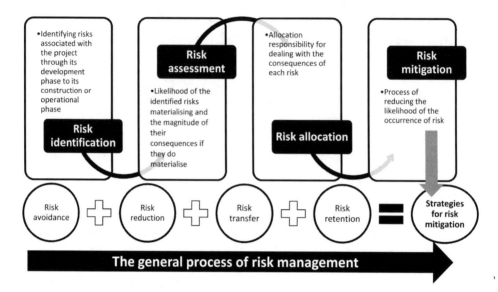

Figure 11.1 The general process of risk management.

Table 11.1 Total contracting revenue of the ENR top 250 international contractors

Year	Contracting revenue (US$ billion)
2013	543.97
2014	521.55
2015	501.14
2016	468.12

Source: *Engineering News-Record.*

The top 250 international contractors reported $468.12 billion in contracting revenue in 2016[2] from projects outside their home countries, down from $543.97 billion in 2013 (Table 11.1). Even if the overall reduction in demand is continuing and exchange rate pressures are persisting in the international construction market, some recovery from the great recession of 2008 can be seen. The international construction market has started growing by the end of 2018. A decline in margins is compelling the international construction industry to find innovative ways to increase its efficiency by adopting mechanisation to tackle skills shortages and improve productivity.[3]

In 2013–14, Indian project export contracts supported by the EXIM bank – the developmental financial organisation set up by the Government of India specifically to promote exports– amounted to Rs. 34.131 crore which were secured by 40 companies in 35 countries. As of 31 March 2020, 319 project export contracts valued at Rs. 1,40,326 crore and supported by the bank were under execution in 74 countries across Asia, Africa and the Commonwealth of Independent States (CIS) by 99 Indian companies.[4] During 2015–16, the bank funded 95 project export contracts in 39 countries by 50 exporters, aggregating to Rs. 22,551 crore, while under the buyer's credit it sanctioned $2.19 billion for 22 projects valued at $2.49 billion.[5]

11.1.1 India's export contracts

During FY 2015–16, 95 contracts amounting to Rs. 225.51 billion covering 39 countries were secured by 50 Indian exporters, as against 105 contracts worth Rs. 497.81 billion covering 40 countries, secured by 56 Indian exporters during FY 2014–15 (Table 11.2). The contracts secured during the year comprised 55 turnkey contracts valued at Rs. 114.12 billion, 18 construction contracts valued at Rs. 97.87 billion, 10 supply contracts valued at Rs. 12.37 billion and 12 technical consultancy and services contracts valued at Rs. 1.15 billion. The export of construction and real estate industry services is presented in Table 11.3.

11.1.2 Complexities in international construction contracting business

The construction process is considered a complex activity since it involves a number of tasks and objectives, the most significant of which is realising value for money (Antonioua et al., 2013). Generally, large construction projects requiring complex engineering technologies, complex management and operation issues (especially for international construction joint ventures [ICJVs]), huge capital investment, narrow profit margins, stretched contract periods, complex processes, environmental factors, multiplicity of laws framed by international bodies and host countries, unfamiliar socioeconomic conditions and different language and

Table 11.2 Contracts secured

Year	No. of contracts secured	Amount in Rs. billion	No. of countries	No. of contractors
2011–12	53	229.75	23	28
2012–13	85	242.55	38	47
2013–14	75	341.31	35	40
2014–15	105	497.81	40	56
2015–16	95	225.51	39	50

Source: Compiled from various annual reports of Exim Bank.

Table 11.3 Export of services of the construction and real estate industry

Year	Export of services in Rs. million
2010–11	133,649
2011–12	189,735
2012–13	214,716
2013–14	253,044
2014–15	232,970
2015–16	229,792
2016–17	98,260

Source: CMIE.

culture host countries make international construction a high-risk endeavour. International construction projects involving interactions between individuals, organisations and agencies from different national backgrounds and cultural contexts make them more complex. High transaction costs, friction between project participants along with coordination and communication difficulties contribute to poor performances and disputes between parties. Indeed, international dispute settlement bodies such as the International Centre for Settlement of Investment Disputes (ICSID), the United Nations Commission on International Trade Law (UNCITRAL), the International Federation of Consulting Engineers (FIDIC) and the New York Convention and Bilateral Investment Treaties (BITs) ensure a proper legal framework and disputes settlement mechanisms. The most popular arbitration rules for international construction projects in South Asia are: the International Chamber of Commerce (ICC) or the United Nations Commission on International Trade Law. Other country-specific rules that have proved useful to international contractors are: the Singapore International Arbitration Center Rules (SIAC) and the Philippines Construction Industry Arbitration Commission Rules (CIAC).[6] Inadequately defined contract clauses in projects can lead to severe risk.

Payment delays disrupt and expose the whole supply chain to considerable hardship, the insolvency of construction firms and the inability of clients to make payments. The continuous flow of projects and project delivery risks are common in many countries. Obtaining third-party guarantees issued by a substantial international economic institution can secure payment on a construction project to an extent. Such third-party guarantees can be in the form of a letter of credit, bonds or bank guarantees. Construction projects will generally face five main groups of risks: (i) preliminary design; (ii) tender; (iii) detailed design; (iv) construction works; and (v) financing the investment (see Figure 11.2).

11.2 Types of risk

There are three types of risk: (i) factors within the control of the project participants; (ii) factors in the control of others, e.g. planning requirements, building regulations, government taxation, banks – rate of interest; and (iv) acts of God – outside the control of the project participants, e.g. weather. For the purposes of our discussion, we classify these risks as controllable and uncontrollable risks.

11.2.1 Controllable and uncontrollable risks

Risks are classified as controllable and uncontrollable risks. Risks arise as a result of actions or events that are either within or outside the project's control. Thus, they are termed controllable and uncontrollable risks. A risk outcome that is directly or at least partially within the control of the decision-maker or can be controlled by the project participants is called a controllable risk. For instance, the price of a specific decorative facing brick increases. The outcome of this occurrence is that the project team can explore cheaper substitutes for the decorative facing brick. These types of situations are within the control of the project team. Whereas uncontrollable risks are those risks that cannot be influenced by the project participants because such risks emanate from the external environment or socio-political, economic or climatic spheres.

Suppose the fire code establishes a new requirement that must be incorporated into the fabric of a building before a fire certificate can be granted. The project has no choice but to conform. The specific type of risk encountered and who, if anyone, has control over it will decide both the nature of the response to it and also who is allocated the responsibility for managing that risk.

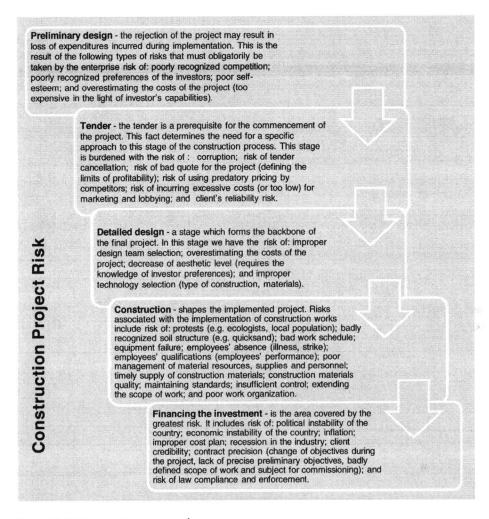

Construction Project Risk

Preliminary design - the rejection of the project may result in loss of expenditures incurred during implementation. This is the result of the following types of risks that must obligatorily be taken by the enterprise risk of: poorly recognized competition; poorly recognized preferences of the investors; poor self-esteem; and overestimating the costs of the project (too expensive in the light of investor's capabilities).

Tender - the tender is a prerequisite for the commencement of the project. This fact determines the need for a specific approach to this stage of the construction process. This stage is burdened with the risk of : corruption; risk of tender cancellation; risk of bad quote for the project (defining the limits of profitability); risk of using predatory pricing by competitors; risk of incurring excessive costs (or too low) for marketing and lobbying; and client's reliability risk.

Detailed design - a stage which forms the backbone of the final project. In this stage we have the risk of: improper design team selection; overestimating the costs of the project; decrease of aesthetic level (requires the knowledge of investor preferences); and improper technology selection (type of construction, materials).

Construction - shapes the implemented project. Risks associated with the implementation of construction works include risk of: protests (e.g. ecologists, local population); badly recognized soil structure (e.g. quicksand); bad work schedule; equipment failure; employees' absence (illness, strike); employees' qualifications (employees' performance); poor management of material resources, supplies and personnel; timely supply of construction materials; construction materials quality; maintaining standards; insufficient control; extending the scope of work; and poor work organization.

Financing the investment - is the area covered by the greatest risk. It includes risk of: political instability of the country; economic instability of the country; inflation; improper cost plan; recession in the industry; client credibility; contract precision (change of objectives during the project, lack of precise preliminary objectives, badly defined scope of work and subject for commissioning); and risk of law compliance and enforcement.

Figure 11.2 Construction project risk.

11.2.2 Common risks encountered by international construction projects

Risk management, from the project management perspective, attempts to identify, prevent, contain and reduce negative impacts and maximise opportunities and positive outcomes in the interests of projects and stakeholders. It is a systematic approach that allows risks to be embraced, avoided, reduced or eliminated through a logical, comprehensive and documented strategy.

11.2.2.1 Economic risks

Changes in the economic environment of a host country, e.g. changes in inflation, exchange rates, tax rates and tax regimes, have direct effects on the profitability of project exporters. High inflation rates reduce the attractiveness of foreign investment owing to a country's currency depreciation on the foreign exchange market and have an effect on cost overruns in construction projects (Zhi, 1995; Gunhan and Arditi, 2005) and can cause financial and payment-related risks of currency exposure for foreign investors (Han and Diekmann, 2001; Hastak and Shaked, 2000). Tax rate changes have a direct and immediate financial impact. Firms often have to pay taxes both to the host country and to their parent country owing to the tax regimes of the host country (Kapila and Hendrickson, 2001). Sharply decreasing GDP causes a crisis for the local economy which affects the performance of international contracting companies (Zhi, 1995).

11.2.2.2 Political risks

Political risk is mostly pertinent in emerging markets. It is the risk that an investment's returns could suffer as a result of political changes or instability in the host country. Expropriation, war, riots and breach of contract by a government agency are referred to as political risks. International financial institutions generally include political risk in their direct lending conditions for the projects they finance since they are hesitant to assume political risks. The guarantees of the Multilateral Investment Guarantee Agency (MIGA) protect investments against political risks. The MIGA helps investors access funding sources with improved financial terms and conditions.

Political instability in a host country may affect the profitability and other goals of an international project (Kapila and Hendrickson, 2001). Common sources of political risk include wars, internal and external conflicts, territorial disputes, government changes and terrorist attacks around the world (Hoti and McAleer, 2004). The strength of the legal system in a host country affects the formation and operation of ICJVs, since conflicts between partners due to contract-related problems are taken into account according to the legal system in the host country (Ozorhon, Dikmen, and Birgonul, 2007).

11.2.2.3 Currency risk

International construction projects are subject to currency risks as they are likely to involve multiple currencies. Currency risks arise whenever foreign exchange funds are used to finance a project or a project generates revenue only in the local currency. Hard currency loans can create currency risks if revenues are in the local currency. Take the case of a power plant in India financed in dollars. If electricity tariffs are in rupees, it can create an asset–liability currency mismatch. For example, if the Indian rupee depreciates against the dollar by 10 per cent, revenues remain unchanged but liabilities are now 10 per cent higher.[7] Therefore, currency risks can create windfall gains or losses for the contracting firms. The major question that arises in such a situation is determining who should assume this currency risk. International contracting firms or lenders have no control over the exchange rate and will therefore adopt risk management strategies or price the exchange rate risk into their rates/tariffs. Lenders will not accept any significant currency risk and expect the contracting firms to ensure that any currency risk the project may assume will not affect its debt service.

11.2.2.3.1 CURRENCY RISK MANAGEMENT STRATEGIES

Foreign exchange risk can be an important concern if the project generates revenue only in the local currency. There are different ways of managing currency risk. The most popular currency risk strategies are: natural hedge; local currency swap; exchange rate – indexed contracts/index output prices to the exchange rate; currency swap; foreign currency loan under peg; mix of local currency and international hard currency loans; and try to obtain government guarantee of foreign exchange.

11.2.2.3.2 NATURAL HEDGE

Natural hedges are an alternative way of reducing exchange rate exposure. Generally, sponsors structure the financing to provide for a natural hedge during construction and operation to limit foreign exchange risk, with the significant volume of US dollars required being funded by equity in US dollars. The Nam Theun 2 (NT2), a hydropower dam project located on the Nam Theun River in Laos People's Democratic Republic (Laos PDR), is an example of a natural hedge. The NT2 project involves the development, construction and operation of a trans-basin electric power-producing plant that is expected to generate $1.9 billion in foreign exchange earnings over a 25-year period through the export of electricity to Thailand. The US$1.45 billion NT2 project is partially financed by Thai banks through Thai baht-denominated loans. It exports a significant proportion of its energy production to Thailand. Consequently, the Thai baht-denominated loans are not exposed to currency risk. Similarly, the power purchase agreement for Bhutanese hydroelectric projects that export their production to India, is in Indian rupees.

11.2.2.3.3 LOCAL CURRENCY SWAP

The purpose of local currency swaps is to hedge against risk exposure associated with exchange rate fluctuations. Under this system of swapping, two parties agree to exchange the principal and/or interest payments on a loan in one currency for an equivalent loan in another currency. Such a swap permits lenders, borrowers and investors to hedge (a part of) their loans or investments. However, for several emerging markets, currency swaps are not commercially offered. The International Finance Corporation (IFC) provides currency swaps for a number of these markets. Furthermore, the Currency Exchange Fund (TCX) is a special-purpose fund that provides currency hedge products for local borrowers in frontier and less liquid emerging markets. The TCX is a fund managed by TCX Investment Management Company BV, a private company that was founded in 2007 by a group of (i) development finance institutions, (ii) specialised microfinance investment vehicles (MIVs) and (iii) donors to offer solutions. The TCX was established to manage currency risk in developing and frontier markets following an initiative by the Netherlands Development Finance Company (FMO). The Dutch government is among its investors. The TCX provides risk capital to facilitate the participation of other investors.[8] The global infrastructure funding gap for 2013–2030, as estimated by McKinsey, is $3.2 trillion a year through 2030. Another way is to build a currency swap the contract so that the revenue of the project can be indexed to the exchange rate. Pegging a country's currency to a foreign currency, i.e. fixing the value of an emerging market's currency to that of a sounder currency, is another method that enables a developer to consider taking out a loan in a foreign currency.

11.2.2.4 Socio-cultural risks

International construction projects engage participants from diverse political, legal, socio-economic and cultural backgrounds. Such projects, with varying levels of diverse nationalities and cultures participating, bring about a project culture which is absolutely unique with particular management difficulties that must be overcome to achieve project success. Numerous studies have identified the socio-cultural issues in host countries as one of the variables to explain the project performance of international construction projects. Some studies have identified civil conflicts due to ideological differences, unequal income distribution, religious clashes (Hoti and McAleer, 2004), language barriers, class structure, nationalism and corruption in the host country as other aspects of socio-cultural risks (Isik, Arditi, Dikmen, and Birgonul, 2010).

11.2.2.5 Climate

While not a risk issue in every jurisdiction, climatic issues necessitate consideration. Projects in the Middle East, for example, suffer climatic risk due to the heat and humidity during the summer months which affect labour efficiency and consequently work quality. The legislated labour break during the middle of the day is also a climate-related risk. Countries with monsoonal seasons also require careful climatic risk assessment and work scheduling.

11.3 Risk classification as per different phases of the project

The construction process can be divided into the planning/design/construction phases, the operation phase and the termination phase. Risks are inherent parts of each of these phases. Therefore, this section deals with risks in different phases of construction.

11.3.1 Risks during the planning/design/construction phases

11.3.1.1 Risk of cancellation or change of scope

A project is susceptible to cancellation if a new government sets different priorities from those set by the previous government, or if parliamentary approval is needed before major PPP contracts can proceed. Such a cancellation can harm project exporting firms, as they might have already made considerable investments in preparing their project proposal. Moreover, a decision on the part of the public authorities to change the project scope at a late phase could have costly consequences for private firms delivering the project. Unnecessary circumstances can affect a project, varying from delays, extended costs and alterations, to cancellation.[9] Generally, the clients of larger infrastructure projects such as power plants are the government of the host country and financial contributions come directly from the budget. It is a major risk because budgets are approved for much shorter periods than required to build such projects and are subject to budgetary cuts if required which can result in delays, suspension of work or even cancellation. Consequently, guaranteed financing through other sources is preferable, even if the terms and conditions are less favourable. When cash flow reduces or discontinues, work stops; the delay or cancellation of projects involving large capital investments is very expensive and can have disastrous consequences.[10] Therefore, project developers, financing institutions and their lawyers need to understand this phenomenon and plan strategies and mechanisms to deal with it in advance.

International contracts with longer durations are often renegotiated. A request for the renegotiation of an existing agreement is often accompanied by express or implied threats, including: (i) governmental intervention; (ii) expropriation; (iii) slowdown in performance; or (iv) the complete repudiation or cancellation of the contract itself. Post-deal, intra-deal or extra-deal renegotiation is a constant fact of international business life. While post-deal renegotiations take place when the original deal has reached or is nearing its end, numerous factors discriminate it from negotiation in the first instance, factors that may significantly affect the renegotiation process. Primarily, by virtue of local law, the customs of the particular business concerned or the parties' express or implied contractual commitments to one another, the parties may have a legal obligation to negotiate in good faith with one another despite the fact that the original contract has terminated; as a result, their ability to refuse to engage in post-deal renegotiations may be limited. The existence and precise nature of such a duty will depend on the law governing the contract. Ultimately, the willingness of the participants to reach an amicable settlement will be influenced by their investments in their first relationship and the extent to which they can use those investments advantageously in their second contract.

Intra-deal renegotiations take place in accordance with the original contract because the original contract itself provides that certain parts of the agreement may or will be renegotiated at specified times or in certain circumstances. This is a mechanism of adjustment for mitigating the contractual risk arising from supervening unpredictable events beyond their control or imperfections in the contract. It is an adjustment to an existing contract that is sought by both sides. Such clauses in a contract also enable parties to delay the discussion of certain topics to a later date after the contract has been signed.

The third type of renegotiations refers to extra-deal renegotiations. It encompasses circumstances where one party insists on adapting the terms of a presumed valid contract that does not provide for an express provision authorising renegotiation. The disappointed party can invoke provisions of international instruments, namely the Principles of International Commercial Contracts and, to a certain extent, the United Nations Convention on Contracts for the International Sale of Goods, to justify the extra-deal renegotiations, subject to specific conditions. The United Nations Convention on Contracts for the International Sale of Goods (CISG – Vienna, 1980) provides a modern, uniform and fair regime for contracts for the international sale of goods. The International Institute for the Unification of Private Law (UNIDROIT) is an independent intergovernmental organisation that studies needs and methods for modernising, harmonising and coordinating private and, in particular, commercial law between states and groups of states and formulates uniform law instruments, principles and rules to achieve those objectives. The Principles of International Commercial Contracts of 2010 is a document drawn up by UNIDROIT intended to help harmonise international commercial law contracts.

11.3.1.2 Risk concerning environmental and other permits

Construction projects involve complex and interconnected activities, carried out by different project participants, with large costs and long duration. The complicated and lengthy procedure of obtaining construction permits at different stages of construction delays the project's execution and impacts the project's profitability, as cash flows start later than anticipated. The lengthy procedure of obtaining construction permits often impacts on the profitability of a project (see Figure 11.3).[11] Frequently, project feasibility studies cannot foresee hidden clauses such as restrictions and compensation requirements.

11.3.1.3 Risk of community opposition

Local communities can affect projects in ways that do not just influence permit procedures. Native populations, for example, can have formal or informal veto rights over such projects

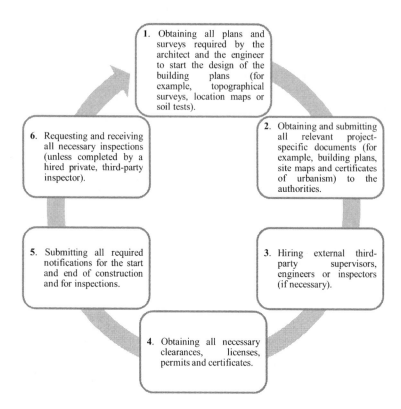

Figure 11.3 Procedures in obtaining construction permits.

within their territories; action groups can organise protests that prompt politicians to withdraw permission; and so on. The forms of community opposition range from lodging formal objections with planning authorities, establishing activist groups, arranging petitions and public protests, and legal proceedings, to lobbying politicians and attempting to attract media interest. Community risk is likely to be high if the project involves land acquisitions or the relocation of local inhabitants. Property market values also play a vital role in the opposition of local communities if they feel that the compensation was insufficient. Key factors that nurture protests are given in Figure 11.4.

11.3.2 Risks during the operation phase

11.3.2.1 Risk of expropriation

One fundamental political risk faced by private infrastructure owners is the risk of outright confiscation or nationalisation of their asset. Expropriation means taking possession, transferring or distributing a property by the state and denial of its use to the owner. If this process extends over a long period of time it is referred to as creeping expropriation. The seizing of land, property or assets is called direct expropriation and indirect expropriations are usually through taking the shares of the firm, appointing an administrator, implementing discriminatory taxes, refusing to grant export or import permits or changing the legislative

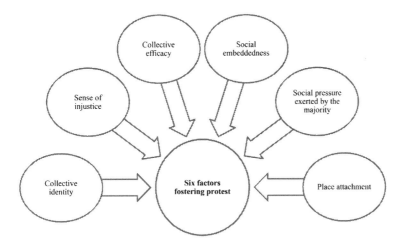

Figure 11.4 Key factors that nurture protest.

landscape which affect the ability of the international entity to undertake its business. More subtly, a series of renegotiations or regulatory changes can result in de facto expropriation, or 'creeping expropriation'. A joint venture was established between a US firm and an Iranian firm to execute a building project in Iran. Afterwards, the Iranian government appointed an administrator to the project and pushed the US firm out of the project's management.[12]This is an example of indirect expropriation. Expropriations are in breach of investment agreements because investment agreements between states and foreign investors are liable to be governed by international investment agreements, bilateral and multi-lateral investment treaties or preferential trade agreements and not by the existing domestic laws of the host government. States have a sovereign right under international law to take property held by nationals or aliens through nationalisation or expropriation for economic, political, social or other reasons. To be lawful under international law, the exercise of the sovereign right to expropriate must fulfil four conditions: (a) property has to be taken for a public purpose; (b) it must be non-discriminatory; (c) it should be in accordance with due process of the law; and (d) accompanied by due compensation (United Nations, 2012).

> In January 2012, the GRZ reversed the June 2010 sale of the SOE Zambia Telecommunications Company (Zamtel) to LAP Green Networks of Libya, which acquired a 75 per cent shareholding in Zamtel for USD 257 million. The GRZ unilaterally reversed the sale and re-appropriated the telecom company, citing corruption and flaws in the privatization process. LAP Green Networks has since challenged the decision in the courts of law. In September 2012, the GRZ terminated and reacquired its concession agreement with the country's largest railway operator, Railway Systems of Zambia (RSZ). The GRZ said termination of the concession, which had been expected to last until 2023, was necessitated by RSZ's inefficiencies, including high levels of derailments and the loss of life and property. The concession was returned to Zambia Railway, the parastatal former operator of Zambia's railway networks.
>
> (U.S. Department of State, 2013, February)

Currently, no agreement has been reached between LAP and the government and the two parties continue to be embroiled in litigation over inter alia, the legality of the expropriation and the applicable compensation (Matambo, 2014).

11.3.2.2 Risk of breach of contract

Generally, a breach of contract occurs when: (i) a party to a contract repudiates or fails to perform their contractual obligations by the time fixed for performance under the contract; (ii) one party indicates that it will not perform its promises; or (iii) a party carries out some act which disables it from performing its obligation.[13]Sometimes, in a PPP concession arrangement, the government might breach its contractual obligations on the grounds of safety, health or other public concerns. Often, irrespective of whether these concerns are justified or not, the value of the asset would be adversely affected.

11.3.2.3 Risk of asset-specific regulation

The operating regulations of mega projects such as airports, power stations and dams are stringent and very specific due to social and environmental considerations. Minute variations in regulations or contract conditions can also have detrimental effects on revenues or cost.

11.3.3 Risks during the termination phase

Many types of risks can occur during the termination phase, such as those relating to: (i) the duration or renewal of the concession; (ii) the transfer of the asset and transfer price; and (iii) the decommissioning of the asset, i.e. tightening standards during the operation phase can increase decommissioning costs. Occasionally, concessions will be terminated before time. Often, asset transfer to a state or to a fresh concessionaire may invite disputes over the transfer price.

11.4 Risk management

Risks stem from uncertainty, which is mainly caused by lack of detailed information at the time a decision is made. Uncertainty can be defined as a situation about which there is no historical data or previous experience. An example could be a new building that utilises an innovative construction material that has not previously been used. Therefore, no historical data exist on which to make fundamental decisions on methods of working. Ultimately, most of the risks found in a construction project will have an adverse consequence on one or more of the following attributes of a project: (i) time – additional time for design, construction and subsequently the occupation of the building; (ii) cost – additional costs; (iii) quality – failure to meet the required quality. Project risk management is the culture, processes and structures adopted by an organisation and directed towards the effective management of risk in projects.

Risk may be defined as the likelihood/probability of the occurrence of an undesirable event that will have a positive or negative impact on a project's objectives. Risk management systems are employed in business firms to control risks in the business process. A thorough knowledge of the probability of different kinds of risks contributes to a more effective risk management system and, therefore, an enhanced project output and better value for both clients and contractors. The common sources of risk in construction projects are: (i) misunderstanding of contract terms and conditions; (ii) design changes and errors; (iii) poorly coordinated work; (iv) poor estimates; (v) poorly defined roles and responsibilities; (vi) unskilled staff; (vii) natural hazards; and (viii) political and legal problems. The general steps in the risk management process are risk planning, risk identification, risk analysis, risk response and risk monitoring and control.

The impact of risk events are frequently a form of loss to the client or project owner because it would mean additional expenditure and therefore a lower return on investment. The impact on the contractor will take the form of a loss of revenue. Consultants on a project have a professional responsibility of concern to their clients and in the present economic/legal environment, clients have shown an inclination towards legal action when they have suffered loss as a direct result of poor advice. Therefore, the risk will impact the client, contractors and the consultants, among others. It is very important to identify possible risk sources and underlying condition that can generate a risk event to alleviate any consequence of a risk occurring.

Each party in a construction contract has varying interests that they attempt to satisfy with favourable contract conditions. In the case of owners, the project is to be completed on time, within budget and it should be functional in every respect. They will try to ensure the contractor's maximum accountability, risk assumption and indemnity responsibilities and will seek the right to terminate the construction contract for convenience (Merwin, Linley and Steedman, 2004). Exposure to risk will be a major concern of the prime contractor and he, in turn, will try to transfer the risk to subcontractors. A subcontractor, while negotiating a deal, can assess a risk and try to manage it carefully by defining appropriate roles, responsibilities and scope as well as charging additional compensation for supplementary risk exposure. The project's architect will be made accountable for design errors and omissions.

The critical risk factors identified by many researchers in different types of international construction projects, such as residential, industrial, commercial and infrastructure, in various foreign countries are: (i) scope and design changes; (ii) technology implementation; (iii) site conditions and unknown geological conditions; (iv) inflation; (v) economic condition of the country (political and economic stability, legal system maturity, socio-cultural differences, international relations, bureaucracy, significance of the project for the country, geography and climate conditions, government attitude towards foreign investors) and the rules and regulation; (vi) the country's requirements (import–export rules, customs procedures, social security law, requirements from foreign firms); (vii) financial failure; (viii) inadequate managerial skills; (ix) improper coordination between teams; (x) lack of availability of resources; (xi) weather and climatic conditions; (xii) statutory clearance and approvals (delay in approvals, delay in progress payments); (xiii) poor safety procedures; (xiv) country's market conditions (labour, material, equipment, local supplier, local subcontractor, poor infrastructure); (xv) contract clauses (rights and obligation of parties, payment method, escalation, taxation, warranty, default of owner, force majeure, cost compensation, time extension, liquidated damages, change orders, variation of work, valuation of variations, disputes, codes and standards, etc.); (xvi) project conditions (design maturity, constructability, geo-technical conditions, location, site conditions, contract clarity, scope clarity, size, duration, payment type, project delivery system); (vii) project requirements (technical, technological, managerial, environmental impact, etc.); (xviii)poor performance of partner; (xix) company resources (financial and technical resources), staff (poor competency of staff), managerial capability, experience, relations with client, etc.; (xx) company conditions – objectives, management capability, risk response strategy, workload, business style, management style, top management support, location of management; (xxi) delay in logistics; and (xxii) delays in execution. In addition, inappropriate project formulation; host government's political continuity; attitude towards foreign investors and their profits; nationalisation/expropriation; enforceability of contracts; bureaucratic delays; communication and transportation; hostilities with neighbouring country or region; fractionalisation by language, ethnic and regional groups; corruption and dishonesty; social conflicts; foreign exchange reverses; debt and debt servicing capacity of the country also play a vital role in international construction project execution (Figure 11.5). A detailed description is given in the following parts. The risk factors as given by Qing-fu et al., (2013) are shown in Figure 11.6.

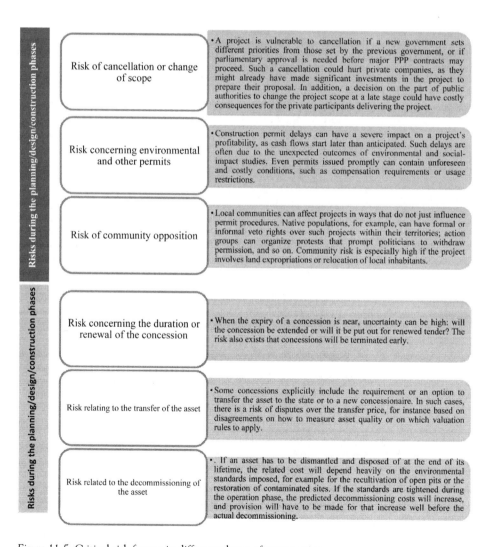

Figure 11.5 Critical risk factors in different phases of construction.

Internal Risk Factors	
Personnel risk	**Construction technology risk**
• Quality of the general worker (Including psychological diathesis, moral integrity, operation technique and efficiency) • Quality of the technical personnel (Including psychological diathesis, moral integrity and technology level) • Quality of the management personnel (Including psychological diathesis, moral integrity and management level) • Quality of the supervising personnel • Instability of the staff • Noncooperation of the employer and supervising engineer	• Backward construction technology • Unreasonable construction technology and scheme • Unsuitable protection measures of construction safety • Failure of the application of new technology and method • Half-baked consideration on the actual condition of the construction site • Unfamiliar with the design drawings and design intention • Construction not according to the drawing • Violating construction standard • Insufficient site information and unforeseeable circumstances underground • Unreasonable personnel organization and arrangement • Unreasonable materials and unreasonable equipment allocation
Design technology risk	**Contract risk**
• Adverse effect of the grade of the design department • Unsatisfactory quality of the designing materials • Validity and legality of the designing materials • Deviation of the design from construction • Design problems of new structures and new type of bridges • Insufficient understanding of the structure characteristics and immature design theory • Delay of the examining and approving of the design alteration	• Errors of omission of the bill of quantities • Errors of the unit price or total price of the project • Indeterminate or defective terms of the contract • Default of the partner
Materials and equipment risk	
• Raw materials, finished products and semi-manufactured products being in short supply • Wrong types and quantity of raw materials, finished products and semi-manufactured products • Disqualification of the quality of raw materials, finished products and semi-manufactured products • Consumption in the course of transportation, storage and construction • Restriction of the local transportation • Delay of supplying and entering the construction site of the construction equipments	• Problems of using special and new materials • Disqualification of the construction equipments • Insufficient production capacity of the construction equipments • Insufficient accessories and fuel of the construction equipments • Construction machinery breakdown and the power fault • Installation errors and debugging errors of the construction equipments • Inadequacy of the equipment maintenance or overloading operations of the construction equipments • Instability of the construction equipments and unsafe operation
External Risk Factors	
Political legal risk	**Natural risk**
• Variation of the macro policy • Discontinuity of the laws and regulations • Problems of the construction examination and approval procedure • Effect of the local regulations and specifications associated with construction • Regional protection policy • Injustice of the arbitration for disputing • Too much intervention of the government or the department in charge • Variation of the relations between the countries • Domestic conflict or unrest	• Bad weather conditions and environment (wind, extreme temperatures, flood, debris flow, earthquake and so on) • Undesirable condition of the construction site (instability of water supply, power supply and gas supply) • Adverse geographical location
	Social risk
	• Disordered social public order • Corrupt social morality • Too low cultural quality
Economic risk	
• Adverse situation of macro economy • Severe currency inflation • Bad credit of the insurance companies and bank	• Variation of the local and national tax policy • Adjustment of the national interest • Increase of the wages and welfare of the staff

Figure 11.6 Risk factors.

11.5 Risk management process in construction projects

Risk management includes maximising the opportunities and the impact of positive events and, at the same time, minimising the probability and the impact of negative events in order to meet the project's objectives (Tipili and Ibrahim, 2015). The risk management process consists of: identifying, assessing and analysing, and responding (Smith, Mernaand Jobbling, 2006). The risk management process, as per Cooper et al. (2005), involves the systematic application of management policies, processes and procedures to the task of establishing the context and identifying, analysing, assessing, treating, monitoring and communicating risks. One of the earliest studies (Chapmanand Ward, 1997) outlined a generic risk management process comprising nine phases: (i) define the key aspects of the project; (ii) focus on

Figure 11.7 Risk management process in construction projects.

a strategic approach to risk management; (iii) identify where risks may arise; (iv) structure the information about risk assumption and relationships; (v) assign ownership of risks and responses; (vi) estimate the extent of uncertainty; (vii) evaluate the relative magnitude of the various risks; (viii) plan a response; and (ix) manage by monitoring and controlling execution (Figure 11.7).

11.5.1 Major processes of project risk management

The construction industry can be subject to an exceptionally broad range of risks and uncertainty. Procuring a building from inception to commission may involve a great number of people all with vastly different specialist skills and responsibilities. Each new project entails unique design and construction problems because most construction projects are bespoke.

Literature on different risk management processes, in general, involves: (i) identifying/determining risks that are likely to affect the project's objectives and documenting the characteristics of each; (ii) evaluating risks and risk interactions to assess the range of possible project outcomes; (iii) defining enhancement steps for opportunities and responses to threats; and (iv) responding to changes in risk over the course of a project (Figure 11.8). Risks have to be identified, quantified and prioritised; then, a plan must be developed to eradicate or reduce them.

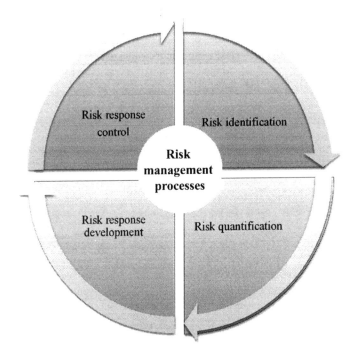

Figure 11.8 Major processes of project risk management.

11.5.2 Risk identification

Risk identification is the primary stage of the risk management process. Without first iden-tifying what the potential risks are, it cannot be ascertained: (i) if they will arise; (ii) what effect they might have if they do arise; and (iii) what measures need to be taken to prevent their occurrence. Clearly, the identification of risks may be considered the most important stage in the risk management process. When identifying risks, previous experience should be considered an invaluable asset. All project team members should be encouraged to share any relevant knowledge gained from previous circumstances of a similar nature.

Frequently identified risks include owner's delayed payment to the contractor; owner's unreasonably imposed tight schedule; change of design required by the owner; lack of work definition by the owner; low productivity of labour and equipment; delay in material supply by suppliers; and changes in the laws and regulation. The delayed payment to the contractor can lead to late payment of salaries, time overrun of the project, cash flow problems, slow-down in progress of the work till receipt of payment, difficulty in procuring materials and services, difficulty in tendering for new projects, subcontractor's refusal to continue works on the project, bad reputation of the contractor and high interest rates due to loans.

Identifying risks is the most important phase of the risk management process as no action can be taken on a risk if it has not been recognised. It is aimed at determining potential risks, i.e. those that may affect the project. Identifying and recording all risks are essential since risks that are not identified at this stage may perhaps be excluded from further analysis. Therefore, the risk identification process should incorporate all risks regardless of whether or not the risks are within the direct control of the organisation, covering the whole lifecycle. Risks that are not within the direct control or external risks (political, economic, social risk or weather related) are those that are outside the control of the project management team. Risks that are within the direct control or internal risks (resource related, project member or team related, designer, contractor, subcontractor, suppliers related, construction site related, documents and information-related risk) may be separated in relation to the party who might be the originator of the risk events such as stakeholders, designer or contractor. A project risk register should be prepared and reviewed periodically. The approaches used to identify risks could include the use of checklists, judgements based on experience and records, flow charts, brainstorming, systems analysis, scenario analysis and system engineering techniques.

11.6 Risk assessment

Risk assessment/analysis, the next step in the risk management process, involves critically evaluating prospective risks, arranging them as per their importance and permitting the management team to select the important ones. During risk assessment, identified risks are evaluated and ranked. The goal is to prioritise risks requiring management. Following their identification, the risks should be assessed. Empirical research on risk assessment studies confirms that construction companies use both qualitative and quantitative techniques for assessing project risks (Baker, Ponniahand Smith, 1998). Therefore, the risk assessment phase normally falls into two distinct steps: qualitative assessment and quantitative assessment. A study has revealed that qualitative analysis is the most widely applied technique in the Chinese construction industry, while the use of quantitative methods less so (Tanget al., 2007). Dikmen and Birgonul (2006) uses an analytic hierarchy process to calculate the risk and opportunity ratings for both risk and opportunity assessment of international projects. Zeng et al. (2007) has put forward a risk assessment methodology rooted in fuzzy reasoning techniques and aimed at managing risks in complex projects. In their study, Osman et al. (2014) used the relative importance index (RII) method to prioritise the project risks and subsequently categorise projects by fuzzy AHP and fuzzy TOPSIS methods and found that these techniques are able to assess the overall risks of construction projects. Quantitative methods are applied to determine the probability and impact of the risks identified (Winch, 2002). Generally, the quantitative methods used are: scenario techniques– Monte Carlo simulation; modelling techniques– sensitivity analysis; and diagramming techniques– decision tree analysis (fault tree analysis and event tree analysis). 'Qualitative risk assessment provides a means to categorize potential risks in terms of their priority, allowing project managers to decide whether to proceed with quantitative assessment or move directly to risk response planning' (El-Sayegh and Mansour, 2015). The majority of the existing approaches provide risk rating; in reality only a few quantify risk.

11.6.1 Project risk analysis techniques

The process of project risk analysis requires appropriate and professional tools/techniques. These techniques are classified as qualitative and quantitative techniques. The qualitative risk analysis technique presents results in the form of qualitative descriptions, where risk assessment is connected with qualitative scales for the probability and the impact of the consequences of risk (De Marco and Muhammad J. T., 2014). It can be used to prioritise risks for further analysis by assessing their probability of occurrence and impact. Quantitative tools for risk analysis rely on the application of numerical measures to express the level of risk. Figures 11.9 and 11.10 present a brief description of the qualitative and quantitative risk analysis techniques generally used.

Even though diverse tools and methods are used for identifying risks, the common techniques used are brainstorming, the Delphi method, interview and cause analysis. Strengths, weaknesses, opportunities and threats (SWOT) analysis and presumption analysis are used to investigate the larger scope of possible events in risk identification. Figure 11.11 shows the common techniques used to identify risks in construction projects in developing countries (Bahamid and Doh, 2017).

Once the project team has been briefed on the use of risk management for the project, a list of risk categories is sent to each member as a guide to identify risk issues and to generate initial discussions. Risks are usually pinpointed through one-to-one interviews, by questionnaire, during brainstorming sessions or in risk workshops at which all members of the project team are present. The client is a key participant in these sessions and, since he is the main stakeholder, will have a prominent view on possible exposures. His attitude to risk will ultimately determine the direction for further investigation and action.

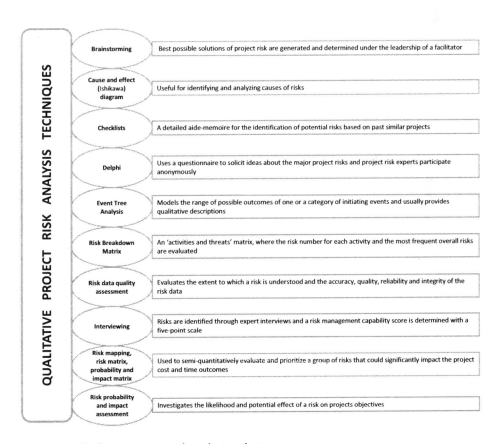

Figure 11.9 Qualitative project risk analysis techniques.

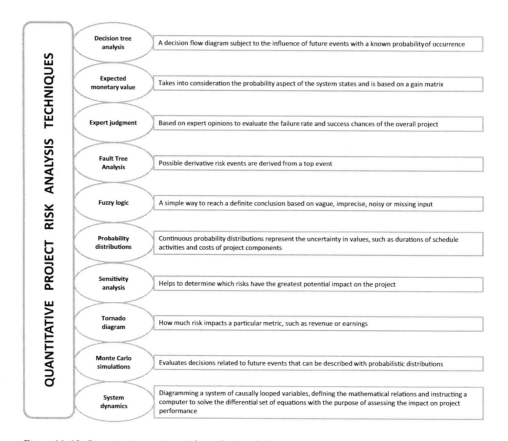

Figure 11.10 Quantitative project risk analysis techniques.

The identified risks are listed in a risk register. Risk registers can be used constructively to record, develop and review the process. Results can then be incorporated into a database which provides a basis for identifying and statistically quantifying future projects.

This information base should also record details of all incidents occurring on the project, even if they did not result in losses or claims. These near misses are indicative of the areas where risks may present themselves in the future and will enable action to be taken to mitigate them.

11.6.1.1 Brainstorming

This is a very effective procedure for identifying risks and the subsequent response/mitigation thereof. Brainstorming provides a select group with the opportunity to imagine every likely risk. Brainstorming involves open, frank and in-depth discussion with the project participants about their concerns, the areas of uncertainty/hazards, the likely risks, the likelihood of any risks occurring, the potential impact of risks and project participants' initial response to identified risks. Brainstorming is based on a synergy effect, where group thinking is more productive than individual thinking, where ideas can be combined or further built on by others. Brainstorming is also based on the avoidance of criticism which improves the production of ideas; where more ideas produce a higher chance of more feasible ideas and

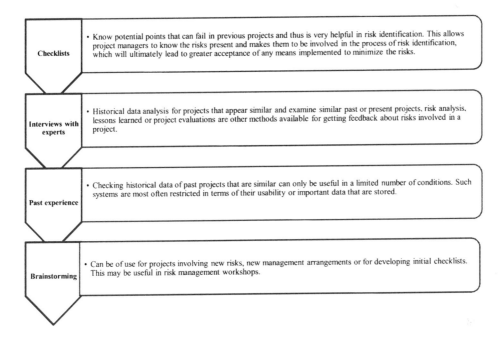

Figure 11.11 Common techniques used to identify risks in developing countries.

without criticism people are encouraged to think more outside the box, which may provide new discoveries. The identified risk will be explained in terms of its course of impact and its effect on the project, and later will form part of a risk log containing the risk description, its probability of occurrence and its potential impact and risk response/mitigation.

Brainstorming takes the form of an informal meeting with no set agenda and no definite time scale. These informal brainstorming sessions should normally be attended by the client, project manager, risk manager and, if applicable, appropriate members of the design team and the end users of the development. It will be more meaningful with the inclusion of the prime contractor and any specialist suppliers/subcontractors. On some occasions, it may be necessary to promote discussion through the use of checklists (a sample checklist is given in Figure 11.12). An experienced risk manager can facilitate meaningful discussion and document the proceedings. Identifying risks at an early stage and including cost contingencies for them could become the basis for a financial claim at a later date.

11.6.1.2 The Delphi technique

The Delphi technique is a method of systematic collection, through carefully designed questionnaires, and collation of judgements from isolated suitable individuals on a specific topic. It attempts to produce objective results from subjective discussions. This method may be applicable in identifying risks but it is more suited to assessing the likelihood of an occurrence and the potential impacts of previously identified risk events. This method essentially involves the sequence of events presented in Figure 11.13.

The main advantage of this method is that all participants act independently; no strong personalities dominate nor is there any peer pressure or bullying. An external expert can be

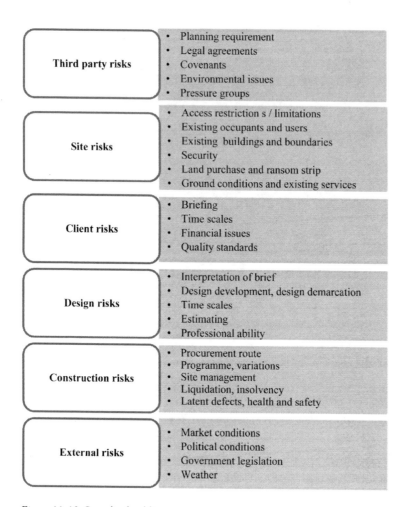

Figure 11.12 Sample checklist.

used in any location. Its main problem is that it is a very repetitive technique and thus time-consuming. Conflicting or incompatible ideas can result from the feedback. In addition, while a consensus opinion may be reached, this often means an opinion that no one person offered in the first instance.

Following vigorous examination, this identification stage should eventually produce a schedule of risks drawn from a consensus of views of the project members. However, it should be realised that risks can be identified all through a project's life. To facilitate the inclusion of these risks in the risk management process, a method of documenting new risks must be developed.

11.6.1.3 *Qualitative assessment*

Qualitative assessment entails the correct registration of the identified risks. An appropriate method of formalising this process is to utilise a qualitative assessment pro forma accompanied by a risk register, which may include, but is not limited to: classification and

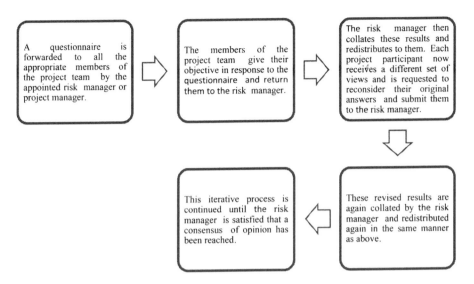

Figure 11.13 The Delphi technique – sequence of events.

reference; description of risk; relationship of risk to other risks (interdependencies); cause of risk, potential impact; likelihood of occurrence (probability); prioritisation of risk, response/mitigation strategy; and risk responsibility.

11.6.1.4 Risk categorisation

Risk categorisation is a way of classifying project threats along with their sources to facilitate identification of areas or stages of the project that are most exposed to those risks and uncertainties. Risk categorisation is an aid to identifying the source of risk and could include: environmental: site conditions, health and safety; contractual: client, contractor, third party (subcontractor), information, protection; and design: planning permission, fire officer's requirements. A risk breakdown structure can be constructed to prioritise risks according to how quick a response they require.

11.7 Risk response

Risk responses are actions taken to avoid, transfer, reduce or eliminate the potential negative impact or threat to a project and its objectives in a systematic manner. Risk response can be defined as the identification, selection, evaluation and action to implement a project (Zhang and Fan, 2014). Risk identification and assessment enable risk managers to decide on suitable risk response strategies. Risk response is the action that the risk management team usually takes against risks. This is the action required to reduce, eradicate or avoid the potential impact of risks on a project. For each risk identified, the risk or project manager must consider alternative methods to mitigate the risk by using the most appropriate process.

There are two types of response to risk: (i) an immediate change or alteration to the project which usually results in the elimination of the risk; and (ii) a contingency plan that will only be implemented if an identified risk materialises.

In order to mitigate the potential impact of any risk, the project manager or his designated risk manager must consider alternative options for action, evaluate the consequences and decide whether or not that action should be taken. As a component of the risk management process, the major intention of any response and mitigation strategy is to initiate and implement proper action to prevent risks from occurring or, at minimum, limit the potential damage they may cause. Furthermore, through the use of adequate and appropriate contingency plans, if the occurrence of a risk is unavoidable, its impact should be limited to the contingency levels contained within the overall project allowances. This should ensure that the overall project objectives of time, cost and quality are not jeopardised.

Mitigation strategies may be formulated to respond to individual or groups of risk as deemed appropriate by the project/risk manager. However, strategies must be agreed, recorded and documented with all responsibilities clearly stated. Specifically, it is obvious that high probability/high impact risks will require rigorous and thorough discussion and examination to ensure the formulation of an appropriate and credible response and mitigation strategy.

11.7.1 Options for responding to risk

The options for dealing with risks are:(i) avoidance; (ii) reduction; (iii) transfer; (iv) sharing; and (v) retention. Each option should be assessed as one or more will apply in every circumstance (Figure 11.14). A number of questions must first be asked to facilitate identification of which route(s) should be adopted: (i) is the risk controllable or uncontrollable; (ii) who is best placed to influence/deal with the source and outcome of the risk; (iii) what secondary or resultant risks arise as a result of the action taken; and (iv) is the cost of mitigating the risk acceptable when compared to the potential impact of the risk itself?

11.7.1.1 Risk avoidance

Risk avoidance may include a review of the overall project objectives leading to a reappraisal of the project as a whole. Risk avoidance is often perceived as the ultimate mitigation strategy in that it implies that the project may be aborted (Ashworth and Perera, 2015). The solution for risks with a significant impact on a project is either to avoid them by changing the scope of the project or cancel the project in the worst case scenario.

> By avoiding a risk exposure, the contractor knows that he will not experience the potential losses that the risk exposure may generate. On the other hand, however, the contractor loses the potential gains (opportunity) that may have been derived from

Figure 11.14 Options for dealing with the risks.

assuming that exposure. A contractor may avoid the political and financial risks associated with a project in a particular unstable country by not bidding on projects in this country.

(Al-Bahar and Keith, 1990)

This method of mitigation involves the removal of the cause of the risk and therefore the risk itself. Another instance of risk avoidance comprises the use of exemption clauses in contracts, either to avoid certain risks or to avoid certain consequences following from the risks so that the potential liability can be avoided. Risk avoidance is most likely to take place where the level of risk is at a point where the project is potentially unviable.

11.7.1.2 Risk reduction

Several risks which exceed acceptable levels are not pressurising enough to be avoided, transferred or shared with other project participants, however, they are still too big enough to retain or ignore. In such cases, procedures are undertaken to diminish risk exposure, either by decreasing the probability of the event or mitigating its potential impact on a project (Jones and Saad, 2003). This method adopts an approach whereby potential exposure to risks and their impact or the severity of the financial loss is alleviated or reduced. Methods of risk reduction may require some initial investment which should then reduce the likelihood of the risk occurring. Risk reduction occurs where the level of risk is unacceptable and alternative action is available. Typical actions to reduce risk are:

- Detailed site survey where adverse ground conditions are identified but the severity is not established. Conducting a detailed ground investigation can lead to preparation of a proper estimate.
- An alternative contract management strategy can reduce risks by allocating the risks between project participants in a different way. Risk can be reduced by outsourcing or appointing a subcontractor to execute the project in some cases.
- Changes in design to accommodate the findings of the risk identification process.

Risk reduction exercises are always worthwhile because they can lead to greater knowledge about a project, thereby reducing the potential impact of risks and the level of uncertainty. Risk reduction invariably leads to greater confidence in a project's outcome. Even though risk reduction gives rise to an increase in the base cost (i.e. the estimate of certain items), it should offer a significantly greater reduction in the level of contingency required. Therefore, risk reduction should only be adopted where the resultant increase in costs is less than the potential loss that could be caused by the risk being mitigated.

It is worth noting that some of the literature on the subject views insurance as risk reduction while others consider it as the transference of risk. Neither is erroneous, the most significant issue being that the risk is dealt with through an appropriate response/mitigation.

11.7.1.3 Risk transfer

One of the risk handling options is risk transfer. Risk transfer plays a significant role in infrastructure development projects and entails the entire or partial transfer of risks among the different parties involved in the execution of a project, i.e. through moving the responsibility to another party or risk sharing through a contract, insurance, partnering or joint venture. Commonly, risks are transferred through the placement of contracts, the appointment of specialist subcontractors or suppliers or by taking out an insurance policy. The client transfers the risk to the contractor or designer, the contractor transfers to his subcontractors, the contractor and subcontractors to their guarantee a bank by means of a warranty, bond or guarantee (Rogers and Duffy, 2012), or the entire project team can transfer to the insurance company (Knecht, 2002).As with risk reduction, whenever a risk is transferred a premium is usually paid. Again, this premium should be less than the potential impact of the risk.

The transference of risk should comprise the passing of risks to those better placed or more capable to maintain control and influence the outcome of the risk. Transference should never be viewed as a negative risk response. Its intention is not to pass the buck by making someone else responsible. Further, it should not be used in a penal or punitive manner as a protective mechanism for other project participants. For risks to be managed properly, an incentive may be required.

When transferring risk, it is important to differentiate between the transference of the risk itself and the allocation of risk responsibility. Where a risk is transferred, the intention should be to transfer the whole of the risk including its potential impact. Where the responsibility for the risk is allocated to a project participant, time, cost and quality repercussions remain which may still adversely affect the project's outcome.

11.7.1.4 Risk sharing

Risk sharing is when a portion of a risk is transferred while some of the risk is retained. This approach may be adopted where the risk exposure is beyond the control of one party. In such instances, it is imperative that each party appreciates the value of the portion of the risk for which it is responsible.

The dilemma of risk sharing is who should bear the loss when a risk occurs (Miceli, 1997). Another study has proposed that risk sharing in contract law gives rise to the problem of who or which party would have borne the loss if they could have foreseen that contingency (Posner and Rosenfield, 1977). The complication here is who can prevent or control the risk more efficiently. Similarly, if the risk cannot be prevented or controlled, which party is in a better position to protect against the loss? In their study, Kobayashi, Omoto and Onishi (2006) stated two risk-sharing principles: (i) the party who can assess and control the risk should bear it; and (ii) if none of the parties can assess or control the risk, the party who can bear it easier or procure the insurance from the market should bear it. Sometimes, in practice, if a risk, such as adverse ground conditions, is allocated to a contractor but it is having greater financial implications, the contractor bears the cost up to an agreed amount and the employer bears the balance, after negotiations.

11.7.1.5 Risk retention

Many risks will continue after exploring all possibilities for response and mitigation of such risks. Risk retention, commonly known as risk acceptance, refers to facing a risk by retaining any potential loss rather than transferring that risk to an insurer or other party. Risk retention can be divided into active risk retention and passive risk retention (Figure 11.15). The former denotes that the risk is identified consciously in a planned manner, while the latter is risk which managers do not realise throughout the planning of project (ignorance) or acknowledges the existence of risk or are not prepared to deal with a project risk because it is a minor risk. Both these methods have been utilised extensively in the construction

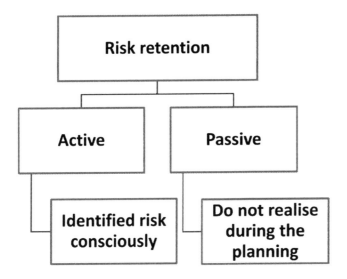

Figure 11.15 Risk retention.

industry. The reduction of these types of risks is an approach employed to reduce the probability and impact of the risk below an adequate threshold and risk sharing is primarily realised through a contractual mechanism to develop a sense of joint responsibility among the project stakeholders (Loosemore, Raftery, Reilly and Higgon, 2006).

It is generally believed that several firms in the construction industry in developing countries use brainstorming and checklists in risk management but do not practice formal risk management principles. Risks must be identified before a project starts and it is a cyclical process. Continuous attempts are required to assess the probability and adverse effects of new risks, in different phases of construction, in relation to existing risks.

11.8 Conclusion

Project risk is any possible event that can negatively affect a project's objectives. The risk can be managed, minimised, shared, transferred or accepted. Project risk should be identified, analysed and managed throughout the life of a project. The project risk management process generally consists of risk management planning, the identification of risk events, a qualitative and quantitative risk analysis and risk response and risk monitoring and control.

Notes

1 *See* https://www.ice.org.uk/ICEDevelopmentWebPortal/media/Documents/News/ICE%20News/ Global-Construction-press-release.pdf Accessedon 2 March 2018.
2 *See* https://www.enr.com/toplists/2017-Top-250-International-Contractors-3 Accessed on 2 March 2018.
3 Some of the data and other information given in the initial parts of this chapter may be similar to Chapter 10 since this chapter also includes the risksininternational projects.
4 *Economic Times.* (2014, May) *Project Exports to Get a Boost: Exim Bank.* Retrieved from https://ec onomictimes.indiatimes.com/industry/transportation/shipping-/-transport/project-exports-to-get-a-boost-exim-bank/articleshow/35010726.cms Accessed on 2 March 2018.

5 *Economic Times.* (2016, May) *Exim Bank Net Profit Plunges 56.5% to Rs 316 crore in FY16.* Retrieved from
    https://economictimes.indiatimes.com/markets/stocks/earnings/exim-bank-net-profit-plunges-56
    -5-to-rs-316-crore-in-fy16/articleshow/52365229.cms Accessed on 2 March 2018.
6 FIDIC. (2006, January) *Managing Payment Risks on International Construction Projects.* Retrieved from   http://fidic.org/sites/default/files/11%20Managing%20Payment%20Risks%20on%20Inte
    rnational%20Construction%20Projects.pdf Accessed on 5 March 2018.
7 IISD. (2015, August) Currency risk in project finance. [Discussion paper]International Institute
    for Sustainable Development. Retrieved from https://www.iisd.org/sites/default/files/publications/c
    urrency-risk-project-finance-discussion-paper.pdf Accessed on 5 March 2018.
8 *See* https://www.oecd.org/dac/peer-reviews/Currency-Exchange-Fund.pdf Accessed on 5 March
    2018.
9 *See*   https://ac.els-cdn.com/S1877042813004552/1-s2.0-S1877042813004552-main.pdf?_tid=
    78770138-e805-11e7-a9fc-00000aab0f6c&acdnat=1514049765_d08a228a8639e482d65e73
    7702dcc35e Accessed on 2 May 2018.
10 *See*   http://www.iaea.org/inis/collection/NCLCollectionStore/_Public/31/007/31007028.pdf
    Accessed on 2 May 2018.
11 Borrowed from World Bank's methodology for *Doing Business* –Dealing with Construction Permits.
    Retrieved   from   http://www.doingbusiness.org/Methodology/dealing-with-construction-permits
    Accessed on 2 May 2018.
12 *See*   https://books.google.co.in/books?id=i19ac3wnJ88C&pg=PA371&lpg=PA371&dq=expro-
    priation+of+assets&source=bl&ots=   o39ksPi6DB&sig=TD8if9XC84RZInqxF86X26_5yoM&
    hl=en&sa=X&ved=0ahUKEwjO1YmXy4zYAhWIu48KHRtaCmw4ChDoAQhJMAc#v=onep
    age&q=expropriation%20of%20assets&f=false Accessed on 13 December 2017.
13 *See* https://www.hkis.org.hk/hkis/general/events/cpd-2015054.pdf Accessed on 15 December
    2017.

References

Al-Bahar, Jamal F. and Crandall, Keith C. (1990) Systematic risk management approach for
    construction projects. *Journal of Construction Engineering and Management, 116*(3), pp. 533–546.
    Retrieved from: https://ascelibrary.org/doi/pdf/10.1061/(ASCE)0733-9364(1990)116%3A3(533)
    Accessed on 12 March 2018.
Antonioua , Fani, Aretoulisb, Georgios N.,Konstantinidisc, Dimitrios and Kalfakakou, Glykeria P.
    (2013) Complexity in the evaluation of contract types employed for the construction of highway
    projects. *Procedia – Social and Behavioral Sciences, 74*, pp. 448–458.
Ashworth, Allan and Perera, Srinath (2015) *Cost Studies of Buildings*, 6th edition. Routledge, London.
Bahamid, R.A. and Doh, S.I. (2017) A review of risk management process in construction projects
    of developing countries. *IOP Conference Series: Materials Science and Engineering, 271*(1), 012042.
    Retrieved from: http://iopscience.iop.org/article/10.1088/1757-899X/271/1/012042/pdf Accessed
    on 22 February 2018.
Baker, S., Ponniah, D. and Smith, S. (1998) Techniques for the analysis of risks in major projects. *The
    Journal of the Operational Research Society, 49*(6), pp. 567–572.
Chapman, C. and Ward, S. (1997) *Project Risk Management: Processes, Techniques and Insights*.John
    Wiley, Chichester.
Cooper, D., Grey, S., Raymond, G. and Walker, P. (2005) *Project Risk Management Guidelines:
    Managing Risk in Large Projects and Complex Procurements.* John Wiley & Sons Ltd., Chichester.
Dikmen, Irem and Birgonul, M. Talat. (2006) An analytic hierarchy process based model for risk and
    opportunity assessment of international construction projects. *Canadian Journal of Civil Engineering,
    33*(1), pp. 58–68.
Economic Times (2014, May) Project exports to get a boost: Exim Bank. https://economictimes.ind
    iatimes.com/industry/transportation/shipping-/-transport/project-exports-to-get-a-boost-exim-ba
    nk/articleshow/35010726.cms Retrieved on March 2, 2018.

Economic Times. (2016, May) Exim Bank net profit plunges 56.5% to Rs 316 crore in FY16. https:/
/economictimes.indiatimes.com/markets/stocks/earnings/exim-bank-net-profit-plunges-56-5-to-rs
-316-crore-in-fy16/articleshow/52365229.cms Retrieved on March 2, 2018.

El-Sayegh, S.M. and Mansour, M.H. (2015) Risk assessment and allocation in highway construction
projects in the UAE. *Journal of Management in Engineering*, 31(6), pp. 04015004–04015011.
Retrieved from: https://ascelibrary.org/doi/pdf/10.1061/%28ASCE%29ME.1943-5479.0000365
Accessed on 22 February 2018.

FIDIC. (2006, March) *FIDIC Conditions of Contract for Construction: For Building and Engineering
Works Designed by the Employer*, Multilateral Development Bank Harmonised Edition,
FédérationInternationale des Ingénieurs-Conseils. Retrieved from: https:// fidic.org/sites/default/f
iles/cons_mdb_gc_v2_unprotected.pdf Accessed on 2 June 2017.

Gunhan, S. and Arditi, D. (2005) Factors affecting international construction. *Journal of Construction
Engineering and Management*, 131(3), pp. 273–282.

Han, S.H. and Diekmann, J.E. (2001) Approaches for making risk-based go/no-go decisions for
international projects. *Journal of Construction Engineering and Management*, 127(4), pp. 300–308.

Hastak, M. and Shaked, A. (2000) ICRAM-1: Model for international construction risk assessment.
Journal of Management in Engineering, 16(1), pp. 59–69.

Hoti, S. and McAleer, M. (2004) An empirical assessment of country risk ratings and associated
models. *Journal of Economic Surveys*, 18(4), pp. 539–587.

Isık, Z., Arditi, D., Dikmen, I. and Birgonul, M.T. (2010) The role of exogenous factors in the strategic
performance of construction companies. *Engineering, Construction and Architectural Management*,
17(2), pp. 119–134.

Jones, M. and Saad, M. (2003) *Managing Innovation in Construction*. Telford, London.

Kapila, P. and Hendrickson, C. (2001) Exchange rate risk management in international construction
ventures. *Journal of Management in Engineering*, 17(4), pp. 186–191.

Knecht, B. (2002) Fast-track construction becomes the norm: Client, architect, and construction
manager must perform delicate balancing act to shrink the construction process and save time and
money. *Architect Record*, 190(2), p. 123.

Kobayashi, Kiyoshi, Omoto, Toshihiko and Onishi, Masamitsu (2006) Risk-sharing rule in project
contracts. [Paper presentation]. In: *23rd International Symposium on Automation and Robotics in
Construction ISARC 2006 Proceedings October 3–5, 2006*, Tokyo, Japan. Retrieved from: https://ww
w.iaarc.org/publications/fulltext/isarc2006-00181_200606131122.pdf Accessed on 12 March 2018.

Loosemore, M., Raftery, J., Reilly, C. and Higgon, D. (2006) *Risk Management in Projects*, 2nd edition.
Taylor &Francis, Abingdon.

De Marco, Alberto and Muhammad Jamaluddin Thaheem (2014) Risk analysis in construction
projects: A practical selection methodology. *American Journal of Applied Sciences*, 11(1), pp. 74–84.
Retrieved from: http://thescipub.com/pdf/10.3844/ajassp.2014.74.84 Accessed on 20 March 2018.

Matambo, N.F. (2014) The threat of expropriation in commercial contracts entered into with states:
Lessons from the case of lap green networks of Libya and the Zambian government (Dissertation),
The University of Cape Town, South Africa. Retrieved from: https://open.uct.ac.za/bitstream/item
/9347/thesis_law:2014_matambo_nf.pdf?sequence=1 Accessed on 15 December 2017.

Merwin, Bruce W., Linley, Joanne and Steedman, Tracy L. (2004) Critical construction contract
clauses: Owner, contractor, and subcontractor perspectives. *Probate and Property*, pp. 1–11, The
American Bar Association. Retrieved from: https://www.pecklaw.com/wp-content/uploads/2014
/09/ABA_PP_v028n05__critical_construction_contract_clauses.pdf Accessed on 14 March 2018.

Miceli, T.J. (1997) *Economics of the Law: Torts, Contracts, Property, Litigation*.Oxford University Press,
New York.

Osman, Taylan, Bafail, Abdallah O., Abdulaal, Reda M.S. and Kabli, Mohammed R. (2014)
Construction projects selection and risk assessment by fuzzy AHP and fuzzy TOPSIS methodologies.
Applied Soft Computing, 17, pp. 105–116.

Ozorhon, B., Dikmen, I. and Birgonul, M.T. (2007) Performance of international joint ventures in
construction. *Journal of Management in Engineering*, 23(3), pp. 156–163.

Posner, R. and Rosenfield, A. (1977) Impossibility and related doctrines in contract law: An economic analysis. *The Journal of Legal Studies*, 6(1), pp. 83–118.

Qing-Fu, Li, Zhang, Peng and Fu, Yan-Chao (2013) Risk identification for the construction phases of the large bridge based on WBS-RBS. *Research Journal of Applied Sciences, Engineering and Technology*, 6(9), pp. 1527–1528. Retrieved from: https://maxwellsci.com/print/rjaset/v6-1523-1 530.pdf Accessed on 17 December 2017.

Rogers, Martin and Aidan Duffy. (2012) Engineering Project Appraisal, 2nd edition. John Wiley & Sons, Chichester.

Smith, N.J., Merna, T. and Jobbling, P. (2006) *Managing Risk in Construction Projects*, 2nd edition. Blackwell Publishing, Oxford.

Tang, W., Qiang, M., Duffield, C., Young, D.M. and Lu, Y. (2007) Risk management in the Chinese construction industry. *Journal of Construction Engineering and Management*, 133(12), pp. 944–956.

Tipili, L.G. and Ibrahim, Y. (2015) Identification and assessment of key risk factors affecting public construction projects in Nigeria: Stakeholders' perspectives. [Paper presentation]. In: *Proceedings 2nd Nigerian Institute of Quantity Surveyors Research Conference*, pp. 707–721, Federal University of Technology (Akure), The Nigerian Institute of Quantity Surveyors, Nigeria.

U.S. Department of State. (2013, February) *2013 Investment Climate Statement*.Bureau of Economic and Business Affairs, Zambia. Retrieved from: https://2009-2017.state.gov/e/eb/rls/othr/ics/2013 /204763.htm Accessed on 15 December 2017.

United Nations. (2012) *Expropriation*. UNCTAD Series on Issues in International Investment Agreements II, New York and Geneva. Retrieved from: http://unctad.org/en/Docs/unctaddiaeia20 11d7_en.pdf Accessed on 15 December 2017.

Winch, G. (2002) *Managing Construction Projects, an Information Processing Approach*. Blackwell Publishing, Oxford.

World Economic Forum and the Boston Consulting Group. (2014) Strategic infrastructure mitigation of political and regulatory risk in infrastructure projects, p. 12, World Economic Forum, Geneva, Switzerland. Retrieved from: http://www3.weforum.org/docs/WEF_Risk_Mitigation_Report_2015 .pdf Accessed on 12 March 2018.

ZengJiahao, Min An and Smith, Nigel John (2007) Application of a fuzzy based decision making methodology to construction project risk assessment. *International Journal of Project Management*, 25(6), pp. 589–600.

Zhang, Y. and Fan, Z.P. (2014) An optimization method for selecting project risk response strategies. *International Journal of Project Management*, 32(3), pp. 412–422.

Zhi, H. (1995) Risk management for overseas construction projects. *International Journal of Project Management*, 13(4), pp. 231–237.

Bibliography

ADB. (1994) *Framework for the Economic and Financial Appraisal of Urban Development Sector Projects: A Reference Guide for Bank Staff, Consultants and Executing Agencies*, Economics and Development Resource Center, Infrastructure Department, Asian Development Bank, Philippines. Manila.

ADB. (2017) *Guidelines for the Economic Analysis of Projects*, Asian Development Bank, Philippines. Manila.

Aibinu, A. and Jagboro, G. (2002) The effects of construction delays on project delivery in Nigerian construction industry. *International Journal of Project Management, 20*(8), pp. 593–599.

Alinaitwe, H.M., Mwakali, J.A. and Hansson, B. (2007) Factors affecting the productivity of building craftsmen-studies of Uganda. *Journal of Civil Engineering and Management, 13*(3), pp. 169–176.

Arditi, D. and Mochtar, K. (2000) Trends in productivity improvement in the US construction industry. *Construction Management and Economics, 18*(1), pp. 15–27.

Ashworth, A. (2006) *Contractual Procedures in the Construction Industry*, 5th edition. Unitec, New Zealand.

Ashworth, A. (2012) *Contractual Procedures in the Construction Industry*, 4th edition. Pearson, Harlow, England.

Assaf, S.A. and Al-Hejji, S. (2006) Causes of delay in large construction, projects. *International Journal of Project Management, 24*(4), pp. 349–357.

Baccarini, D. and Archer, R. (2001) The risk ranking of projects: A methodology. *International Journal of Project Management, 19*(3), pp. 139–145.

Bentley, J. (1987) *Construction Tendering and Estimating*, Routledge, London. E.& F.N. Spon.

Brent, R.J. (1996) *Applied Cost-Benefit Analysis*, Edward Elgar, Cheltenham (UK).

Brook, M. (2004) *Estimating and Tendering for Construction Work*, 3rd edition. Butterworth Heinemann, Boston, MA.

Chan, D.W.M. and Kumaraswamy, M.M. (2002) Compressing construction duration: Lesson learned from Hong Kong building projects. *International Journal of Project Management, 20*(1), pp. 23–35.

Chan, P.W. and Kaka, A. (2007) Productivity improvements: Understand the workforce perceptions of productivity first. *Personnel Review, 36*(4), pp. 564–584.

Chapman, C. and Ward, S. (2002) *Managing Project Risk and Uncertainty*, Wiley, Chichester.

Chitkara, K.K. (2004) *Construction Project Management*, 4th edition. Tata McGraw Hill, New Delhi.

Choudhry, R.M. (2017) Achieving safety and productivity in construction projects. *Journal of Civil Engineering and Management, 23*(2), pp. 311–318.

Clamp, H. and Cox, S. (1998) *Which Contract: Choosing Appropriate Building Contract*, 5th edition. RIBA Pub., Landon.

Clegg, S. (1992) Contracts cause conflict. In: P. Fenn and R. Gameson (Eds), *Construction Conflict: Management and Resolution*, pp. 128–144, Chapman & Hall, London.

Clough, Richard H., Sears, Glenn A., Keoki Sears, S., Segner, Robert O. and Rounds, Jerald L. (2015) *Construction Contracting: A Practical Guide to Company Management*, 8th edition. John Wiley & Sons, Incorporated. Hoboken, New Jersey.

Cooper, D., Gray, S., Raymond, G. and Walker, P. (2005) *Project Risk Management Guidelines - Managing Risk in Large Projects and Complex Procurements*, Wiley, Chichester.

CPWD. (2019) *CPWD Works Manual 2019*, Central Public Works Department, Government of India, New Delhi.

Crawley, J. (1998) *Constructive Conflict Management*, 4th edition. Nicholas Brealey, London.

Dai, J., Goodrum, P.M., Maloney, W.F. and Srinivasan, C. (2009) Latent structures of the factors affecting construction labor productivity. *Journal of Construction Engineering and Management, 135*(5), pp. 397–406.

Dinwiddy, C. and Teal, F. (1996) *Principles of Cost-Benefits Analysis for Developing Countries*, Cambridge University Press, Cambridge.

Dixon, J.A., Scura, L.F., Carpenter, R.A. and Sherman, P.B. (1994) *Economic Analysis of Environmental Impact*, Seconda Edizione. Earthsca Publications, London.

EIB. (2013, April) *The Economic Appraisal of Investment Projects at the EIB*, Projects Directorate, European Investment Bank, Luxembourg.

Fenn, P., Lowe, D. and Speck, C. (1997) Conflict and dispute in construction. *Construction Management and Economics, 15*(6), pp. 513–518.

FIDIC. (2006, March) *Conditions of Contracts for Construction Works* (MDB Harmonized Edition). International Federation of Consulting Engineers, Geneva.

Fisher, T. and Ranasinghe, M. (2000) Culture and foreign companies choice of entry mode: The case of Singapore building and construction industry. *Construction Management and Economics, 19*, pp. 343–353.

Flanagan, R. and Norman, G. (1993) *Risk Management and Construction*, Blackwell, Oxford.

Ghoddousi, P., Poorafshar, O., Chileshe, N. and Hosseini, M.R. (2015) Labour productivity in Iranian construction projects: Perceptions of chief executive officers. *International Journal of Productivity and Performance Management, 64*(6), pp. 811–830.

Gould, F.R. (2005) *Managing the Construction Process*, 3rd edition, Prentice-Hall, Upper Saddle River, NJ.

Halpin, Daniel W. and Senior, Bolivar A. (2011) *Construction Management Fundamentals*, 4th edition. John Wiley & Sons, Inc, Hoboken, NJ.

Harris, Frank and McCaffer, Ronald (2013) *Modern Construction Management*, Wiley Blackwell, John Wiley & Sons, West Sussex, UK.

Heldman, K. (2005) *PMP: Project Management Professional Study Guide*, 3rd edition. Wiley Publishing, Hoboken, NJ.

Hughes, W., Champion, R. and Murdoch, J.R. (2015) *Construction Contracts: Law and Management*, 5th edition. Routledge, London.

Keeney, R.L. and Raiffa, H. (1993) *Decisions with Multiple Objectives: Preferences and Value Tradeoffs*, Cambridge University Press, Cambridge.

Kivrak, S., Arslan, G., Tuncan, M. and Birgonul, M.T. (2014) Impact of national culture on knowledge sharing in international construction projects. *Canadian Journal of Civil Engineering, 41*(7), pp. 642–649.

Knight, F.H. (1921) *Risk, Uncertainty and Profit*, Houghton Mifflin, Boston, MA.

Knutson, Kraig, Schexnayder, Clifford J., Fiori, Christine M. and Mayo, Richard E. (2008) *Construction Management Fundamentals*, 2nd edition. McGraw-Hill Education, New York.

Koehn, E. and Brown, G. (1986) International labor productivity factors. *Journal of Construction Engineering and Management, 112*(2), pp. 299–302.

Kohli, K.N. (1993) *Economic Analysis of Investment Projects: A Practical Approach*, Oxford University Press, Oxford for the Asian Development Bank. Manila, Phillipines.

Kululanga, G., Kuotcha, W., McCaffer, R. and Edum-Fotwe, F. (2001) Construction contractors' claim process framework. *Journal of Construction Engineering and Management, 127*(4), pp. 309–314.

Kumaraswamy, M. (1996) *Conflict, Claims and Disputes in Construction*, Engineering and Architectural Management, E&FN Spon, London.

Kumaraswamy, M.M. (1997) Conflicts, claims and disputes in construction. *Engineering, Construction and Architectural Management, 4*(2), pp. 95–111.

Lam, J. (2003) *Enterprise Risk Management*, John Wiley & Sons, Inc, Hoboken, NJ.

Little, I.M.D. and Mirrlees, J.A. (1974) *Project Appraisal and Planning for Developing Countries*, Heinemann, London.

Liu, J., Meng, F. and Fellows, R. (2015) An exploratory study of understanding project risk management from the perspective of national culture. *International Journal of Project Management*, 33(15), pp. 564–575.

Loosemore, M. and Muslmani, H.A. (1999) Construction project management in the Persian Gulf: Inter-cultural communication. *International Journal of Project Management*, 17(2), pp. 95–100.

Mahalingam, A., Levitt, R.E. and Scott, W.R. (2005) Cultural clashes in international infrastructure development projects: Which cultures matter? In: *Proceedings of the CIB W92/T23/W107 International Symposium on Procurement Systems: The Impact of Cultural Differences and Systems on Construction Performance*, pp. 645–653, Las Vegas, NV. Retrieved from: https://gpc.stanford.edu/sites/g/files/sbiybj8226/f/cp012_0.pdf Accessed on February 20, 2020.

Merna, A. and Bower, D. (1997) *Dispute Resolution in Construction & Infrastructure Projects*, Asia Law & Practice Publishing, Hong Kong.

Mishan, E.J. (1994) *Cost Benefit Analysis: An Informal Introduction*, 4th edition. Routledge, New York.

Mitkus, S. and Mitkus, T. (2014) Causes of conflicts in a construction industry: A communicational approach. *Procedia - Social and Behavioral Sciences*, 110, pp. 777–786.

Mojahed, S. and Aghazadeh, F. (2008) Major factors influencing productivity of water and wastewater treatment plant construction: Evidence from the deep south USA. *International Journal of Project Management*, 26(2), pp. 195–202.

Morledge, R., Smith, A. and Kashiwagi, D.T. (2006) *Building Procurement*, Blackwell Publishing Ltd., London.

Murdoch, J. and Hugh, W. (2008) *Construction Contracts Law and Management*, 4th edition. Taylor and Francis, Simultaneously Published in the USA. New York, USA.

O'Reilly, M. (1999) *Civil Engineering Construction Contracts*, 3rd edition. Thomas Telford, London.

Ochieng, E.G. and Price, A.D.F. (2009) Addressing cultural issues when managing multicultural construction project teams. In: *Association of Researchers in Construction Management, ARCOM 2009 Proceedings of the 25th Annual Conference*, pp. 1273–1282. Nottingham, UK.

Ochieng, E.G. and Price, A.D.F. (2010) Factors influencing effective performance of multi-cultural construction project teams. In: *Association of Researchers in Construction Management, ARCOM 2010 Proceedings of the 26th Annual Conference*, pp. 1159–1167. Leeds, UK.

Ofori, G. (2003) Frameworks for analyzing international construction. *Construction Management and Economics*, 21(4), pp. 379–391.

Olsson, C. (2002) *Risk Management in Emerging Markets*, Pearson. London.

Overholt, W. (1982) *Political Risk*, Euromoney, New York.

Ozorhon, B., Arditi, D., Dikmen, I. and Birgonul, M.T. (2007) Effect of host country and project conditions in international construction joint ventures. *International Journal of Project Management*, 25(8), pp. 799–806.

Ozorhon, B., Arditi, D., Dikmen, I. and Birgonul, M.T. (2008) Implications of culture in the performance of international construction joint ventures. *Journal of Construction Engineering and Management*, 134(5), pp. 361–370.

Perry, J. and Hayes, R. (1985, June) Risk and its management in construction projects. *Proceedings of the Institution of Civil Engineers*, 78(Part I), pp. 499–521.

Pheng, L.S. and Leong, D.H.Y. (2000) Cross-cultural project management for international construction in China. *International Journal of Project Management*, 18(5), pp. 307–316.

Powell-Smith, Vincent, Sims, John H.M. and Dancaster, Christopher (1998) *Construction Arbitrations: A Practical Guide*, 2nd edition. Blackwell Science, Oxford.

Powell-Smith, V. and Stephenson, D. (1993) *Civil Engineering Claims*, 2nd edition. Blackwell Scientific, Oxford.

Radosavljevic, Milan and Bennett, John (2012) *Construction Management Strategies: A Theory of Construction Management*, John Wiley & Sons, West Sussex, UK.

Radujkovic, M. (1997) *Risk Sources and Drivers in Construction Projects*, E & FN Spon, London.

Rees-Caldwell, K. and Pinnington, A.H. (2013) National culture differences in project management: Comparing British and Arab project managers' perceptions of different planning areas. *International Journal of Project Management*, 31(2), pp. 212–227.

Root, F. (1994) *Entry Strategies for International Markets*, Lexington Books, New York.

Sawhinney, M. (2001) *International Construction*, Blackwell Science, London.

Semple, C., Hartman, F.T. and Jergeas, G. (1994) Construction claims and disputes: Causes and cost/time overruns. *Journal of Construction Engineering and Management*, 120(4), pp. 785–795.

Squire, Lyn and van der Tak, Herman G. (1975) *Economic Analysis of Projects*, The Johns Hopkins University Press, Baltimore and London for The International Bank for Reconstruction and Development.

Sykes, J. and Sheridan, P. (1996) Claims and disputes in construction. *Construction Law Journal*, 12(1), pp. 3–13.

Totterdill, B.W. (2006) *FIDIC User's Guide: A Particular Guide to the 1999 Red and Yellow Books*, 1st edition. Thomas Telford, London.

UNIDO. (1986) *Manual for Evaluation of Industrial Projects*, Nations Industrial Development Organization and the Industrial Development Centre for Arab States, UNIDO, Vienna.

Vogl, B. and Abdel-Wahab, M. (2015) Measuring the construction industry's productivity performance: Critique of international productivity comparisons at industry level. *Journal of Construction Engineering and Management*, 141(4), p. 04014085.

Waziri, F. and Khalfan, S. (2014) Cross-cultural communication in construction industry: How do Chinese firms cross the barriers in Tanzania? *European Journal of Business and Management*, 6(13), pp. 118–122.

Wearne, Stephen H. (1989) *Civil Engineering Contracts: An Introduction to Construction Contracts and the ICE Model Form of Contract*, Thomas Telford, London.

World Bank. (2010) *Guidelines Procurement under IBRD Loans and IDA Credits*, The World Bank, Washington, DC.

World Bank. (2010, August) *Standard Bidding Documents: Procurement of Works & User's Guide*, The World Bank, Washington, DC.

World Bank. (2018) *Contract Management Practice*, The World Bank, Washington, DC.

Zhi, H. (1995) Risk management for overseas construction projects. *International Journal of Project Management*, 13(4), pp. 231–237.

Index

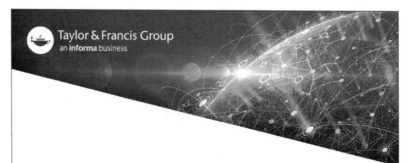